高等学校专业英语教材

测控技术与仪器专业英语教程
（第 4 版）

刘曙光　主编

电子工业出版社
Publishing House of Electronics Industry
北京·BEIJING

内 容 简 介

本书旨在使读者掌握测控技术与仪器专业英语术语及用法，培养和提高读者阅读和翻译专业英语文献资料的能力；主要内容包括电子技术、数字系统、信号处理、测试技术、仪器仪表、可编程序控制器、遥感通信、无线电能传输等。本书由16篇课文和16篇阅读材料组成，并附有所有课文的参考译文。本书配有电子教案、课后习题答案、音频材料，读者可登录华信教育资源网 www.hxedu.com.cn 免费下载。

本书可以作为测控技术与仪器专业的专业英语课程教材，也可供从事相关专业的工程技术人员学习参考。

未经许可，不得以任何方式复制或抄袭本书之部分或全部内容。
版权所有，侵权必究。

图书在版编目(CIP)数据

测控技术与仪器专业英语教程/刘曙光主编. —4版. —北京：电子工业出版社，2019.6
高等学校专业英语教材
ISBN 978-7-121-36820-2

I. ①测… II. ①刘… III. ①测量系统-控制系统-英语-高等学校-教材 ②电子测量设备-英语-高等学校-教材 IV. ①TM93

中国版本图书馆 CIP 数据核字(2019)第 115868 号

策划编辑：秦淑灵
责任编辑：靳　平
印　　刷：北京虎彩文化传播有限公司
装　　订：北京虎彩文化传播有限公司
出版发行：电子工业出版社
　　　　　北京市海淀区万寿路173信箱　邮编100036
开　　本：787×1092　1/16　印张：18.5　字数：636千字
版　　次：2004年2月第1版
　　　　　2019年6月第4版
印　　次：2025年1月第7次印刷
定　　价：49.00元

凡所购买电子工业出版社图书有缺损问题，请向购买书店调换。若书店售缺，请与本社发行部联系，联系及邮购电话：(010)88254888，88258888。
质量投诉请发邮件至 zlts@phei.com.cn，盗版侵权举报请发邮件至 dbqq@phei.com.cn。
本书咨询联系方式：qinshl@phei.com.cn。

前　言

教育部高等学校大学外语教学指导委员会于2016年颁布了《大学英语教学指南》，这给大学英语教学改革提供了依据和参考。《大学英语教学指南》明确了课程定位与性质、教学目标和教学要求、课程设置、评价与测试、教学方法与手段、教学资源、教学管理及教师发展。《大学英语教学指南》指出："大学英语在注重发展学生通用语言能力的同时，应进一步增强其学术英语或职业英语交流能力和跨文化交际能力，以使学生在日常生活、专业学习和职业岗位等不同领域或语境中能够用英语有效地进行交流。"从这里我们感知到明显课改信号：在我国开展了40年的以培养通用英语为主要目标的大学英语教学模式，势必会让位于通用英语和专门用途英语并重的教学模式。

随着全球经济一体化进程不断加快，我们要借鉴西方先进技术的地方越来越多，接触西方专业领域知识的机会也更加频繁。作为了解西方的必要工具——英语的地位也变得举足轻重。英语中的专门术语大多与日常含义不同，能将英语和专业知识融会贯通，是对专业人才适应时代需要的一种必然和迫切的要求。这一任务自然主要落在了各高等院校的专门用途英语教学改革中。根据这一精神，我们编写了本书，以满足测控技术与仪器专业学生学习的迫切需要，也可供从事上述专业的工程技术人员学习参考。

《测控技术与仪器专业英语教程》一书自2004年出版以来，已经出版了3版。本书被国内许多高等院校选为教材，得到了广大学者和读者的认可。本书由16篇课文和16篇阅读材料组成，主要内容包括电子技术、数字系统、信号处理、测试技术、仪器仪表、可编程序控制器、遥感通信、无线电能传输等。与第3版相比，在以下方面做了进一步改进：

原Lesson16在选材上与测控专业学科体系不太一致，同时为反映新技术的发展，要选择贴合度强的英文素材予以替换；

将世界顶级期刊论文引入本书，以反映世界科技先进水平。2007年，美国麻省理工学院以Soljacic M教授为首的研究小组在《Science》上发表论文，他们发现了一种全新的无线供电模式。紧接着，他们又在《Annals of Physics》上发表论文，提出了电能传输的磁共振感应耦合技术，称为"WiTricity"（Wireless electricity transfer）技术。为此，我们将这两篇世界顶级期刊论文编入了本书；

对全书勘误进行了全面修正。

本书内容新颖、问题规范、难度适中，为了适应专业英语的教学要求，书中所设计的内容既对学生所学过的课程进行了必要的覆盖，又有所拓宽和延伸，力求反映测控技术与仪器的现状和发展趋势；既可提高读者英语阅读水平，又能使读者了解学科前沿。为了方便教学，本书配有电子教案、课后习题答案、音频材料，读者可登录华信教育资源网www.hxedu.com.cn免费下载。

本书由黄山学院刘曙光教授主编并统稿，参加编写的有西安交通大学的曹军义教授、西安电子科技大学的费佩燕副教授、西安工业大学的侯志敏副教授、中北大学的孙兴丽副教授、西安工程大学的屈萍鸽副教授、郭亚青副教授、张莉副教授。

本书的出版也得到了多项基金项目的大力支持，如 2017 年度安徽省教育厅教学研究重大项目（2017jyxm0454）、2017 年度安徽省教育厅新工科研究与实践项目（2017xgkxm47）、2018 年度安徽省教育厅校企合作实践教育基地项目（2018sjjd079）、2018 年度安徽省高校自然科学研究重点项目（KJ2018A0408）等，在此对基金项目的帮助表示感谢。

鉴于水平有限，书中难免有不足和欠妥之处，恳请广大读者批评指正。

作者联系方式：Liushuguang@hsu.edu.cn。

作　者

Contents

Lesson 1　Periodic Signals ··· (1)
  1.1　Time-Domain Description ··· (1)
  1.2　Frequency-Domain Description ··· (2)
  1.3　Orthogonal Functions ··· (3)
  1.4　The Fourier Series ··· (5)
  Exercises ··· (10)
  Reading Material: Underwater Acoustic Signal ··· (11)

Lesson 2　Aperiodic Signals ·· (13)
  2.1　Introduction ··· (13)
  2.2　The Exponential Form of the Fourier Series ·· (13)
  2.3　The Fourier Transform ·· (14)
  2.4　The Laplace Transform ·· (20)
  Exercises ··· (26)
  Reading Material: Properties of Signal and Noise ··· (27)

Lesson 3　Sampled-Data Signals ·· (29)
  3.1　Introduction ··· (29)
  3.2　Mathematical Description Using the Dirac Function ······························· (29)
  3.3　Spectra of Sampled-data Signals ··· (30)
  3.4　The z-transform ·· (32)
  Exercises ··· (38)
  Reading Material: Signal Sampling ··· (40)

Lesson 4　Random Signals ··· (43)
  4.1　Introduction ··· (43)
  4.2　Elements of Probability Theory ·· (45)
  4.3　Amplitude Distribution and Moments ··· (48)
  4.4　The Autocorrelation and Power Spectral Density ·································· (51)
  Exercises ··· (56)
  Reading Material: The Processing of Random Signal ···································· (57)

Lesson 5　Static Performance ·· (61)
  5.1　The Ideal Measuring System ·· (61)
  5.2　Sensitivity ·· (61)
  5.3　Accuracy and Precision ··· (62)
  5.4　Possible and Probable Errors ··· (64)
  5.5　Other Static-Performance Terms ·· (64)

Ⅴ

  Exercises ··· (65)

  Reading Material: Noncontact Temperature Measurement ···································· (67)

Lesson 6 Dynamic Performance ·· (70)

  6.1 Zero-order Systems ·· (71)

  6.2 First-order Systems ·· (71)

  6.3 Second-order Systems ·· (74)

  6.4 Step-response Specification ·· (78)

  6.5 Frequency-response Specification ·· (79)

  Exercises ··· (81)

  Reading Material: Eddy Current ··· (82)

Lesson 7 Basic Knowledge of Transducers and Resistance Transducers ······················· (85)

  7.1 Transducer Elements ·· (85)

  7.2 Transducer Sensitivity ·· (85)

  7.3 Characteristics of an Ideal Transducer ··· (86)

  7.4 Electrical Transducers ·· (86)

  7.5 Resistance Transducers ·· (87)

  Exercises ··· (95)

  Reading Material: Laser ·· (96)

Lesson 8 Capacitance, Inductance Transducers and Some Others ······························· (98)

  8.1 Capacitive Transducers ·· (98)

  8.2 Inductive Transducers ·· (99)

  8.3 Linear Variable-differential Transformer (l. v. d. t.) ·· (100)

  8.4 Piezo-electric Transducers ··· (102)

  8.5 Electromagnetic Transducers ·· (102)

  8.6 Thermoelectric Transducers ··· (103)

  8.7 Photoelectric Cells (self-generating) ·· (104)

  8.8 Mechanical Transducers and Sensing Elements ··· (104)

  Exercises ··· (108)

  Reading Material: Video Camera ·· (109)

Lesson 9 Analog Instruments ·· (112)

  9.1 Meter Basics ·· (112)

  9.2 Ammeters ·· (114)

  9.3 Current Measuring Errors ··· (114)

  9.4 DC Voltmeters ··· (116)

  9.5 Voltage Measuring Errors ··· (117)

  9.6 Ohmmeter and Resistance Measurements ·· (119)

  9.7 Series Ohmmeter ·· (119)

  9.8 Shunt Ohmmeter ·· (121)

  9.9 Ohmmeter Accuracy ·· (121)

  9.10 Volt-Ohm-Milliammeters ··· (122)

  9.11 Electronic Voltmeters ··· (124)

9.12	Transistorized Electronic Voltmeters	(124)
9.13	FET Voltmeters	(126)
9.14	Operational Amplifier Electronic Voltmeters	(127)
9.15	Electronic Current Measurements	(129)
9.16	Resistance Measurements	(130)
	Exercises	(134)
	Reading Material: Rectifier Meters	(135)

Lesson 10 Digital Instruments (139)

10.1	Digital Displays	(139)
10.2	Electronic Digital Counter	(141)
10.3	Input Signal Conditioning	(141)
10.4	Timerbase Oscillator	(144)
10.5	Timerbase Dividers	(145)
10.6	Counting Register	(145)
10.7	Digital Voltmeter	(150)
	Exercises	(153)
	Reading Material: Rise-time Measurements	(154)

Lesson 11 Computer-based Test Instruments (158)

11.1	Internal Adapters	(158)
11.2	External PC Instruments	(160)
11.3	Analog-to-Digital Conversion	(162)
11.4	Computer Interface	(165)
11.5	General-purpose Interface BUS	(166)
	Exercises	(170)
	Reading Material: PC-based Workstation	(171)

Lesson 12 Industrial Bus (175)

12.1	What is An Industrial Bus	(175)
12.2	Data Line Isolation Theory	(176)
	Exercises	(179)
	Reading Material: Serial Communications Systems	(180)

Lesson 13 Programmable Logic Controller (183)

13.1	Basic PLC Operation	(183)
13.2	Hard-Wired Control	(183)
13.3	Advantages of PLCs	(184)
13.4	Siemens PLCs	(185)
13.5	CPU	(185)
13.6	Programming Devices	(185)
13.7	Software	(185)
13.8	Connector Cables PPI (Point-to-Point Interface)	(186)
	Exercises	(187)
	Reading Material: Terminology	(188)

Lesson 14　Remote Sensing （191）
　14.1　Sensors （191）
　14.2　Satellites （192）
　14.3　Outlook （192）
　Exercises （193）
　Reading Material：GPS （194）

Lesson 15　Multi-sensor Data Fusion （197）
　15.1　Introduction （197）
　15.2　Technical Background （197）
　15.3　Method （198）
　15.4　Application （199）
　Exercises （205）
　Reading Material：Multi-sensor Image Fusion （206）

Lesson 16　Wireless Power Transfer via Coupled Magnetic Resonances （209）
　16.1　Overview of the Formalism （209）
　16.2　Theoretical Model for Self−resonant Coils （210）
　16.3　Comparison with Experimentally Determined Parameters （212）
　16.4　Measurement of the Efficiency （213）
　Exercises （215）
　Reading Material：Wireless Charging System for Electric Vehicle （216）

参考译文 （220）
　第1课　周期信号 （220）
　第2课　非周期信号 （224）
　第3课　数据采样信号 （231）
　第4课　随机信号 （235）
　第5课　静态性能 （242）
　第6课　动态性能 （244）
　第7课　传感器和电阻式传感器的基础知识 （249）
　第8课　电容式、电感式传感器及其他器件 （254）
　第9课　模拟仪器 （258）
　第10课　数字仪器 （267）
　第11课　基于计算机的测试仪器 （272）
　第12课　工业总线 （276）
　第13课　可编程序控制器 （278）
　第14课　遥感 （280）
　第15课　多传感器数据融合 （281）
　第16课　磁耦合共振式无线电能传输技术 （283）

参考文献 （287）

Lesson 1 Periodic Signals

1.1 Time-Domain Description

The fact that great majority of functions which may usefully be considered as signals are functions of time lends justification to the treatment of signal theory in terms of time and of frequency. A periodic signal will therefore be considered to be one which repeats itself exactly every T seconds, where T is called the period of the signal waveform; the theoretical treatment of periodic waveforms assumes that this exact repetition is extended throughout all time, both past and future. In practice, of course, signals do not repeat themselves indefinitely. Nevertheless, a waveform such as the output voltage of a main rectifier prior to smoothing does repeat itself very many times, and it analysis as a strictly periodic signal yields valuable results. [1] In other cases, such as the electrocardiogram, the waveform is quasi-periodic and may usefully be treated as truly periodic for some purpose. It is worth nothing that a truly repetitive signal is of very little interest in a communication channel, since no further information is conveyed after the first cycle of the waveform has been received. One of the main reasons for discussing periodic signals is that a clear understanding of their analysis is a great help when dealing with periodic and random ones.

A complete time-domain description of such a signal involves specifying its value precise at every instant of time. In some cases this may be done very simply using mathematical notation. Fortunately, it is in many cases useful to describe only certain aspects of a signal waveform, or to represent it by a mathematical formula which is only approximate. The following aspects might be relevant in particular cases.

(1) the average value of the signal.

(2) the peak value reached by the signal.

(3) the proportion of the total time spent between value a and b.

(4) the period of the signal.

If it is desired to approximate the waveform by a mathematical expression, such as a polynomial expansion, a Taylor series, or a Fourier series may be used. A polynomial of order n having the form

$$f(t) = a_0 + a_1 t + a_2 t^2 + a_3 t^3 + \cdots + a_n t^n \tag{1-1}$$

may be used to fit the actual curve at $(n+1)$ arbitrary points. The accuracy of fit will generally improve as number of polynomial terms increases. It should also be noted that the error figure between the true signal waveform and the polynomial will normally become very large away from the region of the fitted points, and that the polynomial itself cannot be periodic. Whereas a polynomial approximation fits the actual waveform at a number of arbitrary points, the alternative Taylor series approximation provides a good fit to a smooth continuous waveform in the vicinity of one selected point. The coefficients of the Taylor series are chosen to make the series and its derivatives agree with the actual waveform at this point. The number of terms in the series determines to what order of derivative this agreement will extend, and hence the accuracy with which series and actual waveform agree in the region of point chosen. The general form of the Taylor series for approximating a function in the region of the point is given by

$$f(t) = f(a) + (t-a) \cdot \frac{\mathrm{d}f(a)}{\mathrm{d}t} + \frac{(t-a)^2}{2!} \cdot \frac{\mathrm{d}^2 f(a)}{\mathrm{d}t^2} + \cdots + \frac{(t-a)^n}{n!} \cdot \frac{\mathrm{d}^n f(a)}{\mathrm{d}t^n} \quad (1\text{-}2)$$

Generally speaking, the fit to the actual waveform is good in the region of the point chosen, but rapidly deteriorates to either side. The polynomial and Taylor series descriptions of a signal waveform are therefore only to be recommended when one is concerned to achieve accuracy over a limited region of the waveform. The accuracy usually decreases rapidly outside this region, although it may be improved by including additional terms (so long as t lies within the region of convergence of the series). [2] The approximations provided by such methods are never periodic in form and cannot therefore be considered ideal for the description of repetitive signals.

By contrast the Fourier series approximation is well suited to the representation of a signal waveform over an extended interval. When the signal is periodic, the accuracy of the Fourier series description is maintained for all time, since the signal is represented as the sum of a number of sinusoidal functions, which are themselves periodic. Before examining in detail the Fourier series method of representing a signal, the background to what is known as the 'frequency-domain' approach will be introduced.

1.2 Frequency-Domain Description

The basic conception of frequency-domain analysis is that a waveform of any complexity may be considered as the sum of a number of sinusoidal waveforms of suitable amplitude, periodicity, and relative phase. [3] A continuous sinusoidal function ($\sin \omega t$) is thought of as a 'single frequency' wave of frequency radians per second, and the frequency-domain description of a signal involves its breakdown into a number of such basic functions. This is the method of Fourier analysis.

There are a number of reasons why signal representation in terms of a set of component sinusoidal waves occupies such a central role in signal analysis. The suitability of a set of periodic functions for approximating a signal waveform over an extended interval has already been mentioned, and it will be shown later that the use of such techniques causes the error between the actual signal and its approximation to be minimized in a certain important sense. A further reason why sinusoidal functions are so important in signal analysis is that they occur widely in the physical world and are very susceptible to mathematical treatment; a large and extremely important class of electrical and mechanical systems, known as 'linear systems', responds sinusoidally when driven by a sinusoidal disturbing function of any frequency. All these manifestations of sinusoidal function in the physical world suggest that signal analysis in sinusoidal terms will simplify the problem of relating a signal to underlying physical causes, or to the physical properties of a system or device through which it has passed. Finally, sinusoidal functions form a set of what are called 'orthogonal function', the rather special properties and advantage of which will now be discussed.

1.3 Orthogonal Functions

1.3.1 Vectors and Signals

A discussion of orthogonal functions and of their value for the description of signals may be conveniently introduced by considering the analogy between vectors and signals. A vector is specified both by its magnitude and direction, familiar examples being force and velocity. Suppose we have two $\boldsymbol{V}_1$ and $\boldsymbol{V}_2$; geometrically, we define the component of vector $\boldsymbol{V}_1$ along vector $\boldsymbol{V}_2$ by constructing the perpendicular form the end of $\boldsymbol{V}_1$ onto $\boldsymbol{V}_2$. We then have

$$\boldsymbol{V}_1 = C_{12}\boldsymbol{V}_2 + \boldsymbol{V}_e \tag{1-3}$$

where vector $\boldsymbol{V}_e$ is the error in the approximation. Clearly, this error vector is of minimum length when it is drawn perpendicular to the direction of $\boldsymbol{V}_2$. Thus we say that the component of vector $\boldsymbol{V}_1$ along vector $\boldsymbol{V}_2$ is given by $C_{12}\boldsymbol{V}_2$, where C_{12} is chosen such as to make the error vector as small as possible. A familiar case of an orthogonal vector system is the use of three mutually perpendicular axes in co-ordinate geometry.

There basic ideas about the comparison of vectors may be extended to signals. Suppose we wish to approximate a signal $f_1(t)$ by another signal or function $f_2(t)$ over a certain interval $t_1 < t < t_2$; in other words,

$$f_1(t) \approx C_{12} f_2(t) \qquad \text{for} \quad t_1 < t < t_2$$

We wish to choose C_{12} to achieve the best approximation. If we define the error function

$$f_e(t) = f_1(t) - C_{12} f_2(t) \tag{1-4}$$

it might appear at first sight that we should choose C_{12} so as to minimize the average value of $f_e(t)$ over the chosen interval. The disadvantage of such an error criterion is that large positive and negative errors occurring at different instants would tend to cancel each other out. This difficulty is avoided if we choose to minimize the average squared-error, rather than the error itself (this is equivalent to minimizing the square root of the mean-squared error, or 'r. m. s' error). Denoting the average of $f_e^2(t)$ by ε, we have

$$\varepsilon = \frac{1}{(t_2 - t_1)} \int_{t_1}^{t_2} f_e^2(t) dt = \frac{1}{(t_2 - t_1)} \int_{t_1}^{t_2} [f_1(t) - C_{12} f_2(t)]^2 dt \tag{1-5}$$

Differentiating with respect to C_{12} and putting the resulting expression equal to zero gives the value of C_{12} for which is a minimum.[4] Thus

$$\frac{d}{dC_{12}} \left\{ \frac{1}{(t_2 - t_1)} \int_{t_1}^{t_2} [f_1(t) - C_{12} f_2(t)]^2 dt \right\} = 0$$

Expanding the bracket and changing the order of integration and differentiating gives

$$C_{12} = \int_{t_1}^{t_2} f_1(t) f_2(t) dt \bigg/ \int_{t_1}^{t_2} f_2^2(t) dt \tag{1-6}$$

1.3.2 Signal Description by Sets of Orthogonal Function

Suppose that we have approximated a signal $f_1(t)$ over a certain interval by the function

$f_2(t)$ so that the mean square error is minimized, but that we now wish to improve the approximation. It will be demonstrated that a very attractive approach is to express the signal in terms of a set of function $f_2(t)$, $f_3(t)$, $f_4(t)$, etc., which are mutually orthogonal. Suppose the initial approximation is

$$f_1(t) \approx C_{12} f_2(t) \qquad (1\text{-}7)$$

and that the error is further reduced by putting

$$f_1(t) \approx C_{12} f_2(t) + C_{13} f_3(t) \qquad (1\text{-}8)$$

where $f_2(t)$ and $f_3(t)$ are orthogonal over the interval of interest. Now that we have incorporated the additional term $C_{13} f_3(t)$, it is interesting to find what the new value of must be in order that the mean square error is again minimized. We now have

$$f_e(t) = f_1(t) - C_{12} f_2(t) - C_{13} f_3(t) \qquad (1\text{-}9)$$

and the mean square error in the interval $t_1 < t < t_2$ is therefore

$$\varepsilon = \frac{1}{t_2 - t_1} \cdot \int_{t_1}^{t_2} [f_1(t) - C_{12}(t) f_2(t) - C_{13}(t) f_3(t)]^2 \mathrm{d}t \qquad (1\text{-}10)$$

Differentiating partially with respect to C_{12} to find the value of C_{12} for which the mean square error is again minimized, and changing the order of differentiation and integration, we have again [5]

$$C_{12} = \int_{t_1}^{t_2} f_1(t) f_2(t) \mathrm{d}t \Big/ \int_{t_1}^{t_2} f_2^2(t) \mathrm{d}t \qquad (1\text{-}11)$$

In order words, the decision to improve the approximation by incorporating an additional term in does not require us to modify the coefficient, provided that $f_3(t)$ is orthogonal to $f_2(t)$ in the chosen time interval. [6] By precisely similar arguments we could show that the value of C_{13} would be unchanged if the signal was to be approximated by $f_3(t)$ alone.

This important result may be extended to cover the representation of a signal in terms of a whole set of orthogonal functions. The value of any coefficient does not depend upon how many functions from the complete set are used in the approximation, and is thus unaltered when further terms are included. [7] The use of a set of orthogonal functions for signal description is analogous to the use of three mutually perpendicular (that is, orthogonal) axes for the description of a vector in three-dimensional space, and gives rise to the notion of a 'signal space'. [8] Accurate signal representation will often require the use of many more than three orthogonal functions, so that we must think of a signal within some interval $t_1 < t < t_2$ as being represented by a point in a multidimensional space.

To summarize, there are a number of sets of orthogonal functions available such as the so-called Legendre polynomials and Walsh functions for the approximate description of signal waveform, of which the sinusoidal set is the most widely used. [9] Sets involving polynomials in t are not by their very nature periodic, but may sensibly be used to describe one cycle (or less) of a periodic waveform; outside the chosen interval, errors between the true signal and its approximation will normally increase rapidly. A description of one cycle of a periodic signal in terms of sinusoidal function will, however, be equally valid for all time because of the every member of the orthogonal.

1.4 The Fourier Series

The basis of the Fourier series is that complex periodic waveform may be analyzed into a number of harmonically related sinusoidal waves which constitute an orthogonal set. If we have a periodic signal $f(t)$ with a period equal to T, then $f(t)$ may be represented by the series

$$f(t) = A_0 + \sum_{n=1}^{\infty} A_n \cos n\omega_1 t + \sum_{n=1}^{\infty} B_n \sin n\omega_1 t \qquad (1\text{-}12)$$

where $\omega_1 = 2\pi/T$. Thus $f(t)$ is considered to be made up by the addition of a steady level A_0 to a number of sinusoidal and cosinusoidal waves of different frequencies. The lowest of these frequencies is ω_1 (radians per second) and is called the 'fundamental'; waves of this frequency have a period equal to that of the signal. Frequency $2\omega_1$ is called the 'second harmonic', $3\omega_1$ is the 'third harmonic', and so on. Certain restrictions, known as the Dirichlet conditions, must be placed upon $f(t)$ for the above series to be valid. The integral $\int |f(t)| dt$ over a complete period must be finite, and may not have more than a finite number of discontinuities in any finite interval. Fortunately, these conditions do not exclude any signal waveform of practical interest.

1.4.1 Evaluation of the Coefficients

We now turn to the question of evaluating the coefficients A_0, A_n and B_n. Using the minimum square error criterion described in foregoing text, and writing for the sake of convenience, we have

$$\left. \begin{aligned} A_0 &= \frac{1}{2\pi} \int_{-\pi}^{\pi} f(x) dx \\ A_n &= \frac{1}{\pi} \int_{-\pi}^{\pi} f(x) \cos nx \, dx \\ B_n &= \frac{1}{\pi} \int_{-\pi}^{\pi} f(x) \sin nx \, dx \end{aligned} \right\} \qquad (1\text{-}13)$$

Although in the majority of cases it is convenient for the interval of integration to be symmetrical about the origin, any interval equal in length to one period of the signal waveform may be chosen.

Many waveforms of practical interest are either even or odd functions of time. If $f(t)$ is even then by definition $f(t)=f(-t)$, whereas if it is odd $f(-t)=-f(t)$. If $f(t)$ is even and we multiply it by the odd function $\sin n\omega_1 t$ the result is also odd. Thus the integrand for every B_n is odd. Now when an odd function is integrated over an interval symmetrical about $t=0$, the result is always zero. Hence all the B coefficients are zero and we are left with a series containing only cosines. By similar arguments, if $f(t)$ is odd the A coefficients must be zero and we are left with a sine series. It is indeed intuitively clear that an even function can only be built up from a number of other functions which are themselves even, and vice versa.

We have already seen how the Fourier series is simplified in the case of an even or odd function, by losing either its sine or its cosine terms. A different type of simplification occurs in the case of a waveform possessing what is know as 'half-wave symmetry'. In mathematical terms, half-wave symmetry exists when

$$f(t) = -f(t+T/2) \tag{1-14}$$

In other words, any two values of the waveform separated by $T/2$ will be equal in magnitude and opposite in sign. Generalizing, only odd harmonics exhibit half-wave symmetry, and therefore a waveform of any complexity which has such symmetry cannot contain even harmonic components. Conversely, a waveform know to contain any second, fourth, or other harmonic components cannot display half-wave symmetry.

Usually, we have always integrated over a complete cycle to derive the coefficients. However in the case of an odd or even function it is sufficient, and often simpler, to integrate over only one half of the cycle and multiply the result by 2. Furthermore if the wave is not only even or odd but also display half-wave symmetry, it is enough to integrate over one quarter of a cycle and multiply by 4. These closer limits are adequate in such cases the function that is being integrated is repetitive, repeating twice within one period when it also exhibit half-wave symmetry.

1.4.2 Choice of Time Origin, and Waveform Power

The amount of work involved in calculating the Fourier series coefficients for a particular waveform shape is reduced if the waveform is either even or odd, and this may often be arranged by a judicious choice of time origin (that is, shift of time origin).[10] This shift has therefore merely had the effect of converting a Fourier series containing only sine terms into one containing only cosine terms; the amplitude of a component at any one frequency is, as we would expect, unaltered. For a complicated waveform which is neither even nor odd, it must be expected to include both sine and cosine terms in its Fourier series.

As the time origin of a waveform is shifted, the various sine and cosine coefficients of its Fourier series will change, but the sum of the squares of any two coefficients A_n and B_n will remain constant, which means that the average power of the waveform, a concept familiar to electrical engineers, is unaltered.

The above ideas lead naturally to an alternative trigonometric of the Fourier series. If the two fundamental components of a waveform are

$$A_1 \cos \omega_1 t \quad \text{and} \quad B_1 \sin \omega_1 t$$

their sum may be expressed in an alternative form using trigonometric identities

$$A_1 \cos \omega_1 t + B_1 \sin \omega_1 t = \sqrt{(A_1^2 + B_1^2)} \cos\left(\omega_1 t - \arctan \frac{B_1}{A_1}\right)$$
$$= \sqrt{(A_1^2 + B_1^2)} \sin\left(\omega_1 t + \arctan \frac{B_1}{A_1}\right) \tag{1-15}$$

Thus the sine and cosine components at a particular frequency are expressed as a single cosine or sine wave together with a phase shift. If this procedure is applied to all harmonic components of the Fourier series, we get the alternative forms

$$f(t) = A_0 + \sum_{n=1}^{\infty} C_n \cos(n\omega_1 t - \phi_n) \quad \text{or} \quad f(t) = A_0 + \sum_{n=1}^{\infty} C_n \sin(n\omega_1 t + \theta_n) \quad (1\text{-}16)$$

where

$$C_n = \sqrt{A_n^2 + B_n^2}, \phi_n = \arctan(B_n/A_n), \theta_n = \arctan(A_n/B_n) \quad (1\text{-}17)$$

Finally, we note that sine the mean power represented by any component wave is

$$(A_n^2 + B_n^2)/2 = C_n^2/2 \quad (1\text{-}18)$$

and the power represented by the term A_0 is simply A_0^2, the total average waveform power is equal to

$$P = A_0^2 + \frac{1}{2} \sum_{n=1}^{\infty} C_n^2 \quad (1\text{-}19)$$

But P may be expressed as the average value over one period of $[f(t)]^2$, using again the convention that is considered to represent a voltage waveform applied across a ohm resistor. Hence

$$P = A_0^2 + \frac{1}{2} \sum_{n=1}^{\infty} C_n^2 = \frac{1}{T} \int_{-T/2}^{T/2} [f(t)]^2 dt \quad (1\text{-}20)$$

This result is a version of a more general one known as Parseval's theorem, and shows that the total waveform power is equal to the sum of the powers represented by its individual Fourier components. It is, however, important to note that this is only true because the various component waves are drawn from an orthogonal set.

Words and Expressions

accuracy ['ækjurəsi] *n.* 精确性，准确度，精度
amplitude ['æmplitju:d] *n.* 振幅，幅度
aperiodic ['eipiəri'ɔdik] *adj.* 非周期的
approach [ə'prəutʃ] *n.* 接（逼）近；近似法（值）；途径，方法
approximation [ə,prɔksi'meiʃən] *n.* 近似值
arbitrary ['ɑ:bitrəri] *adj.* 任意的
channel ['tʃænl] *n.* 信道，频道
coefficient [,kəui'fiʃənt] *n.* 系数
convergence [kən'və:dʒəns] *n.* 收敛
conversely [kən'və:sli] *adj.* 相反的，逆的
coordinate [kəu'ɔ:dinit] *n.* 坐标（系）
criterion [krai'tiəriən] *n.* 标准，规范
deteriorate [di'tiəriəreit] *vi.* 恶化，变坏，退化
differentiate [,difə'renʃieit] *vt.* 求……的微分
dimension [di'menʃən] *n.* 维数
Dirichlet conditions 狄利克雷条件
discontinuity ['dis,kɔnti'nju(:)iti] *n.* 心电图，心动电流图
even ['i:vən] *adj.* 偶数的

expansion [iks'pænʃən] *n.* 展开（式）
foregoing [fɔː'gəuiŋ] *adj.* 前述的，在前的
geometrical [dʒiə'metrikəl] *adj.* 几何学的
half-wave symmetry 半波对称
harmonical [hɑː'mɔnikəl] *adj.* 谐波的
identity [ai'dentiti] *n.* 恒等式
instant ['instənt] *n.* 瞬时，瞬间
integrand ['intigrænd] *n.* 被积函数
integrate ['intigreit] *vt.* 求……的积分
intuitively [in'tjuː(ː)itivli] *adv.* 直观地，直觉地
geometry [dʒi'ɔmitri] *n.* 几何学
Legendre polynomials 勒让德多项式
linear ['liniə] *adj.* 线性的
main [mein] *n.* 电源，电力网
manifestation [ˌmænifes'teiʃən] *n.* 表现
minimum ['miniməm] *n.* 最小值，最小化
mutually ['mjuːtʃuəli] *adv.* 相互地
notation [nəu'teiʃən] *n.* 符号，记号
odd [ɔd] *adj.* 奇数的，单数的
ohm [əum] *n.* 欧姆
order ['ɔːdə] *n.* 次序，阶
origin ['ɔridʒin] *n.* 原点
orthogonal [ɔː'θɔgənl] *adj.* 正交的，直角的
peak [piːk] *adj.* 最高的；*n.* 最高峰
periodicity [ˌpiəriə'disiti] *n.* 周期
perpendicular [ˌpəːpən'dikjulə] *adj.* 垂直的，正交的
phase [feiz] *n.* 相位
polynomial [ˌpɔli'nəumjəl] *adj.* 多项式的；*n.* 多项式
quasi-periodic 准周期的
radian ['reidjən] *n.* 弧度
rectifier ['rektifaiə] *n.* 整流器
resistor [ri'zistə] *n.* 电阻器
series ['siəriːz] *vt.* 展成级数；*n.* 级数
set [set] *n.* 集合
sinusoidal [ˌsainə'sɔidəl] *adj.* 正弦的
susceptible [sə'septəbl] *adj.* 敏感的，易受影响的
symmetry ['simitri] *n.* 对称
symmetrical [si'metrikəl] *adj.* 对称的，均匀的
term [təːm] *n.* 术语，（数）项

theorem [ˈθiərəm] *n.* 定理，法则
time domain　时域
trigonometric [ˌtrigənəˈmetrik] *adj.* 三角法的
trigonometric identities　三角恒等式
vector [ˈvektə] *n.* 矢量，向量
vice versa　反之亦然
vicinity [viˈsiniti] *n.* 附近
Walsh function　沃尔什函数

Notes

1. Nevertheless, a waveform such as the output voltage of a main rectifier prior to smoothing does repeat itself very many times, and it analysis as a strictly periodic signal yields valuable results.
   不过，像电源整流器输出电压这样的波形，在平滑之前，还是重复本身很多次的，将其作为严格的周期信号进行分析，会产生颇有价值的结果。

2. The accuracy usually decreases rapidly away from this region, although it may be improved by including additional terms (so long as t lies within the region of convergence of the series).
   在所选择区域之外，精度通常会迅速降低，尽管可以通过补充一些项，使之有所改善（只要 t 位于序列的收敛域内）。

3. The basic conception of frequency-domain analysis is that a waveform of any complexity may be considered as the sum of a number of sinusoidal waveforms of suitable amplitude, periodicity, and relative phase.
   频域分析的基本概念是：任何复杂波形都可以看成许多具有适当振幅、周期和相对相位的正弦波之和。

4. Differentiating with respect to C_{12} and putting the resulting expression equal to zero gives the value of C_{12} for which ε is a minimum.
   对 C_{12} 求微分，然后令所得表达式为 0，就可以得到使 ε 最小的 C_{12} 值。

5. Differentiating partially with respect to C_{12} to find the value of C_{12} for which the mean square error is again minimized, and changing the order of differentiation and integration, we have again aquation (1-11).
   为了求出使均方误差仍保持最小的 C_{12} 值，先对 C_{12} 求偏微分，再交换微分与积分次序，我们再次得到式（1-11）。

6. In order words, the decision to improve the approximation by incorporating an additional term in does not require us to modify the coefficient C_{12}, provided that $f_3(t)$ is orthogonal to $f_2(t)$ in the chosen time interval.
   换言之，如果 $f_2(t)$ 与 $f_3(t)$ 在所选择的时间区间内正交，在并入用 $f_3(t)$ 表示的附加项来改善逼近程度时，系数 C_{12} 无须修正。

7. This important result may be extended to cover the representation of a signal in terms of a whole set of orthogonal functions. The value of any coefficient does not depend upon how many functions from the complete set are used in the approximation, and is thus unaltered when further terms are included.

    这个重要结论可以推广到用整个正交函数集表示信号的情况。在进行逼近时，任何系数值与完备集合中用了多少函数没有关系，因此函数包含更多的项时，这些系数值不会改变。

8. The use of a set of orthogonal functions for signal description is analogous to the use of three mutually perpendicular (that is, orthogonal) axes for the description of a vector in three-dimensional space, and gives rise to the notion of a 'signal space'.

    利用一个正交函数集描述信号，类似于在三维空间中利用三个互相垂直的轴描述矢量，这就引出了"信号空间"的概念。

9. To summarize, there are a number of sets of orthogonal functions available such as the so-called Legendre polynomials and Walsh functions for the approximate description of signal waveform, of which the sinusoidal set is the most widely used.

    总之，有许多正交函数集可用来近似描述信号波形，如所谓的勒让德多项式和沃尔什函数等，正弦函数集是其中最常用的。

10. The amount of work involved in calculating the Fourier series coefficients for a particular waveform shape is reduced if the waveform is either even or odd, and this may often be arranged by a judicious choice of time origin (that is, shift of time origin).

    对于一个具体的波形而言，如果它是偶函数或奇函数，那么计算该波形傅里叶级数系数时，可以通过适当的选择时间原点来减小其计算工作量。

Exercises

Translate the following passages into English or Chinese.

1. 简谐信号是最简单和最重要的周期信号。任意一个周期信号都可以用简谐信号来表达，两者之间联系的桥梁是傅里叶级数，所以傅里叶级数是周期信号分析的理论基础。
2. 一个在时域上显得很复杂的信号，将其变换或映射到频域（包括 s 和 z 域），就能够分解为非常简单的基本信号形式以进行分析和求解。
3. 频谱是由不连续的谱线组成的，每条谱线代表一个谐波分量，这种频谱称为离散频谱。谱线之间的间隔等于基波频率 ω_0 的整数倍。即频谱中的每一条谱线只能出现在基波频率 ω_0 的整数倍上，各谐波的频率 $n\omega_0$ 都是基波频率的整数倍。
4. The Fourier series is a particular type of orthogonal series representation that is very useful in solving engineering problems, especial communication problems. The orthogonal functions that are used are either sinusoids, or, equivalently, complex exponential functions.
5. For periodic waveforms, the Fourier series representations are valid over all time. Consequently, the (two-side) spectrum, which depends on the waveshape from $t = -\infty$ to $t = \infty$, may be evaluated in terms of Fourier coefficients.

Reading Material

Underwater Acoustic Signals

In the operation of a sonar system the operator is repeatedly faced with the problem of detecting a signal which is obscured by noise. This signal may be an echo resulting from a transmitted signal over which the operator has some control, or it may have its origin in some external source. These two modes of operation are commonly distinguished as active and passive sonar, respectively. Similar situations arise in radar surveillance and in disciplines for techniques and for illustrations of the basic principles.

Since there are many ways in which one can think about signal detection, it is desirable to define a term to denote special cases. The word detection will be used when the question to be answered is, 'Are one or more signals present?' when the system is designed to provide an answer to this question, either deterministic or probabilistic, one speaks of hypothesis testing. The case of a single signal occurs so often that many systems are designed to provide only two answers, 'Yes, a signal is present,' or 'No, there is no signal.' One can make the problem more complicated by endeavoring to classify the signal into categories. Decisions of this latter kind will be referred to as target classification.

Normally a piece of detection equipment is designed to operate in a fixed mode and the parameters such as integrating time of rectifier circuits or persistence of the oscilloscope tube for visual detection cannot be changed readily. There will always be some uncertain signals, which the observer will be hesitant to reject or accept. In these cases the operator might have the feeling that if the integrating time of the detector or the persistence of the oscilloscope tube were longer, he could reach a decision about the existence of the signal. Wald (1950) has formulated this intuitive feeling into a theory of detection. When one is able to vary deliberately the interval over which one stores data in the reception system in order to achieve a certain level of certainty, one speaks of sequential detection.

Frequently it is desirable to determine not only the presence or absence of the signal but also one or more parameters associated with the signal. The parameters of interest can vary widely from a simple quantity such as time of arrival or target bearing to the recovery of the complete waveform. When a system is designed to recover one or more parameters associated with the signal, one speaks of signal extraction.

The word signal was not defined and it was assumed that the reader had an intuitive felling for the word. Some elaboration may be in order since the definition of signal as subjective and depends on the application. One may say that 'signal' is what one wants to observe and noise is anything that obscures the observation. Thus, a tuna fisherman who is searching the ocean with the aid of sonar equipment will be overjoyed with sounds that are impairing the performance of a nearby sonar system engaged in tracking a submarine. Quite literally, one man's signal is another man's noise.

Signals come in all shapes and forms. In active sonar system one may use simple sinusoidal signals of fixed duration and modulations thereof. There are impulsive signals such as those made with explosions or thumpers. At the other extreme one may make use of pseudorandom signals. In passive systems, the signals whose detection is sought may be noise in the conventional meaning of the word; noise produced by propellers or underwater swimmers, for example. It should be evident that one of our problem will be the formulation of mathematical techniques that can be used to describe the signal.

Although the source in an active sonar search system may be designed to transmit a signal known shape, there is no guarantee that the return signal whose detection is sought will be similar. In fact, there are many factors to change the signal. The amplitude loss associated with inverse spherical spreading is most unfortunate for the detection system nut it does not entail any distortion of the wave shape. (Incidentally, this happy state of affairs does not apply to two-dimensional waves except in the far field where the wave can be approximated locally as a plane wave.) The acoustic medium has an attenuation factor, which depends on the frequency. This produces a slight distortion of the wave shape and a corresponding change in the energy spectrum of the pulse. The major changes in the waveform result from acoustic boundaries and inhomogeneities in the medium.

When echoes are produced by extended targets such as submarines, there are two distinct ways in which the echo structure is affected. First, there is the interference between reflections from the different leads to a target strength that fluctuates rapidly with changes in the aspect. Secondly, there is the elongation of the composite echo due to the distribution of reflecting features along the submarines. This means that the duration of the composite echo is dependent in a simple manner on the aspect angle. If T is the duration of the echo from a point scatterer, and L is the length of the submarine, the duration of the returned echo will be $T+(2L/C)\cos\theta$, where θ is the acute angle between the major axis of the submarine and the line joining the source and the submarine. C is the velocity of sound in the water. Of course, $L\cos\theta$ must be replaced by the beam width of the submarine when θ is near.

A final source of pulse distortion is the Doppler shifts produced by the relative motions between the source, the medium, the bottom, and the targets. Since the source, the medium, and the target (or detector in passive listening) may each have a different vector velocity relative to the bottom, the variety of effects may be quite large.

Lesson 2 Aperiodic Signals

2.1 Introduction

In the previous chapter we have seen how a periodic signal may be expressed as the sum of a set of sinusoidal waves which are harmonically related. The spectrum of such a signal consists of a number of discrete frequencies and is known as a 'line' spectrum. Although the analysis of periodic signals gives results which can be of great practical interest, the great majority of signals are not of this type. Firstly, even signals which repeat themselves a very large number of times are generally turned 'on' and 'off'. In other words they may not generally be assumed to exist for all time past, present, and future, and it is important to understand the effects which time-limitation has upon their frequency spectra. Secondly, and quite apart from any question of time limitation, there is an important class of signal waveforms (amongst which are included random signals) which are simply not repetitive in nature and which cannot therefore represented by Fourier series containing a number of harmonically-related frequencies. Fortunately, however, it is possible to derive frequency spectra for such signals using as a starting point of the work we have already done on the Fourier series.

2.2 The Exponential Form of the Fourier Series

It has already been seen (section 1.4.2) that the Fourier series of a periodic signal may be expressed in two ways: either as a set of sine and cosine waves of appropriate amplitude and frequency, or as a set of waves of sinusoidal form which are defined by their amplitudes and relative phase angles. A third form of the Fourier series, the exponential form, will now be discussed, since it is particularly helpful for the derivation of frequency spectra of aperiodic signals.

Instead of using the simple trigonometric form

$$f(t) = A_0 + \sum_{n=1}^{\infty} A_n \cos n\omega_1 t + \sum_{n=1}^{\infty} B_n \sin n\omega_1 t \tag{2-1}$$

we may write the exponential form

$$\begin{aligned} f &= \cdots + a_{-2} \exp(-j2\omega_1 t) + a_{-1} \exp(-j\omega_1 t) + a_0 + a_1 \exp(j\omega_1 t) + \\ &\quad a_2 \exp(j2\omega_1 t) + \cdots \\ &= \sum_{m=-\infty}^{\infty} a_m \exp(jm\omega_1 t) \end{aligned} \tag{2-2}$$

where m is any integer. Although these two forms look rather different, that they are in fact the same may be shown by using the identities

$$\cos x = (e^{jx} + e^{-jx})/2 \tag{2-3}$$

and

$$\sin x = -j(e^{jx} + e^{-jx})/2 \qquad (2\text{-}4)$$

Substitution into the first equation followed by rearrangement of terms yields the exponential form without difficulty. The coefficients of the two forms are related as follows

$$a_0 = A_0; a_m = (A_m - jB_m)/2, \text{when } m \text{ is positive} \qquad (2\text{-}5)$$

and

$$a_m = (A_m + jB_m)/2, \text{when } m \text{ is negative} \qquad (2\text{-}6)$$

These results show that the coefficients of the exponential series are in general complex, and that they occur in conjugate pairs (that is, the imaginary part of a coefficient a_n is equal but opposite in sign to that of coefficient a_{-n}). Although the introduction of complex coefficients is at first difficult to understand, it should be remembered that the real part of a pair of coefficients denotes the magnitude of the cosine wave of the relevant frequency, and that the imaginary part denotes the magnitude of the sine wave. If a particular pair of coefficients a_n and a_{-n} are real, then the component at the frequency $n\omega_1$ is simply a cosine; if a_n and a_{-n} are purely imaginary, the component is just a sine; and if, as is the general case, a_n and a_{-n} are complex, both a cosine and a sine term are present.

2.3 The Fourier Transform

2.3.1 Derivation

The Fourier transform (also called the Fourier integral) does for the non-repetitive signal waveform while the Fourier series does for the repetitive one. We have just seen how the line spectrum of a recurrent pulse waveform is modified as the pulse duration decreases, assuming the period of the waveform (and hence its fundamental component) remains uncharged.[1] Suppose now that the duration of the pulses remains fixed but the separation between them increases, giving rise to an increasing period. In the limit, we will be left with a single rectangular pulse, its neighbours having moved away on either side towards infinite. In this case the fundamental frequency tends towards zero and the harmonics become extremely closely spaced and of vanishingly small amplitudes. Once again, we are left with a continuous spectrum.

Mathematically, this situation may be expressed by modifications to the exponential form of the Fourier series already derived

$$f(t) = \sum_{m=-\infty}^{\infty} a_m \exp(-jm\omega_1 t) \qquad (2\text{-}7)$$

where the complex coefficients a_m are found by $f(t)$ multiplying by $\exp(jm\omega_1 t)$ and taking the average of the result over a complete period

$$\begin{aligned} a_m &= \frac{1}{2\pi}\int_{-\pi}^{\pi} f(x)\exp(-jmx)\,dx \\ &= \frac{1}{T}\int_{-\pi/2}^{\pi/2} f(t)\exp(-jm\omega_1 t)\,dt \end{aligned} \qquad (2\text{-}8)$$

In the new situation where we let the tend of T to infinity, each individual coefficient becomes vanishingly small, and it might seem that the above formular is no longer useful. However the product $a_m \cdot T$ does not vanish as $T \to \infty$, so we now choose to write this as a new variable G. Furthermore, as $T \to \infty$, $\omega_1 \to 0$ and the term tends to a continuous rather than a discrete variable which we will denote by ω. Since the variable G is a function of this continuous frequency variable ω, we now rewrite the second of the above equations as

$$G(\omega) = \int_{-\infty}^{\infty} f(t) e^{-j\omega t} dt \qquad (2\text{-}9)$$

Returning to the first equation which expresses as a sum of an infinite set of harmonic components, we now have

$$f(t) = \sum_{m=-\infty}^{\infty} \frac{G(\omega)}{T} \exp(jm\omega_1 t)$$
$$= \sum_{m=-\infty}^{\infty} G(\omega) \frac{\omega_1}{2\pi} \exp(jm\omega_1 t) \qquad (2\text{-}10)$$

Once again the term is replaced by the continuous variable ω, and the fundamental frequency ω_1 (which is now vanishingly small) is written as $d\omega$. Our summation becomes in the limit an integration and the equation is thus rewritten as

$$f(t) = \frac{1}{2\pi} \int_{-\infty}^{\infty} G(\omega) e^{j\omega t} d\omega \qquad (2\text{-}11)$$

These two derived equations, which show how a non-repetitive time-domain waveform $f(t)$ is related to its continuous spectrum, are known as the Fourier integral equations.

It is very important to grasp the significance of these two equations. The first tell us how the energy of the waveform is continuously distributed in the frequency range between $\omega = \pm \infty$, whereas the second shows how, in effect, the waveform may be synthesized an infinite set of exponential functions the form $e^{-j\omega t}$, each weighted by the relevant value of $G(\omega)^2$. [2]

It is perhaps worth exploring a little further the physical meaning of the continuous frequency variable $G(\omega)$. It is indeed hard to visualize a wave such as an isolated pulse being composed of an infinite set of waves of infinitely small amplitudes, or the energy of such a waveform being continuously distributed in the frequency domain. The task is perhaps made easier by referring to the more familiar situations as follows: there are two beams, one of them is a simply-supported beam loaded at a number of distinct points, whereas the other is continuously loaded along its length by, say, gravel or concrete. [3] In the first case it is easy to say that the loading is applied only at discrete points, just as a repetitive signal waveform contains only certain discrete frequencies. [4] However, if one asked what the load on the continuously loaded beam is at a point, the answer must be that at that point (or any other) the applied load is vanishingly small. The sensible approach is to ask how much the average loading is over a small distance, and to give the answer in kilograms per meter. In the same way a continuous frequency spectrum implies that the component at any point-frequency is vanishingly small and that it is only sensible to ask about the energy contained in a small band of frequencies centered around that point. Therefore the variable $G(\omega)$ is best thought of as a frequency density function.

2.3.2 Examples of Continuous Spectra

To illustrate the use of the Fourier integral, we now formally evaluate the spectrum of the isolated pulse waveform of Figure 2.1. The limits of integration are clearly reduced so that

$$G(\omega) = \int_{-\tau}^{\tau} 1 \times e^{-j\omega t} dt = -\frac{1}{j\omega}[e^{-j\omega t}]_{-\tau}^{\tau} = 2\tau\left(\frac{\sin \omega\tau}{\omega\tau}\right) \qquad (2\text{-}12)$$

As would be expected from our earlier discussion, this function is of $(\sin x)/x$ form and is illustrated in Figure 2.1(b). It passed through zero whenever $\sin\omega\tau=0$, which occurs when ω is an integer multiple of π/τ radians/second. It may seem strange that the pulse contains no energy at such frequencies, but it is not hard to demonstrate. Consider, for example, the frequency $\omega = \pi/\tau$, or $f = 1/2\tau$. If we wish to find out how much of such a frequency is contained in the pulse, the rule is to multiply the pulse by the appropriate sinusoidal waveform and to integrate over the interval of 2τ limit. It is clear that the result must be zero because the integral of the sinusoidal waveform over any interval of 2τ is always zero.

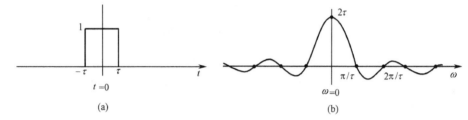

Figure 2.1 (a) A signal isolated pulse and (b) its frequency spectrum

The spectrum of this pulse waveform illustrates a number of important points about time-limited signals. If the pulse duration 2τ is very large, its spectral energy is centered around $\omega = 0$, and in the limit the $(\sin x)/x$ function becomes just a line at zero frequency; the pulse waveform has become, in other words, just a steady level of infinite duration.[5] Conversely, if the pulse is made extremely short, higher frequencies are increasingly represented in its spectrum, and in the limiting case of an infinitely narrow pulse the spectrum becomes flat and extends throughout the frequency band. These results are shown in Figure 2.2, and underline the important principle that a waveform which is very time-limited will occupy a wide band of frequencies, and vice versa. Finally these results suggest what to expect when a nominally continuous or repetitive signal is switched on or off. The above pulse waveforms may indeed be regarded as steady 'signals' which are turned on and off. The sudden changes which occur at the moments of switching introduce, in effect, new frequencies, which cause the single line spectrum of a steady level to be broadened. It is important to appreciate these properties of a time-limited signal, whether it represents an electrical waveform in a communications channel or data recorded or observed over a finite interval.

Having defined the Fourier transform equations it would of course be possible to evaluate the spectrum of various non-repetitive signals. The difficulty of the task depends upon the form of the integral to be evaluated and therefore varies greatly according to the signal waveform chosen. Here we must content ourselves with further example which aids an understanding of continuous spectra, the exponentially decaying sinusoidal waveform of Figure 2.3. Analytically, this waveform is described by $f(t) = e^{-\alpha t}\sin\omega_0 t, t \geqslant 0$.

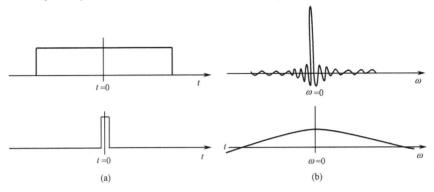

Figure 2.2 (a) Two isolated pulse, one of long and the other of short duration, and (b) their corresponding spectra. Only a small part of the $(\sin x)/x$ spectrum of the short pulse is shown

Its spectrum is therefore given by

$$g(\omega) = \int_{-\infty}^{\infty} f(t)e^{-j\omega t}\,dt = \int_{0}^{\infty} e^{-\alpha t}\sin\omega_0 t\, e^{j\omega t}\,dt \qquad (2\text{-}13)$$

The integration is quite straightforward if $\sin\omega_0 t$ is replace by

$$\sin\omega_0 t = \frac{1}{2j}[\exp(j\omega_0 t) - \exp(-j\omega_0 t)] \qquad (2\text{-}14)$$

and yields

$$G(\omega) = \frac{\omega_0}{\alpha^2 + \omega_0^2 - \omega^2 + j2\alpha\omega} \qquad (2\text{-}15)$$

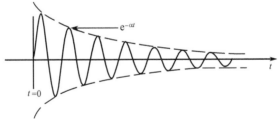

Figure 2.3 The exponentially decaying sinusoidal waveform

The first thing to notice about this that $G(\omega)$ is complex, whereas in the case of the isolated pulse of Figure 2.1 it is purely real. The reason for this difference is that the pulse waveform is drawn symmetrical about $t = 0$, so that its spectrum contains only cosine components. The decaying sinusoidal wave, however, is far from symmetrical about $t = 0$ and could therefore only be synthesized from both sines and cosines, and this fact is reflected by its complex spectrum.

2.3.3 Symmetry of the Fourier Integral Equations

These is an obvious symmetry in the two Fourier integral equations

$$G(\omega) = \int_{-\infty}^{\infty} f(t) e^{-j\omega t} dt \qquad (2\text{-}16)$$

$$f(t) = \frac{1}{2\pi} \int_{-\infty}^{\infty} G(\omega) e^{j\omega t} d\omega \qquad (2\text{-}17)$$

Indeed, apart from the $(1/2\pi)$ multiplier in the second equation and the change of sign in the exponential index, the equations are identical in form. The symmetry between time and frequency domains becomes perfect if we consider an even time function such as the isolated pulse waveform already discussed, which has an even spectrum containing only cosines. In such a case, if t' is substituted for $(-t)$ the first equation becomes

$$G(\omega) = \int_{-\infty}^{\infty} f(-t') e^{j\omega t'} (-dt') = -\int_{-\infty}^{\infty} f(-t') e^{j\omega t'} dt' \qquad (2\text{-}18)$$

But if $f(t)$ is an even function $f(-t') = f(t')$, and therefore

$$G(\omega) = \int_{-\infty}^{\infty} f(t') e^{j\omega t'} dt' \qquad (2\text{-}19)$$

which is now identical in form to the second equation apart from the $(1/2\pi)$ multiplier. The corollary is that, since the spectrum of the square pulse waveform is of $(\sin x)/x$ form, a time function with a shape has spectral energy evenly distributed in a certain range $\pm \omega'$, and none outside it. This symmetry between time and frequency domains is illustrated in Figure 2.4, and accounts for the common description of the Fourier integral equations as the 'Fourier transform pair'.

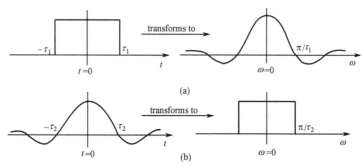

Figure 2.4 Symmetry between time and frequency domains (a) just as a rectangular pulse waveform has a $(\sin x)/x$ spectrum, so (b) a time function of $(\sin x)/x$ from has a rectangular spectrum, with energy equally distributed in the range $-\pi/\tau_2 < \omega < \pi/\tau_2$

2.3.4 Limitations of the Fourier transform

So far it has been implied that the Fourier transform may be successfully used for any norepetitive signal, but there are in fact limitations and difficulties in its applications. An important limitation may be inferred from the equation

$$G(\omega) = \int_{-\infty}^{\infty} f(t) e^{-j\omega t} dt \qquad (2\text{-}20)$$

The Fourier transform will clearly only exist if the right-hand side of the equation is finite. Furthermore, since the magnitude of $e^{-j\omega t}$ is always unity it is sufficient for the transform to exist if

$$\int_{-\infty}^{\infty} |f(t)| \, dt < \infty \tag{2-21}$$

Various waveforms of practical interest, such as continuous sine and cosine function and the so-called unit step function illustrated in Figure 2.5 do not meet this latter condition, and strictly speaking do not therefore possess Fourier transform.[6] Fortunately, however, it is possible to consider them as limiting cases of waveforms which do possess transform.[7] For example, we may evaluate the Fourier transform of sine or cosine wave which exists only in the interval $-T < t < T$; we then allow T to become very large and in the limit approach the transform of the continuous function itself. In the case of the unit step function, it is convenient first to evaluate the Fourier transform of the exponentially decaying step shown in Figure 2.5, and then to allow T to become very small; in the limit, we are left with the spectrum of the unite step function of Figure 2.5(a).[8] For the exponentially decaying step, we have

$$G(\omega) = \int_0^{\infty} e^{\alpha t} e^{-j\omega t} \, dt = \frac{1}{-(\alpha + j\omega)} [e^{-\alpha t} e^{-\omega t}]_0^{\infty} \tag{2-22}$$

When $t=\infty$, the term $e^{-\alpha t}$ is zero if $\alpha > 0$, and therefore

$$G(\omega) = \frac{1}{-(\alpha + j\omega)}(0-1) = \frac{1}{\alpha + j\omega}, \text{ for } \alpha > 0 \tag{2-23}$$

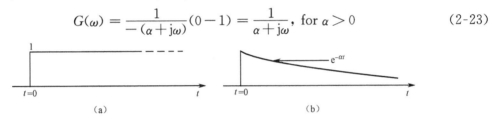

Figure 2.5 (a) The unit step function, and (b) an expotentially decaying step function

Before deriving the transform of the step function itself by letting tend to zero, it is appropriate to mention the problem of singularities. It quite often occurs that the value of a function is undefined at a certain point; such a point is referred to as a singularity. If we let $\alpha \to 0$ in the above example, the value of $G(\omega) \to \infty$ at $\omega = 0$ and there is therefore a singularity at $\omega = 0$. The step function waveform itself is not precisely defined at $t=0$, since there is a discontinuity at this point.

Singularities therefore occur quite often in the context of signal theory. From a mathematical point of view, the presence of such singularities may or may not give rise to special difficulties, but in any case caution should be exercised, particularly when considering the behaviour of the function in the region of a singularity.

Returning to the question of the Fourier transform of the unit step function, we now let $\alpha \to 0$ in the above expression. At first sight $G(\omega)$ would appear to tend to $(1/j\omega)$, but special care must be taken because of the singularity at $\omega = 0$. Suppose we make α small but not zero. Over the range of frequencies for which $\omega \gg \alpha$, $G(\omega) \approx (1/j\omega)$; but when ω is virtually zero so that $\omega \ll \alpha$, we have $G(\omega) \approx (1/\alpha)$. In the limit as α becomes smaller and smaller, we are left with a continuous spectrum equal to $(1/j\omega)$, together with a very large spike of amplitude $1/\alpha$ centered on $\omega=0$. The latter is really a spectral line representing the 'DC' or average value of the step waveform. Thus the unit step function has a spectrum consisting of an infinite set of sinusoidal components of amplitude inversely proportional to frequency, together with a zero-frequency component representing the average value of the waveform.

In effect, we have derived the spectrum of the unit step function by multiplying it by a factor $e^{-\alpha t}$, which makes the Fourier integral convergent. This useful technique may be applied to other waveforms, although such a 'convergence factor' will not work for a signal waveform which extends throughout all time, because the value of $e^{-\alpha t}$, with α positive, increase without limit as t becomes more and more negative. Therefore the method may only be successfully applied to a waveform which is switched on at some definite instant and may be considered zero beforehand; normally the instant at which the signal first assumes a nonzero value is designated $t=0$. [9] If this convergence factor is applied to the general case of a signal $f(t)$ which is zero before $t=0$, we may define a modified version of the Fourier transform.

$$G_1(\omega) = \int_0^\infty f(t) e^{-\alpha t} e^{-j\omega t} dt = \int_0^\infty f(t) e^{-(\alpha+j\omega)t} dt \qquad (2-24)$$

Strictly speaking, of course, G_1 is a function not only of ω, but also of the factor α, which must be chosen so that the integral converges in any particular case. For convenience we introduce the new variable $s=(\alpha+j\omega)$, and we therefore write G_1 as a function of s

$$G_1(s) = \int_0^\infty f(t) e^{-st} dt \qquad (2-25)$$

In doing so, we have defined the Laplace transform of the signal $f(t)$.

2.4 The Laplace Transform

2.4.1 Relationship with the Fourier Transform

The Laplace and Fourier transform are closely related. As we have seen, the Fourier transform allows a signal to be expressed as sum of sinusoidal and cosinusoidal component being represented by a pair of imaginary exponential terms of the form $e^{j\omega t}$. By introducing a so-called convergence factor it is possible to derive the frequency spectra of certain signals for which the Fourier integral may not otherwise be evaluated. So far we have thought of this convergence factor $e^{-\alpha t}$ as being applied as a multiplier to an awkward signal, so that the integral may be evaluated, and we then let $e^{-\alpha t}$ tend to zero. But just as the form of the Fourier integral implies that is being analyzed as an infinite set of exponential terms of the form $e^{j\omega t}$, so the form of the Laplace transform equation suggests that we should think of the Laplace transform as representing $f(t)$ by an infinite set of term of the form $e^{-\alpha t}$, where s is in general a complex number known as the 'complex frequency'. Such terms give rise to not only the sine and cosine waves of the Fourier method, but also growing and decaying sine and cosine waves and growing and decaying exponentials.

Since the Lasplace transform in effect analyses a signal into both oscillatory and non-oscillatory functions which expand and contract with time, it allows the signal $f(t)$ to be less restricted than in the case of the Fourier integral. On the other hand, there are still restrictions because the real part α of the variable must always be sufficient to provide convergence. If for example $f(t)$ contains a growing exponential component e^{7t}, the integral will only converge if the term $e^{-\alpha t}$ at least counteracts its growth, which means that α must be at least 7. But a term in $f(t)$ equal to, for example, $\exp(t^2)$ must dominate $e^{-\alpha t}$ at sufficiently large values of t regardless of the value α, and hence the Laplace transform could not be used. The other restriction, already mentioned, is that the Laplace transform cannot cope with signal which extend into the infinite past, and the integral is only evaluated in the interval $0 < t < \infty$. The reason for this is that values of α which ensure convergence in the interval $0 < t < \infty$ do not also give it in the interval $-\infty < t < 0$.

2.4.2 Use of the Laplace Transform

As a first example, consider the decaying exponential waveform shown in Figure 2.6. Its Laplace transform is given by

$$G(s) = \int_0^\infty f(t) e^{-st} dt = \int e^{-(s+\alpha)t} dt = \frac{1}{s+\alpha} [e^{-(s+\alpha)t}]_0^\infty = \frac{1}{(s+\alpha)} \qquad (2\text{-}26)$$

For our second case we consider the sinusoidal wave of Figure 2.6, for which the Laplace transform is

$$G(s) = \int_0^\infty \sin \omega_0 t \exp(-st) dt = \frac{\omega_0}{(s+j\omega_0)(s-j\omega_0)} \qquad (2\text{-}27)$$

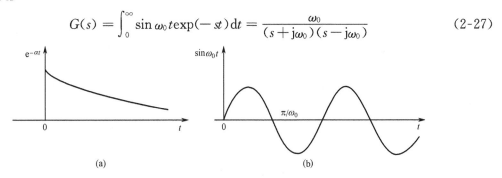

Figure 2.6 The decaying exponential waveform. (a) an exponential decaying function, and (b) the sinusoidal function

It is not easy to appreciate these results, because although one can fairly easily imagine a signal being synthesized from a number of continuous sinusoidal wave, it is more difficult to consider the range of possible wave shapes suggested by Figure 2.6. Of course, the variable $s = (\alpha + j\omega)$ can be real, imaginary, or complex, and if we make it imaginary ($s = j\omega$) we are in effect considering the signal to be composed of sinusoidal waves, as in the Fourier transform. In the case of the decaying exponential wave of Figure 2.6, putting $s = j\omega$ yields the spectrum

$$G(\omega) = 1/(\alpha + j\omega) \qquad (2\text{-}28)$$

the magnitude and phase of which are plotted in Figure 2.7. This result shows the relative amplitudes and phases of the sinusoidal waves needed to synthesize the exponential curve. In the region $-\infty < t < 0$, they all cancel to produce zero resultant, but in the region $0 < t < \infty$ they add together to give the required waveform. If we put $s = j\omega$ in order to investigate the spectrum of the waveform of Figure 2.6 we get the follow expression

$$G(j\omega) = \frac{\omega_0}{(j\omega + j\omega_0)(j\omega - j\omega_0)} \tag{2-29}$$

In this case $G(j\omega)$ displays singularities which imply that the magnitude of the frequency spectrum is infinite at $\omega = \pm \omega_0$, whereas we have in fact chosen a sinusoidal wave of unit amplitude as our time function. This difficulty arises because by putting $s = j\omega$ we are effectively trying to evaluate the Fourier transform of a wave which continues for ever, for which the integral is not convergent. In such cases the results of substituting for s in a Laplace transform expression are not easily interpreted.

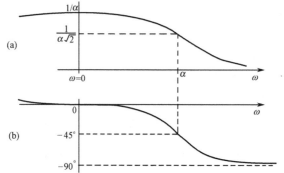

Figure 2.7 (a) Magnitude function, and (b) phase function of the spectrum of the waveform of Figure 2.6

So far we have shown how to derive the Laplace transform of a time function, expressing it as a function of the complex frequency variable s, but no mention has been made of the opposite process of deriving the time function corresponding to a given function of s. One of the main difficulties of the Laplace transform method is that this reverse process of 'inverse transformation' is not a straightforward mathematical operation. Formally, the inverse transform is defined as

$$f(t) = \frac{1}{2\pi j} \int_{\sigma - j\omega}^{\sigma + j\omega} G(s) e^{st} ds \tag{2-30}$$

and while the general form of this equation is similar to the corresponding Fourier integral equation, the limits of $(\sigma \pm j\omega)$ give rise to considerable difficulty in practice. The evaluation if its integral requires familiarity with functions of complex variables and the calculus of residues. The approach adopted here is the practical one commonly used in texts which deal with the Laplace transform, namely to provide a table of some of the more common transform pairs for ready reference.

Even an extensive table of Laplace transform can hardly hope to cover all the functions likely to be encountered in practice, and while it is beyond the scope of this book to investigate the various techniques of inverse transformations, one simple procedure which is quite often useful will now be described. It relies upon the fact that the inverse Laplace transform of the sum of number of function is equal to the sum of their inverse transform, and is called the method of partial fractions. In a great many cases, it is found that a function consists of products of simple linear factors, and the method of partial fractions allows us to express such a function as the sum of a number simpler function, each of which appears in our table of transform.

It might be objected that the Laplace transform is hardly worth the trouble, if its ability to cope with a few signal waveform not amenable to the Fourier transform is offset by complex frequencies. It should however be stressed that this elementary introduction to the Laplace transform does little to suggest the full value of the method. This will become clearer later, suffice it to say at this stage that the Laplace transform is a powerful mathematical tool which may be used to solve a great many problems other than those of signal analysis. In particular a problem stated as a set of differential equations may often to be reduced to a set of much simpler algebraic equations if the Laplace transform is used. Since the dynamics of certain types of mechanical and electrical systems may be expressed mathematical in differential equation form, the Laplace transform method can be of great value in exploring their performance.

2.4.3 The Pole-zero Description of a Signal

Consider once again the function

$$G(s) = 1/(s+\alpha) \tag{2-31}$$

Which, as we have already seen in the previous section, represents the Laplace transform of a decaying exponential signal waveform. It is general property of such functions that there are values of the complex frequency variable s which make $G(s)$ tend to infinity and values of s which make $G(s)$ zero; such values are known as the 'pole' and 'zeros' of $G(s)$ respectively, and they give rise to a convenient graphical description of the corresponding signal. In the simple case quoted above, it is clear that $G(s) \to \infty$ if $s = -\alpha$, so the latter value is a pole of $G(s)$; there are no zeros. If we consider a more complicated function of the general form

$$G(s) = \frac{A(s-z_1)(s-z_2)(s-z_3)\cdots}{(s-p_1)(s-p_2)(s-p_3)\cdots} \tag{2-32}$$

where A is a constant, then the values $s = z_1, z_2, z_3$, etc., represent zeros of $G(s)$ and $s = p_1, p_2, p_3$, etc., represent poles. It is indeed always possible to write a Laplace transform in the above way, and therefore to define the signal in terms of a set of poles and zeros. It is found that poles are either real, or they occur in complex conjugate pairs, and the same is true of zeros. Thus a pole at $s = \alpha + j\omega$ is matched by one at $s = \alpha - j\omega$, and this must always be the case when the corresponding signal is a real function of time.

It is common practice to represent the poles and zeros of $G(s)$ graphically by drawing their positions on a so-called Argand diagram (complex plane). In such a diagram the real part of a complex variable is plotted along the abscissa, and the imaginary part along the ordinate. The poles and zeros of function $G(s)$ are in general complex values of s, so the complex plane gives a convenient method of displaying them, in which case it is widely referred to as the 's-plane' diagram. Suppose, for example, we have a time function $f(t)$, the Laplace transform of which is

$$G(s) = \frac{4s(s-2)}{(s+1+j3)(s+1-j3)} \tag{2-33}$$

Apart form the constant multiplier of 4, we may completely represent the function $G(s)$ by drawing zeros at $s=0$ and $s=2$, and poles at $s=-1\pm j3$, in the complex plane.

Words and Expressions

appreciate [ə'priːʃieit] *vt.* 正确评价，估价
Argand diagram (complex plane)　阿根图（复平面）
abscissa [æb'sisə] *n.* 横坐标
awkward ['ɔːkwəd] *adj.* 笨拙的
amenable [ə'miːnəbl] *adj.* 应服从的
algebraic [ældʒi'breiik] *adj.* 代数的
convergent [kən'vəːdʒənt] *adj.* 收敛的
counteract [ˌkauntə'rækt] *vt.* 抵消，中和
conjugate ['kɔndʒugeit] *adj.* 共轭的
corollary [kə'rɔləri] *n.* 推论
decay [di'kei] *n.* ; *vi.* 衰减
derivation [ˌderi'veiʃən] *n.* 引出，来历，出处
duration [djuə'reiʃən] *n.* 持续时间
denote [di'nəut] *vt.* 指示，表示
exponential [ˌekspəu'nenʃəl] *adj.* 指数的
integral ['intigrəl] *n.* 积分; *adj.* 积分的
multiplier ['mʌltiplaiə] *n.* 乘数，乘法器
negative ['negətiv] *adj.* 负的，阴的
oscillatory ['ɔsilətəri] *adj.* 振荡的
ordinate ['ɔːdinit] *n.* 纵坐标
positive ['pɔzətiv] *adj.* 正的，阳的
partial fraction　部分分式
spike [spaik] *n.* 长钉，道钉
symmetrical [si'metrikəl] *adj.* 对称的
significance [sig'nifikəns] *n.* 有效位，意义
synthesize ['sinθisaiz] *vt.* 综合，合成
singularity [ˌsiŋgju'lærəti] *n.* 奇异点
trigonometric [ˌtrigənə'metrik] *adj.* 三角法的

Notes

1. We have just seen how the line spectrum of a recurrent pulse waveform is modified as the pulse duration decreases, assuming the period of the waveform (and hence its fundamental component) remains uncharged.

假定信号波形周期（即它的基频）不变，当脉冲宽度减小时，我们可以看到周期脉冲波形线谱的变化。

2. The first tell us how the energy of the $f(t)$ waveform is continuously distributed in the frequency range between $\omega=\pm\infty$, whereas the second shows how, in effect, the waveform may be synthesized an infinite set of exponential functions the form $e^{-j\omega t}$, each weighted by the relevant value of $G(\omega)^2$.

第一个方程告诉我们 $f(t)$ 波形的能量是在频率范围$(-\infty, \infty)$上连续分布的。第二个方程说明，实际上该波形可以由 $G(\omega)^2$ 的相关值与形如 $e^{-j\omega t}$ 的无限指数函数集加权合成。

3. There are two beams, one of them is a simply-supported beam loaded at a number of distinct points, whereas the other is continuously loaded along its length by, say, gravel or concrete.

有两根梁，其中一根梁仅在几个点上有载荷，而另一根梁是在全长上均匀地加载，如石头或混凝土。

4. In the first case it is easy to say that the loading is applied only at discrete points, just as a repetitive signal waveform contains only certain discrete frequencies.

简言之，第一种情况是在离散点上加载，正如仅包含离散频率的周期信号波形。

5. If the pulse duration 2τ is very large, its spectral energy is concentrate around $\omega=0$, and in the limit the $(\sin x)/x$ function becomes just a line at zero frequency; the pulse waveform has become, in other words, just a steady level of infinite duration.

如果脉冲宽度 2τ 非常大，它的频谱能量分布在以 $\omega=0$ 为中心的区间上，在极限情况下 $(\sin x)/x$ 函数变成了零频率处的一条直线。换言之，脉冲波形已经变成了无限宽的固定电平。

6. Various waveform of practical interest, such as continuous sine and cosine function and the so-called unit step function illustrated in Figure 2.5 do not meet this latter condition, and strictly speaking do not therefore possess Fourier transform.

实际的许多波形，如图 2.5 所示的连续正弦波和余弦波，以及所谓的单位阶跃函数，都不满足后一个条件，严格地说都不能进行傅里叶变换。

7. Fortunately, however, it is possible to consider them as limiting cases of waveforms which do possess transform.

然而，幸运的是，可以把它们作为进行傅里叶变换的有限的几个波形。

8. In the case of the unit step function, it is convenient first to evaluate the Fourier transform of the exponentially decaying step shown in Figure 2.5, and then to allow σ to become very small; in the limit, we left with the spectrum of the unite step function of Figure 2.5a.

在研究单位阶跃函数时，首先考察如图 2.5 所示的指数衰减阶跃信号，然后令 T 变得非常小，在极限情况下，我们得到了如图 2.5（a）所示的单位阶跃函数频谱。

9. Therefore the method may only be successfully applied to a waveform which is switched on at some definite instant and may be considered zero beforehand; normally

the instant at which the signal first assumes a nonzero value is designated $t=0$.
所以，这种方法只能成功地用在某确定时刻有非零值而在此之前可视为零的波形；通常，将信号最先出现非零值的时刻规定为 $t=0$。

Exercises

Translate the following passages into English or Chinese.

1. 信号不是周期性出现的，而只是持续一段时间，不再重复出现，如过渡过程、爆炸产生的冲击波、起落架着陆时的信号等，把这一类信号看成非周期信号。分析非周期信号的思路是：在时域上，当周期 $T_1 \to \infty$ 时，周期信号变为非周期信号；在频域上，周期信号的频谱在 $T_1 \to \infty$ 时的极限，变为非周期信号的频谱，即傅里叶变换。

2. 信号在时域压缩 α 倍（$\alpha>0$）时，则在频域中频带加宽，幅值压缩 $1/\alpha$ 倍；反之信号在时域扩展时（$\alpha<1$），在频域中将引起频带变窄，但幅值增高。

3. It should be clear that the spectrum of a voltage (or current) waveform is obtained by a mathematical calculation and that it does not appear physically in an actual circuit. f is just a parameter (called frequency) that determines which point of the spectral function is to be evaluated.

4. One of particular interest is the consequence of working with real waveforms. In any physical circuit that can be built, the voltage (or current) waveforms are real functions (as opposed to complex functions) of time.

Reading Material

Properties of Signal and Noise

In communication system, the received waveform is usually categorized into the desired part containing the information, and the extraneous or undesired part. The desired part is called the signal and the undesired part is called noise.

The waveforms will be represented by direct mathematical expressions or by the use of orthogonal series representations such as the Fourier series. The waveform of interest may be the voltage as a function of time, $v(t)$, or the current as a function of time, $i(t)$. Often the same mathematical techniques can be used when working with either type of waveform. For generality, waveforms will be denoted simply as $w(t)$ when the analysis applied to either case.

Physically Realizable Waveforms

Practical waveform that is physically realizable (i. e. , measurable in a laboratory) satisfy several conditions:

(1) the waveform has significant nonzero values over a composite time interval that is finite.

(2) The spectrum of the waveform has significant values over a composite frequency interval that is finite.

(3) The waveform is a continuous function of time.

(4) The waveform has a finite peak value.

(5) The waveform has only real values. That is, at any time, it cannot have a complex value $a+jb$ where b is nonzero.

The first condition is necessary because systems (and their waveforms) appear to exist for a finite amount of time. Physically signals also produce only a finite amount of energy. The second condition is necessary because any transmission medium such as wires, coaxial cable, waveguides, or fiber optic cable has a restricted bandwidth. The third condition is a consequence of the second. The forth condition is necessary because physical devices are destroyed if voltage or current of infinite value is present within the device. The fifth condition follows from the fact that only real waveforms can be observed in the real world, although properties of waveforms, such as spectra, may be complex.

Mathematical models that violate some or all of conditions listed previously are often used, and for one main reason to simplify the mathematical analysis. In fact, we often have to use a model that violates some of these conditions in order to calculate any type of answer. However, if we are careful with the mathematical model, the correct result can be obtained when the answer is properly interpreted. For example, consider the digital waveform shown in Figure 2. 8. The mathematical model waveform has discontinuities at the switching times. This situation violates the third condition the physical waveform is continuous. The physical waveform is of finite duration (decays to zero before $t=\pm\infty$), but the duration of the mathematical waveform extends to infinity.

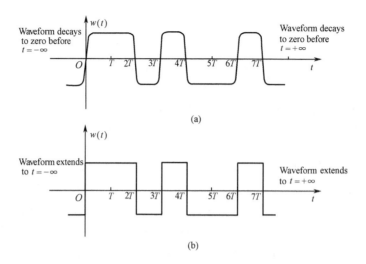

Figure 2.8 (a) A digital signal's physical waveform and (b) math model waveform

In other words, this mathematical model assumes that the physical waveform existed in its steady-state condition for all time. Spectral analysis of the model will approximate the correct results except for the extremely high frequency components. The average power that is calculated from the model will give the correct value for the average power of the physical signal that is measured over an appropriate time interval. The total energy of the mathematical model's signal will be infinity because it extends to infinite time, whereas that of the physical signal will be finite. Consequently, this model will not give the correct value for the total energy of the physical signal without using some additional information. However, the model can be used to evaluate the energy of the physical over some finite time interval of the physical signal. This mathematical model is said to be a power signal because it has the property of finite power (and infinite energy), whereas the physical waveform is said to be an energy signal because it has finite energy. All physical signals are energy signals, although we generally use power signal mathematical models to simplify the analysis.

In summary, waveforms may often be classified as signals or noise, digital or analog, deterministic or nondeterministic, physically realizable or nonphysically realizable, and belonging to the power or energy type.

Lesson 3 Sampled-Data Signals

3.1 Introduction

Sampled-data signals have defined values only at certain instants of time, and arise whenever continuous functions are measured or recorded intermittently. In recent year such signals have come to assure great importance because of developments in digital electronics and computing. Since it is not possible to feed continuous data into a digital computer, any signal data input must be represented as a set of numerical values. In almost every case the numbers represent sampled values of the continuous signal at successive equally-spaced instants.

It should be emphasized from the start that a set of sample values only forms an adequate substitute for the underlying continuous signal waveform if the interval between successive samples is sufficiently small.

3.2 Mathematical Description Using the Dirac Function

The Dirac function, often referred to as the impulse function, is a pulse of extremely short duration and unit area. In other words the product of its duration and its mean height is unity, even though its precise shape is undefined. The physical significance of such a function may be illustrated by an example. Suppose a mechanical impulse is delivered to a golf ball by the head of a golf club. Other things being equal, the momentum imparted to the ball and the distance it travels depend upon the value of the impulse, which is given by the product of the force and the time for which it is exerted. Or, assuming that the force is not constant, it is the area under the force-time graph which determines the impulse, since the Dirac function is defined by its unit area, it may be used to denote a unit mechanical impulse, and we shall use the same function later in this text to describe a sudden electrical disturbance applied to a signal processing device.

The Dirac function is also useful for describing a sampled data signal, which may be considered to consist of a number of equally spaced pulses of extremely short duration. It is convenient in this case to think of all the signal samples as pulses of identical duration, with their heights proportional to the values of the signal at the relevant instants. In practice, this approach proves mathematically sound, provided the pulse duration is negligible compared with the interval between successive samples.

In order to discuss the spectrum of a sampled-data signal, it is necessary to evaluate that of the unit Dirac function. Using the convention that the symbol $\delta(t)$ represents a Dirac pulse occurring at $t=0$, we have

$$G(j\omega) = \int_{-\infty}^{\infty} \delta(t) e^{-j\omega t} dt \qquad (3-1)$$

To evaluate this integral we make use of the so-called 'shifting property' of the Dirac function. Here a function $f(t)$ is multiplied by a unit Dirac function occurring at the instant $t=a$ which is denoted by the symbol $\delta(t-a)$. Since the area of the Dirac pulse is unity, the area under the curve representing the product $[f(t)\delta(t-a)]$ is the value of $f(t)$ at $t=a$. When the product is integrated between $t=-\infty$ and ∞, the result is just this area. Hence the shifting property may be formally stated as follows

$$\int_{-\infty}^{\infty} f(t)\delta(t-a)\,dt = f(a) \tag{3-2}$$

The spectrum $G(j\omega)$ of a Dirac pulse at $t=0$ is therefore simply equal to the value of $e^{-j\omega t}$, at $t=0$, which is unity. Hence

$$G(j\omega) = 1 \tag{3-3}$$

This result tells us that all frequencies are equally represented by cosine components. When a large number of cosines of similar amplitude but different frequencies are added together, they tend to cancel each other out everywhere except at $t=0$, where they all reinforce. Therefore as higher and higher frequencies are include the resultant becomes an extremely narrow pulse centered on $t=0$.

It is also interesting to consider the spectrum of a delayed Dirac pulse which occurs at some instant $t=T$. Clearly, such a pulse must contain the same frequency components as the one occurring at $t=0$, except that each and every one of them is delayed by T seconds. Formally the spectrum is given by

$$G(j\omega) = \int_{-\infty}^{\infty} \delta(t-T)e^{-j\omega t}\,dt = e^{-j\omega T} \tag{3-4}$$

The term $e^{-j\omega t}$ has unit magnitude for any value of ω and phase shift of $-\omega T$ radians. As we would expect, a phase shift $-\omega T$ radians imposed upon a frequency component of radians/second represents a time delay of T seconds. Thus the spectrum $e^{-j\omega T}$ implies that all components are present with equal amplitude and a phase shift proportional to frequency. The same result could be obtained by noting that the Laplace transform of a unit impulse centered on $t=0$ is also unity. Since shifting a time function by T second is equivalent to multiplying its Laplace transform by e^{-st} the transform of a unit impulse occurring at $t=T$ is $1\times e^{-st} = e^{-sT}$. Substitution of $j\omega$ for s gives its frequency spectrum.

3.3 Spectra of Sampled-data Signals

3.3.1 The Discrete Fourier Transform

It is now possible to write down time and frequency-domain expressions for a sampled-data signal. Denoting a sample value x by a unit Dirac pulse weighted by x, the signal is described by

$$f(t) = x_0\delta(t) + x_1\delta(t-T) + x_2\delta(t-2T) + x_3\delta(t-3T) + \cdots \tag{3-5}$$

where T is the sampling interval. Its Laplace transform may be written inspection as

$$G(s) = x_0 + x_1 e^{-sT} + x_2 e^{-s(2T)} + x_3 e^{-s(3T)} + \cdots \tag{3-6}$$

and its Fourier transform as

$$G(j\omega) = x_0 + x_1 e^{-j\omega T} + x_2 e^{-j\omega(2T)} + x_3 e^{-j\omega(3T)} + \cdots \qquad (3\text{-}7)$$

The Fourier transform of a sampled-data signal is generally referred to as a discrete Fourier transform (DFT). There are two reasons for this: firstly, the signal itself is discrete, in the sense of being defined only at discrete instants in time the sampling instant; and second, the common practice of using a digital computer to evaluate the spectrum of a sampled-data signal means that its Fourier transform can only be estimated for a set of discrete values of ω. This is not to imply that $G(j\omega)$ as defined in the above equations is discrete, for it is a continuous function of ω; but with a digital computer we can only estimate it at suitably spaced intervals in ω, and use the values obtained to represent the underlying continuous functions. Fortunately, as we shall show later, such a discrete representation of the spectrum need not involve any loss of essential detail.

Although it is always a simple matter to write down an expression for the spectrum of a sampled-data signal, it is less easy to visualize its form. However, some of the main features of such spectra may be inferred in the following way. Since a sampled-data signal is in effect composed of a set of weighted Dirac pulses each of which contains energy distributed over a very wide band of frequencies, it must be expected that the signal spectrum will be similarly 'wideband'. The second point to notice is that since terms such as $e^{-j\omega T}, e^{-j\omega 2T}, e^{-j\omega 3T}$ are all repetitive in the frequency domain with a period of $2\pi/T$ radians/second, so $G(j\omega)$ also must be the spectrum itself.[1] Finally, it would be reasonable to suppose that the spectrum of sampled version of a waveform would reflect not only the frequency characteristics of the Dirac functions representing the sample pulse, but also components present in the underlying continuous signal.

Figure 3.1 which shows the spectra of two typical continuous signals and those their sampled version, confirms the above arguments. For example, the continuous wave $\cos\omega_0 t$ has two spectral lines at $\omega \pm \omega_0$, its sampled version has a spectrum in which these two spectral lines repeat indefinitely at intervals in ω of $2\pi/T$, where T is the sampling interval. An isolated rectangular pulse has a continuous spectrum of $(\sin x)/x$ form; when sampled, the spectrum repeats itself every $2\pi/T$ radians/second. In both sampled-signal spectra, the low-frequency region between $\omega = -\pi/T$ and $\omega = \pi/T$ represents components present in the underlying continuous signal waveform, the repetitive nature of the spectra being due to the sampling process itself. In fact, there is never any need to evaluate the spectrum of a sampled-data signal at frequencies above $\omega = \pi/T$, since it is bound to be merely a repetition of the lying between $\omega = -\pi/T$ and $\omega = \pi/T$.

3.3.2 The Fast Fourier Transform

In order to find the spectrum of a sampled-data signal, it is necessary to compute values of a function of the form

$$\begin{aligned} G(j\omega) &= x_0 + x_1 e^{-j\omega T} + x_2 e^{-j\omega(2T)} + x_3 e^{-j\omega(3T)} + \cdots \\ &= x_0 + x_1 \cos \omega T + x_2 \cos 2\omega T + x_3 \cos 3\omega T + \cdots \\ &\quad - j(x_1 \cos \omega T + x_2 \cos 2\omega T + x_3 \cos 3\omega T + \cdots) \end{aligned} \qquad (3\text{-}8)$$

The real and imaginary parts of this expression may be evaluated for a suitable set of values of ω, assuming the sampling interval T is known as with continuous functions, an even sampled-data signal has a purely real spectrum and it is therefore only necessary to compute the cosines; conversely, if the time function is odd, it is only necessary to compute the sines.

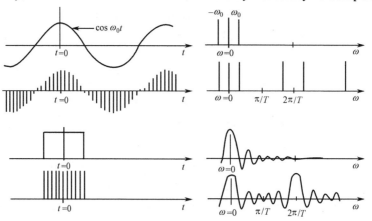

Figure 3.1　Four signals and their spectra. The continuous and sampled version of the cosine and pulse waveforms have similar spectra in the region $-\pi/T<\omega<\pi/T$, but the sampled-data spectra repeats indefinitely at intervals in ω of $2\pi/T$

It is interesting to consider the number of computer operations needed to evaluate the DFT of a signal consisting of N samples. As we have already seen, a signal of this length is defined by $N/2$ harmonics and in the general case each is represented by both a cosine and a sine component. Evaluation of any one term involves multiplying each signal sample value by a factor of the term $\cos n\omega T$ or $\sin n\omega T$ and some N^2 products are therefore involved in calculating the spectrum of an N-sampled signal. Therefore the time taken to calculate the Fourier transform is roughly proportional to the square of the number of sample values.

However, a more careful investigation shows that many of the multiplication operations involved are repeated, and the fast Fourier transform (FFT) aims to eliminate such repetitions as much as possible. This form of the FFT is therefore a recipe, or 'algorithm', which is designed to increase the efficiency of calculation. [2] It may be shown that the computation time for an N-sample transform is approximately proportional to $N\log_2 N$ when the FFT is used, as opposed to N^2 when no special precautions are taken to eliminate redundant calculations. Fast Fourier algorithms become more and more attractive as the number of signal samples increases, and are generally most efficient when this number is an integer power of 2. With signals of this sort of length, reduction of computing time by a factor of 50 or 100 is commonly achieved.

3.4　The z-transform

3.4.1　Introduction

Although it is quite possible to use the Fourier or Laplace transforms to describe the frequency-domain properties of a sampled-data signal, there is another transform which is tailor-made for the purpose. The z-transform provides not only a useful shorthand notation for the Fourier transform of such a signal, but also a very convenient method of defining the signal by a set of poles and zeros.

The z-transform of a sampled-data signal may be simply defined using its Laplace transform as a starting point. In section 3.3.1 it was shown that the Laplace transform of a signal is given by

$$G(s) = x_0 + x_1 e^{-sT} + x_2 e^{-s(2T)} + x_3 e^{-s(3T)} + \cdots \qquad (3\text{-}9)$$

We now define the new variable $z = e^{sT}$ and denote the z-transform of the signal by $G(z)$. Hence

$$G(z) = x_0 + x_1 z^{-1} + x_2 z^{-2} + x_3 z^{-3} + \cdots \qquad (3\text{-}10)$$

z is a new frequency variable which generally has both real and imaginary parts. Thus if

$$z = e^{sT} = e^{\sigma T} e^{j\omega T} = e^{\sigma T} \cos \omega T + j e^{\sigma T} \sin \omega T \qquad (3\text{-}11)$$

We have already seen that the term implies a phase shift proportional to frequency and hence a constant time delay of T seconds. Conversely, the term $e^{j\omega T}$ represents a time advance of T seconds. This accounts for the common description of z as a 'shift operator', denoting a time advance equal to the sampling interval T. [3]

It is clear from the above definition that the z-transform of a sampled signal is a power series in z^{-1} in which the coefficients of the various terms are equal to the sample values; when a transform is expressed in this form, it is possible to regenerate the time function merely by inspection. Suppose, for example, we have the z-transform

$$G(z) = \frac{z}{(z-a)} = \frac{1}{(1-az^{-1})} \qquad (3\text{-}12)$$

We may rewrite this as a power series in z^{-1}

$$G(z) = 1 + az^{-1} + a^2 z^{-2} + a^3 z^{-3} + a^4 z^{-4} + \cdots \qquad (3\text{-}13)$$

3.4.2 z-Plane Poles and Zeros

We have seen that the frequency spectrum of any sampled-data signal is repetitive in form with a period equal to $2\pi/T$ radians/second. Since a signal may be described by a set of s-plane poles and zeros whose positions bear a direct relationship to the signal spectrum, it must be expected that the s-plane pole-zero pattern of a sampled-data signal will also be repetitive in form. It is indeed possible to show that the sampled version of a continuous signal has the same s-plane pole-zero configuration as the original, except that all poles and zeros are repeated indefinitely at intervals of in the direction of the imaginary axis.

One of the advantages of using the s-plane pole-zero description of a continuous signal is that it allows the spectrum to be visualized quite easily, by considering the lengths and phases of vectors drawn from the various poles and zeros to a succession of points on the imaginary axis. In the case of a sampled-data signal, however, this technique becomes almost valueless, since there is an finite set of poles and zeros to be taken into account. This difficulty emphasizes a further advantage of the z-transform, namely that it allows the representation of a sampled-data signal by a finite set of poles and zeros.

In order gain some insight into the relationships between s-plane and z-plane poles and zeros, it is useful to investigate what happens to the complex variable z when s takes on certain typical values. This process is generally referred to as 'mapping' the s-plane into the z-plane.[4] Suppose, for example, s is purely imaginary, then

$$s = j\omega, \quad \text{and} \quad z = e^{j\omega T} = \cos\omega T + \sin\omega T \qquad (3\text{-}14)$$

If we now consider values of z to be plotted on a z-plane Argand diagram, the locus of z will trace out a circle of unit radius as ω varies, starting on the real axis when $\omega=0$ and repeating its trajectory at intervals in ω of $2\pi/T$. Next let $s=\sigma$, where σ is real, giving $z=e^{\sigma}$. If σ is positive number greater than 1; if σ is negative, z is a real positive number less than 1. Suppose finally that $s=\sigma+j\omega$, giving $z=e^{\sigma}e^{j\omega T}$. This value of z is represented by a vector of length e^{σ} which makes an angle ωT radians with the positive real axis. Whenever σ is negative, a point inside the unit circle in the z-plane is specified which corresponds to a point in the left-hand half of the s-plane. Therefore the complete left-hand half of the s-plane maps into the area inside the unit circle in the z-plane.

Actually, this is not quote the whole story. Suppose we have a sampled-data signal described by a single z-plane pole at the point marked B in the right-hand part of Figure 3.2. Its z-transform is then $G(z)=1/(z-r_3)$, and its frequency spectrum is found by substituting $j\omega$ for s, giving

$$G(j\omega) = 1/(e^{j\omega T} - r_3) \qquad (3\text{-}15)$$

The denominator may be represented by a vector drawn from the point B to a point on the unit circle. Every time ω changes by $2\pi/T$ radians/second, the complete unit circle is traced out and the changes in length and phase of this vector -and therefore in the frequency spectrum- are repeated. Hence a single z-plane pole at a point such as B is equivalent to an infinite repetitive set of s-plane poles, adjacent members of the set being separated by an interval of $2\pi/T$ in the direction of the imaginary axis. The s-plane point B in the left-hand half of figure is merely one of this infinite set. This explain why the pole-zero description of a sampled-data signal is only economic if the z-plane is used.

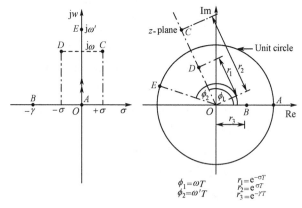

Figure 3.2 Equivalent locations in the s and z planes

The time functions corresponding to a few z-transforms will now be evaluated in order to give some familiarity with the method. Consider first the function

$$G_1(z) = (1-z^{-8}) = (z^8-1)/z^8 \qquad (3\text{-}16)$$

There are eight zeros given by the roots of the equation $z^8=1$, and an eight-order pole at $z=0$. Two of the zeros are clearly $z=1$ and $z=-1$, and the other six are distributed around the unit circle in the z-plane as shown in Figure 3.3. The figure also shows the corresponding time function and its spectrum, the latter having three null points between $\omega=0$ and $\omega=\pi/T$ corresponding to the three zeros labeled A, B, and C. If the zero at $z=1$ is now removed, we have the new z-transform

$$G_2(z) = \frac{1}{z^8}\left[\left(z-\frac{1}{\sqrt{2}}+j\frac{1}{\sqrt{2}}\right)\left(z-\frac{1}{\sqrt{2}}-j\frac{1}{\sqrt{2}}\right)\left(z+\frac{1}{\sqrt{2}}+j\frac{1}{\sqrt{2}}\right)\cdot\right.$$
$$\left.\left(z+\frac{1}{\sqrt{2}}-j\frac{1}{\sqrt{2}}\right)(z+j)(z-j)(z+1)\right] \qquad (3\text{-}17)$$

which multiplies out to give

$$G(z) = z^{-1}+z^{-2}+z^{-3}+z^{-4}+z^{-5}+z^{-6}+z^{-7}+z^{-8} \qquad (3\text{-}18)$$

The corresponding time function consist of a set of eight unit-height samples and is known in Figure 3.4. The spectrum differs form the first example principally be having a finite zero-frequency, or mean term. It is interesting to note that this latter zero pattern may also be achieved by canceling the zero at $z=1$ by a coincident pole. Hence $G_2(z)$ may also be written in the form

$$G_2(z) = (z^8-1)/[z^8(z-1)] \qquad (3\text{-}19)$$

Emphasizing that a given z-transform may often be expressed more neatly as a set of poles and zeros rather than as a set of zeros alone.

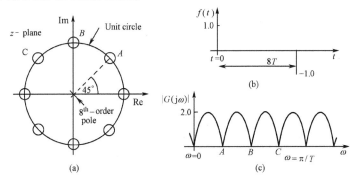

Figure 3.3 (a) A z-plane pole-zero configuration, (b) the corresponding time function, and (c) the magnitude of its spectrum

Any z-transform which is expressible as a finite set of zeros on the unit circle of the z-plane must represent a time function of finite duration which, apart from a mere shift of time origin, is either odd or even. The reason for this becomes clear if it is recalled that a finite set of zeros on the unit circle is equivalent to an infinite repetitive set on the imaginary axis in the s-plane. Any vector drawn from one such s-plane zero to a point on the imaginary axis must always have a phase angle of $\pm\pi/2$ radians and hence the total phase angle due to all such zero vector must always be an integer multiple of $\pi/2$. Thus the spectrum is always either purely real or purely imaginary, representing a time function composed of either cosines or sines. It is interesting to note that the time function of Figure 3.3 is odd, and that of Figure 3.4 is even, apart form a shift in time origin.

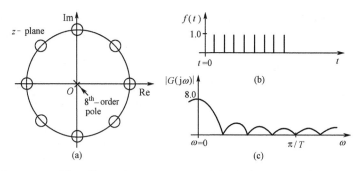

Figure 3.4 The effects of removing one of the z-plane zeros of Figure 3.3

A slight variation on the theme of possible alternative representations of a signal by poles and zeros is provided by the sampled-data signal of Figure 3.5(a) which is a truncated version of the 'infinite' decaying exponential waveform of Figure 3.5(b).[5] Since the truncated signal may be formed by subtracting the waveform of Figure 3.5(c) form from that of Figure 3.5(b), its z-transform is written by inspection as

$$G_3(z) = \frac{z}{(z-a)}(1 - a^k z^{-k}) \tag{3-20}$$

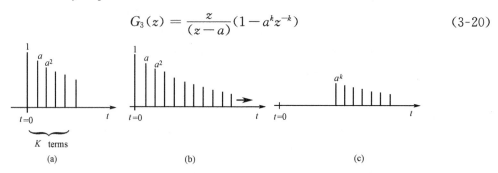

Figure 3.5 Truncated and 'infinite' version of a sampled exponential signal

It therefore has a set of k zeros equally spaced around a circle of radius a, except that the one at $z=a$ is cancelled by a coincident poles, as shown in Figure 3.6 for the case when $k=12$. It is interesting to note that if a and k are chosen so that all terms in the waveform of Figure 3.4 are negligible, then the infinite and truncated version of the signal are effectively the same. And since the infinite version may be represented by a single pole at $z=a$, we conclude that a sufficiently large group of zeros on a circle of radius a is equivalent to a single pole at the same radius.

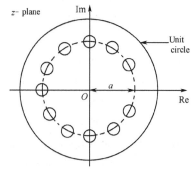

Figure 3.6 z-plane zeros of a sampled exponential signal truncated to 12 terms

Consider next the z-transform

$$G_4(z) = \frac{z(z-1)}{z^2 - 2\alpha z + (\alpha^2 + \beta^2)} \tag{3-21}$$

which has zeros at $z=0$ and $z=1$, and a complex conjugate pole pair at $z=\alpha \pm j\beta$. To take a numerical example suppose that $\alpha=0.81$ and $\beta=0.55$ which gives the pole positions shown in Figure 3.7. These poles are close to the unit circle and therefore represent strong frequency components in the region of $\omega T=\theta$, or $\omega=\theta/T$ radians/second. The zero at $z=1$ implies that the corresponding time function has zero mean, and therefore that all its sample values would sum to zero. It is not so simple to express $G_4(z)$ as a power series in z^{-1} in order to drive the corresponding signal, but the result of doing so is illustrated in Figure 3.7(b). The presence of z-plane poles just in side the unit circle gives rise to a power series with an infinite number of terms which represents a decaying oscillatory function. A single cycle of the dominant component occupies about even sampling intervals and its frequency is therefore

$$f \approx \frac{1}{11T} \text{ hertz, or } \quad \omega = 2\pi f = \frac{2\pi}{11T} = \frac{0.57}{T} \quad \text{radians/second} \tag{3-22}$$

The value is roughly what we would expect since

$$\tan\theta = \frac{\beta}{\alpha} = \frac{0.55}{0.81} \tag{3-23}$$

and hence $\theta=34°$, which is very close to 0.57 radians. The final point to make is that poles or zeros at the z-plane origin correspond to a pure time-shift and do not otherwise affect the form of the time function. Thus the zero at $z=0$ in the above example merely ensures that the first finite sample value occurs at $t=0$; if the zero was not present, it would occur at $t=T$. Conversely, a single pole at the origin would correspond to a first sample value at $t=-T$. Generally speaking, the first nonzero sample occurs at $t=0$ when the highest powers z in the numerator and denominator polynomials of $G(z)$ are equal.

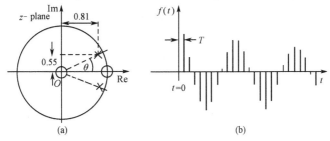

Figure 3.7 (a) A pole-zero configuration which includes a complex conjugate pole pair just inside the unit circle, and (b) the corresponding time function

Words and Expressions

algorithm [ˈælgəriðəm] n. 算法
decaying oscillatory function 衰减振荡函数
harmonic [haːˈmɔnik] n. 谐波
intermittently [ˌintə(ː)ˈmitəntli] adv. 间歇地

locus [ˈləukəs] *n.* 轨迹
mapping [ˈmæpiŋ] *vt.* 映射
momentum [məuˈmentəm] *n.* 要素
negligible [ˈneglidʒəbl] *adj.* 可忽略的
power series   幂级数
reinforce [ˌriːinˈfɔːs] *vt.* 加强
shift operator   移位算子
shifting property   筛选特性
trajectory [ˈtrædʒiktəri] *n.* 轨道
truncate [ˈtrʌŋkeit] *vt.* 截断；*adj.* 截断的
underlying [ˌʌndəˈlaiiŋ] *adj.* 基础的，在下面的

Notes

1. The second point to notice is that since terms such as $e^{-j\omega T}$, $e^{-j\omega(2T)}$, $e^{-j\omega(3T)}$ are all repetitive in the frequency domain with a period of $2\pi/T$ radians/second, so also must be the spectrum itself.
   需要说明的第二个问题是，由于诸如 $e^{-j\omega T}$，$e^{-j\omega(2T)}$，$e^{-j\omega(3T)}$ 这些项都是在频域内以 $2\pi/T$（弧度/每秒）周期重复的，它们的频谱一定也是周期的。

2. This form of the FFT is therefore a recipe, or 'algorithm', which is designed to increase the efficiency of calculation.
   离散傅里叶变换的这种形式是一种以增强计算效率为目的的设计方法或"算法"。

3. This accounts for the common description of z as a 'shift operator', denoting a time advance equal to the sampling interval T.
   这说明在一般描述方法中将 z 作为"移位算子"，表明超前时间等于抽样间隔 T。

4. This process is generally referred to as 'mapping' the s-plane into the z-plane.
   这个过程一般称为从 s 平面到 z 平面的映射。

5. A slight variation on the theme of possible alternative representations of a signal by poles and zeros is provided by the sampled-data signal of Figure 3.5 (a) which is a truncated version of the 'infinite' decaying exponential waveform of Figure 3.5 (b).
   与用零极点表示信号稍有不同的信号表达式如图 3.5（a）所示，它是图 3.5（b）表示的无限衰减指数波形的截断形式。

Exercises

Translate the following passages into English or Chinese.
1. 把信号只在某些不连续的瞬时给出函数值，称为离散时间信号。它可以来自模拟信号的抽样，或者是时间和幅度均不连续的离散信号。在离散信号的分析和处理时，通常把按一定先后次序排列、在时间上不连续的一组数的集合，称为"序列"。

2. 两个不同的序列可以对应相同的 z 变换，而收敛域并不相同，因此，为了使序列和 z 变换一一对应，在给出序列 z 变换的同时，必须指定其收敛域。
3. 在离散时间信号的分析和处理中，常常要对序列进行相加、相乘、延时和卷积等运算，z 变换的特性对于简化运算非常有用。
4. Pulse amplitude modulation (PAM) is an engineering term that is used to describe the conversion of the analog signal to a pulse-type signal where the amplitude of the pulse denotes the analog information.
5. If we assume that each of the digital words has n binary digits, there are $M=2^n$ unique code words that are possible, each code word corresponding to a certain amplitude level. However, each sample value from the analog signal can be any one of an infinite number of levels, so that the digital word that represents the amplitude closest to the actual sampled valued is used. This is called quantizing.

Reading Material

Signal Sampling

Whenever a continuous signal is to be represented by a set of samples, a decision must be made about the sampling rate. If the sampling rate is too low, information about the detailed fluctuations of the continuous waveform will be lost; and if too high, an unnecessarily large number of samples will have to be stored or processed. We shall now show that the clue to an appropriate of a continuous signal and that of its sampled version.

The first point to make about the sampling process is that it is a form of amplitude modulation; from a mathematical point of view, sampling a continuous signal is equivalent to multiplying it by a train of equally-spaced unit Dirac pulse. We may conveniently consider each sample to be a weighted value of the continuous waveform at the relevant instant. Since the sampled signal is obtained by multiplication of the continuous waveform and the Dirac pulse ensures that the result will be simply equal to the value of the train, its spectrum may be found by convoluting their respective spectra.

Our first task is therefore to define the spectrum of an infinite train of unit Dirac pulses, separated from one another by some sampling interval. For convenience, let us assume that one of the pulses occurs at $t = 0$. Such a pulse train is a strictly periodic waveform, symmetrical about $t = 0$, and must therefore have a line spectrum containing only cosine terms; its fundamental frequrncy will clearly be $1/T$ hertz, or $2/T$ radians/second. It remain to define the relative magnitudes of the various harmonics. Actually this is quite simply done, if we recall that the magnitude of any harmonic term in a periodic waveform may be found by multiplying the waveform by a cosine (or sine) of appropriate frequency, and integrating the product over one complete period. Figure 1 shows, as an example, part of a cosine of the third harmonic frequency $(\omega = 6\pi/T)$; if we multiply this by the pulse train waveform above, and integrate over the period $-T/2 < t < T/2$, the shifting property of the Dirac pulse insures that the result will be simply to the value of the cosine wave at $t = 0$, which is unity. Indeed this same result applies to any other harmonic frequency (and for the zero-frequency term) because cosines of all frequencies have a value of unity at $t = 0$. We conclude that the spectrum of all pulse train of Figure 3.8 contains an infinite set of cosine harmonic, all of the same amplitude and separated by $2\pi/T$ radians/second. This result emphasizes a special property of the Dirac pulse train that its time-domain waveform and its frequency spectrum are identical in form. It is perhaps unsurprising that its spectrum extends to infinitely high frequencies, because it made up from Dirac pulses which are themselves completely 'wideband'; and since the time-function is strictly repetitive, the spectrum can contain only discrete harmonic frequencies.

The next problem is to investigate what happens when the spectrum just discussed is convoluted with that of a typical continuous signal, for such a convolution yield the spectrum of the signal's sampled version. Let us denote the spectrum of the continuous signal by $G_1(\omega)$

and the spectrum of the pulse train by $G_2(\omega)$. The result of convoluting these two functions is a further function $G_3(\omega)$, given by

$$G_3(\omega) = \int_{-\infty}^{\infty} G_1(\omega - \Omega) G_2(\Omega) d\Omega$$

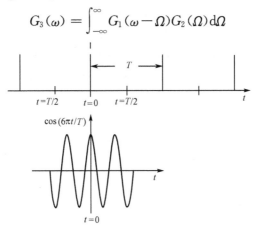

Figure 3.8 The spectrum of the pulse train

where Ω is an auxiliary frequency variable. We should of course remember that $G_1(\omega)$ is generally complex, whereas $G_2(\omega)$ is purely real since it is the spectrum of an even time function. However, the above convolution integral is simpler to visualize if we think of both $G_1(\omega)$ and $G_2(\omega)$ as real functions of ω; and the general validity of the result is not effected by this assumption. Now as we already shown the spectrum $G_2(\omega)$ consists of a series of equally-spaced spectral lines, which may also conveniently be represented by Dirac functions (that is, frequency-domain 'pulse'). Hence

$$G_3(\omega) = \cdots + G_1\left(\omega + \frac{4\pi}{T}\right) + G_1\left(\omega + \frac{2\pi}{T}\right) + G_1(\omega) + G_1\left(\omega - \frac{2\pi}{T}\right) + G_1\left(\omega - \frac{4\pi}{T}\right) + \cdots$$

This important result shows that the spectrum of a sampled signal is a repeated version of that of the underlying continuous waveform, it also gives a vital clue to the minimum sampling rate which may be used if the sampled values are to from an adequate substitute for the original signal. Figure 3.9 has shown spectrum $G_1(\omega)$ with significant frequency components in the frequency range $-\omega$ to ω: sampling causes this spectrum to repeat every $2\pi/T$ radians/second. It is therefore clear that if ω is greater than π/T there will be overlap between adjacent repetitions of $G_1(\omega)$. Without such overlap it would be possible, at least in principle, to recover the original continuous waveform from its sampled version by passing the latter through a linear filter which transmitted equally all components in the range $-\omega$ to ω, and rejected all other. But with such overlap, the spectrum of the sampled signal is no longer simply related to that of the original (especially in the regions of $\pm \omega$), and no linear filtering operation could be expected to recover the original from its sampling rate is too low is often referred to as 'aliasing'.

The above argument specifies, in effect, a minimum sampling rate for the faithful representation of a continuous signal by a sample values contain complete information about fluctuations in the original signal, and that they could therefore be used to reconstitute the original if required. It is at first sight remarkable that a set of samples can ever form a complete substitute for continuous waveform, for it seems inevitable that some details of the latter's fluctuations must be lost in the sampling process. The reason why this is not so is that we have assumed an upper limit ($\hat{\omega}$) to the frequencies contained in the continuous waveform, which is equivalent to specifying a maximum rate at which it can change its values. If its maximum rate of change is limited, them all possible fluctuations can be detected by taking samples at suitably spaced instants.

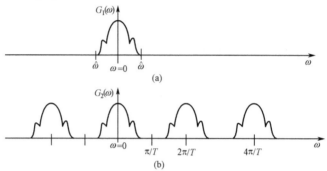

Figure 3.9 (a) The spectrum of a typical continuous signal, and (b) that of its sampled version when the sampling interval is T seconds

The minimum sampling rate for adequate representation of a continuous signal is formally specified in the so called 'sampling theorem'. If $\hat{f}$ is the frequency in hertz corresponding to $\hat{\omega}$ radians/second, then $\hat{f}=\hat{\omega}/2\pi$; and if the 'overlap' condition is to be avoided we have already shown that $\hat{\omega}<\pi/T$ or $\hat{f}=\hat{\omega}/2\pi<1/2T$ and hence $T<\pi/\hat{\omega}=1/2\hat{f}$.

Formally, the sampling theorem may therefore be stated as follows: 'A continuous signal which contains no significant frequency components above f hertz may in principle be recovered from its sampled version, if the sampling interval is less than $1/2\hat{f}$ seconds'.

There we noted that the frequency spectrum if a sampled data signal is always repetitive in form, and we now recognize this to be an essential characteristic of the sampling has been, that portion of the sampled-data signal's spectrum lying in the range $-\pi/T<\omega<\pi/T$ may be used to define both the sampled-data signal itself, and the underlying waveform which it represents.

Lesson 4　Random Signals

4.1　Introduction

So far we have dealt with continuous and sampled signals having defined waveshapes. Such signals are described as 'deterministic', and the frequency spectra we calculate for them specify the magnitudes and relative phases of the sinusoidal wave which, if add together, would exactly resynthesize the original waveforms. By contrast, the value of a random waveform or signal is not specified at every instant of time, nor is it possible to predict its future with certainty on the basis of its past history. The reason for this is generally that we have insufficient understanding of the physical process producing the random signals; on the other occasions, the understanding is present, but the effort involved in predicting the signal is too great to be worthwhile. In such case it is usual to evaluate some average properties of the signal which describe it adequately for the task in hand.

A common initial reaction is that a random signal with ill-defined properties can have little place in any scientific theory of signal analysis, but the opposite is in fact nearer the truth. Suppose, for example, it is desired to send a message along a telegraph link. It is almost valueless to send a known, deterministic, message since the person at the receiving end learns so little by its reception. As a simple example, it would be pointless to transmit a continuous sinusiodal tone, since once its amplitude, frequency and phase characteristics have been determined by the receiver, no further information is conveyed. But if the tone is switched on and off as in Morse-code message, the receiver does not know whether a 'dot' or 'dash' is to be sent next, and it is this very randomness or uncertainty about the future of the signal which allows useful information to be conveyed. On the other hand it is quite possible to say something about the average properties of Morse-code massages, and such knowledge might be very useful to the designer of a telegraph system, even though he does not know what dot-dash sequence is to be sent on any particular occasion. Viewed in this light it is unsurprising that random signal theory has played a major role in the development of modern communications systems.

A signal may be random in a variety of ways, perhaps the most common type of randomness is in the amplitude of a signal waveform, illustrated in Figure 4.1(a). In this case future values of the signal cannot be predicted with certainty, even when its past history is taken into account. Figure 4.1(b) shows another common form of randomness, of the general type displayed by the Morse-code message just mentioned. Here the signal is always at one or other of two definite values but the transitions between these two levels occur at random instants. Signal in which the times of occurrence of some definite even or transition between states are random arise in many diverse fields of study, such as queuing theory, nuclear particle physics and neurophysiology, as well as in electronic communications.[1] Figure

4.1(c) shows a signal possessing both the common types of randomness so far mentioned, and represents an electrocardiogram, or recording of the electrical activity of the heart. The heartbeat is somewhat irregular, so that the timing of successive ECG complexes has a random component; furthermore, the waveshape itself is not exactly repetitive in form.

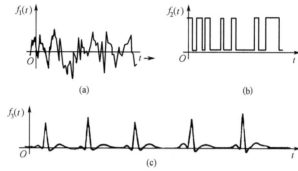

Figure 4.1　Signal displaying (a) random amplitude, (b) random timing of transition between fixed levels, and (c) random amplitude and timing components in an essentially repetitive waveform

In practice a signal quite often contains both random and deterministic components. For example, the ECG complexes of Figure 4.1(c) might usefully be considered to consist of a strictly repetitive signal plus small random amplitude and timing components. Very often such a random component is the result of recording or measurement errors, or arises at some point in the system which is outside the control of the experimenter. Random disturbances are widely encountered in electronic circuits, where they are referred to as electrical 'noise'. However, the methods of analysis of random waveforms described in this chapter apply either a waveform represents a useful signal or an unwanted 'noise'. It is important to be able to describe a noise waveform quantitatively, at least so that the effects on it of signal processing devices may be assessed. And we shall see later that a most important type of signal processing operation is one that attempt is made to extract or enhance a signal waveform in the presence of unwanted disturbances.

As already noted, the method used to describe random signals is to assess some average properties of interest. The branch of mathematics involved is statistics, which concerns itself with the quantitative properties of a population as a whole, rather than of its individual elements. In the present context, we may think of this population as being made up from a very large number of successive value of a random signal waveform. The other branch of mathematics of direct interest for random signal analysis is that of probability, which is closely related to statistics. Probability theory concerns itself with the likelihood of various possible outcomes of random process or phenomenon, whereas statistics seek to summarize the actual outcomes using average measures. In order to allow us to develop useful average measures for random signals, we first examine some of the basic notions of probability theory.

4.2 Elements of Probability Theory

4.2.1 The Probability of an Event

Suppose a die with six faces is thrown repeatedly. As the number of trials increases, and provided the die is fair, we expect that the number of times it lands on any one face to be close to one sixth of the total number of throws. If asked about the chance of the die landing on a particular face in the next trial, we would therefore assess it as one in six. Formally, if the trial is repeated N times and the event A occurs n times, the probability of event A is defined as

$$p(A) = \lim_{N \to \infty} (n/N) \qquad (4-1)$$

The definition of probability as the relative frequency of the event is great intuitive appeal, although it gives rise to considerable difficulty in more rigorous mathematical treatments. The main reason for this is that any actual experiment necessarily involves a finite number of trials and may fail to approache the limit in a convincing way. Even in a simple coin tossing experiment, a vast number of tosses may fail to yield the probabilities of 0.5 expected for both 'heads' and 'tails'. For this reason an alternative approached to the definition of probability, based upon a set of axioms, is sometimes adopted. Whichever definition is used, however, it is clear that a probability of 1 denotes certainty and a probability of 0 implies that the event never occurs.

Next suppose that we define the event (A or B) as occurring whenever either A or B occurs. In a very large number N of trials, suppose A occurs n times and B occurs m times. If A and B are mutually exclusive (so that they never occur together) the event (A or B) occurs $(n+m)$ times. Hence

$$p(A \text{ or } B) = \lim_{N \to \infty}\left(\frac{n+m}{N}\right) = \lim_{N \to \infty}\left(\frac{n}{N}\right) + \lim_{N \to \infty}\left(\frac{m}{N}\right)$$
$$= p(A) + p(B) \qquad (4-2)$$

This is the basic additive law of probability, which may be extended to cover the case of an experiment or trial with any number of mutually exclusive outcomes.

4.2.2 Joint and Conditional Probabilities

Suppose we now conduct an experiment with two sets of possible outcomes. Then the joint probability p (A and B) is the probability that the outcome A from one set occurs together with outcome B from the other set. For example, the experiment might consist of tossing two dice simultaneously, or of drawing two cards from a pack. Suppose that of N experiments performed n produce outcome A; of these, suppose that m also produce outcome B. Then m is the number experiments which give rise to both A and B, so that

$$p(A \text{ and } B) = \lim_{N \to \infty}\left(\frac{m}{N}\right) = \lim_{N \to \infty}\left(\frac{n}{N}\right)\left(\frac{m}{n}\right) \qquad (4-3)$$

Note that the limit of (n/N) as $N \to \infty$ is $p(A)$, and that the limit of (m/n) is the probability that outcome B occurs given that outcome A has already occurred. This latter probability is

called the 'conditional probability of B given A' and will be denoted by symbol $p(B/A)$. Thus
$$p(A \text{ and } B) = p(A)p(B/A) \tag{4-4}$$
By similar arguments, it may be shown that
$$p(A \text{ and } B) = p(B)p(A/B) \tag{4-5}$$
And hence
$$p(A)p(B/A) = p(B)p(A/B) \tag{4-6}$$
or
$$p(A/B) = p(A)p(B/A)/p(B) \tag{4-7}$$
This result, known as Bayes' rule, relates the conditional probability of A given B to that of B given A.

It is obvious that if outcome B is completely unaffected by outcome A, then the conditional probability of B given A will be just the same as the probability of B alone, hence
$$p(B) = p(B/A) \tag{4-8}$$
And
$$p(A \text{ and } B) = p(A)p(B) \tag{4-9}$$
In this case the outcomes A and B are said to be statistically independent.

To illustrate these results, suppose we have a box containing 3 red balls and 5 blue balls. The first experiment consists of taking one ball from the box at random, replacing it and then taking another, and we are interested in the probability that the first ball withdrawn is red (outcome A) and the second one blue (outcome B). In this case, the result of the first part of the experiment in no way affects that of the second, since the first ball is replaced. Hence the probability of the joint event (red followed by blue) is
$$p(A \text{ and } B) = p(A)p(B) = \frac{3}{8} \times \frac{5}{8} = \frac{15}{64}$$

We now repeat the experiment without replacing the first ball before taking out the second, so that the two parts of the experiment are no longer statistically independent. If a red ball is withdrawn first, then 5 blue and 2 red balls remain, giving the conditional probability
$$p(B/A) = \frac{5}{7}$$
The brief discussion of joint and conditional probabilities, which has been applied to the case of an experiment with two sets of possible outcomes, may be extended to cover any number of such sets. Its relevance to signals and signal analysis may be illustrated by reference to Figure 4.2 (a), which shows a portion of a random sampled-data signal in which each sample value takes on one of six possible levels. This is analogous to the die with six faces, and we could estimate the probability of the next sample taking on any particular value by analyzing a sufficiently long portion of the signal's past history (note that, unlike the case of the fair die, the six probabilities may well not be equal). Thus any one sample value is thought of as the result of a trial which has, in this case, six possible outcomes. A rather more elaborate description of the signal could be obtained by asking whether a particular value of the signal

tends to influence following ones. For example, does the occurrence of a sample 3 mean that a 4 or 2 (or any other value) is more likely to follow than would be suggested by the simple probabilities already derived? This is tantamount to asking whether a particular value is constrained by those preceding or following it and could be answered by assessing suitable conditional probabilities. Such matters will be investigated in subsequent sections where so-called correlation functions are discussed.

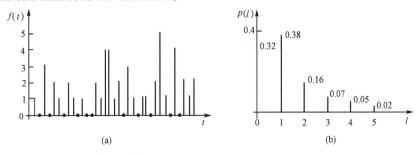

Figure 4.2 (a) A portion of a random sampled-data signal with six discrete levels, and (b) the probabilities associated with each level

So far we have considered experiments having a finite number of possible outcomes, and the analogous situation of a random signal which takes on a finite number of levels. In the case of a signal such as that of Figure 4.2(a), analysis of a very long portion of the signal would allow us to estimate the probabilities associated with the six possible signal levels; those could then be plotted on a graph as in Figure 4.2(b), commonly met in practice, however, are signals which display a continuous range of amplitude levels. The probability of finding such a signal at a particular level then becomes vanishingly small and forced to use a continuous probability variable, the probability density function, to describe it.

4.2.3 The Probability Density Function

The continuous random signal shown in Figure 4.3 takes on an infinite set of amplitude values, and the probability of its assuming some particular value such as y therefore becomes vanishongly small. In this case it is only sensible to talk about the probability that the signal occupies a small range of values such as that between y and $(y+\delta y)$. This is simply equal to the fraction of the total time spent by the signal in this range, and may be estimated by summing all time intervals such as δt_1, δt_2, $\delta t_3 \cdots$ over a long length of record (say T_0 second) and then dividing the result by T_0. Denoting this probability by q, we have

$$q = \lim_{T_0 \to \infty} \left(\frac{\delta t_1 + \delta t_2 + \delta t_3 + \cdots}{T_0} \right) \tag{4-10}$$

Now the value of q clearly depends on the size of the interval δy which tends to zero, so does q. However, the quotient $(q/\delta y)$ tends to a constant value as $\delta y \to 0$, and this value is called the probability density $p(y)$. $p(y)$ gives us a probability measure which is independent of the precise value of δy chosen, the actual numerical probability that the signal falls in the range y to $(y+\delta y)$ is found by multiplying $p(y)$ by the interval size δy; hence

$$p(y) = q/\delta y \text{ or } q = p(y)\delta y \qquad (4\text{-}11)$$

It is pointed out in section 3.3.1 that the spectrum $G(j\omega)$ of an aperiodic time function should be thought of as a density function, because it is only sensible to talk about spectral energy in some narrow band of frequency. The probability density function of a random signal involves a similar concept: it is not sensible to talk about the probability that a continuous signal adopts a particular value y, but only that it falls within some narrow rang of values between y and $(y+\delta y)$.

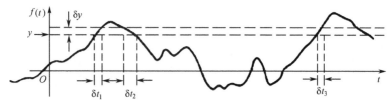

Figure 4.3 A continuous random signal

To take a simple example, suppose we have a continuous random signal $f(t)$ which that $-1 < f(t) < 1$ for all t, with no bias towards any particular range within these limits. In other words $p(y)$ is a 'rectangular' or 'even' distribution having a constant value in the range $-1 < y < 1$, as shown in Figure 4.4. The actual numerical value of $p(y)$ may be found by considering a small range of values such as y_1 to $(y_1 + \delta y_1)$. The probability of finding the signal in this range is

$$\frac{\delta y_1}{1-(-1)} = \frac{y_1}{2} = p(y_1)\delta y_1 \qquad (4\text{-}12)$$

$$p(y_1) = 0.5 \qquad (4\text{-}13)$$

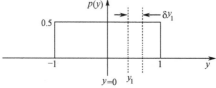

Figure 4.4 A rectangular probability density function

Therefore in this particular case the probability density has a constant value of 0.5 in the range $-1 < y < 1$. Generalizing to any form function $p(y)$, the probability of finding the signal somewhere in the range $-\infty < y < \infty$ must be unity, and hence

$$\int_{-\infty}^{\infty} p(y) \, dt = 1 \qquad (4\text{-}14)$$

In other words, any probability density function must have unit area.

4.3 Amplitude Distribution and Moments

We have already seen two examples of functions which describe the probability that a random signal takes on a particular value, or falls within a certain narrow range of values. Such functions are often referred to as 'amplitude distributions', and, like other long-term or average measures used to describe random signals, they may also be applied to deterministic ones. For example, the sinusoidal wave of Figure 4.5 (a) which spends a relatively large proportion of its time near the values $\pm A$ but never exceeds them, has the amplitude probability density curve shown in Figure 4.5 (b). On the whole, however, such amplitude probability functions are not used for deterministic signals because compact (and complete) analytic descriptions of the functions themselves are normally available.

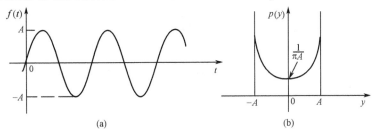

Figure 4.5 (a) Part of a continuous sinusoidal wave of amplitude A, and (b) its amplitude probability density

Another important point is that the amplitude distribution of a signal tells us nothing about its detailed structure or spectrum, nor is it a unique property of the signal. Figure 4.6 shows portions of one random and three deterministic waveform all having identical amplitude distributions but with very different structures and frequency spectra.

It is sometimes of interest to know the probability that a signal has a value below a certain level a. Denoting this probability by $P(a)$, we have

$$P(a) = \int_{-\infty}^{a} p(y) \mathrm{d}y \qquad (4\text{-}15)$$

P is normally referred to as the 'cumulative' distribution function. Alternatively, if the probability of finding the signal above some level b is required, it is only necessary to change the limits of integration to b and $+\infty$.

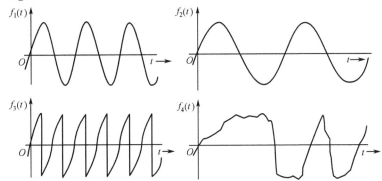

Figure 4.6 Portions of four waveforms with identical amplitude distributions

A further very important set of properties of a random (or deterministic) signal, known as the central moments, may be evaluated if the amplitude distribution function is known. The first central moment, more widely known as the average or mean value, and the second central moment or variance, are the most important, although higher order ones are sometimes also of interest. Consider first the case of a random signal of the general type already shown in Figure 4.2, which can assume n discrete values $y_1, y_2, y_3, \cdots, y_n$. If we consider a very large number ($N \to \infty$) of simple values, the value y_1 is expected to occur Np_1 times, where p_1 is the probability associated with it; value y_2 is expected to occur Np_2 times, and so on. The average, or mean, of all the sample values is found by adding together all the sample values multiplied by the total number of samples. Hence the average ($\bar{y}$) is given by

$$\bar{y} = \sum_{m=1}^{n} p_m y_m \qquad (4\text{-}16)$$

where the capital sigma represents a summation of n terms. This average, or mean, value is also called the 'expected' value of y and is often written as $E(y)$ or $<y>$. When the random signal is continuous and can take up an infinite set of values, the summation is replaced by a probability density. In this case the average value is

$$\bar{y} = \int_{-\infty}^{\infty} y p(y) \mathrm{d}y \qquad (4\text{-}17)$$

The second central moment, or variance, is generally given the symbol σ^2, where σ is referred to as the 'standard deviation'. The variance is defined as the average or expected value of the square of the function's departure from its mean, and is therefore a measure of signal fluctuation. By arguments similar to those used when defining the mean, the second central moment is given by

$$\sigma^2 = \overline{(y-\bar{y})^2} = \sum_{m=1}^{n} (y_m - \bar{y})^2 p_m \qquad (4\text{-}18)$$

for a discrete random signal having n possible levels, and as

$$\sigma^2 = \overline{(y-\bar{y})^2} = \int (y_m - \bar{y})^2 p(y) \mathrm{d}y \qquad (4\text{-}19)$$

for a continuous one.

A distinction must be made between the 'central moments' and the 'moments' of a function. The former are measures of fluctuation about the mean value, whereas the simple moment is calculated without taking the mean into account. Thus the second moment of a discrete random variable with n possible levels is simply

$$\overline{y^2} = \sum_{m=1}^{n} p_m y_m^2 \qquad (4\text{-}20)$$

Not surprisingly, the central moments and moments are closely related and knowledge of one set allows the other to be calculated. In particular, the second moment is related to the second central moment and the mean as follows

$$\overline{y^2} = \sigma^2 + (\bar{y})^2 \qquad (4\text{-}21)$$

This relationship explains why the second central moment or variance is such an important measure. As has already been pointed out in section 2.4.2, the average squared value of a signal (its second moment) equals its average power, adopting the convention that the signal is considered to represent an electrical voltage across or current through a 1-ohm resistor. Now the square of the mean value represents the power in the zero frequency (DC) component of the signal, and therefore the second central moment (σ^2) must represent the power in all other frequency components. it is therefore widely referred to as the 'AC power'.

4.4 The Autocorrelation and Power Spectral Density

4.4.1 The Spectral Properties of Random Signals

The probability functions which we have so far investigated provide no clues to the structure of a random signal in the time-domain, or to its frequency spectrum. At first it may not be obvious that the spectrum of a random signal may be discussed at all, since the signal is unpredictable and is not defined by an analytic function. However, Figure 4.7, which shows portions of two random signals, is sufficient evidence that a spectral measure is likely to be very useful, assuming that one may be adequately defined. Suppose we treated these two portions as deterministic signals, by assuming them to have the waveforms shown in the interval $-T_0 < t < T_0$ and to be zero elsewhere, and then evaluate their frequency spectra. It is obvious that the spectra would be quite different, if only because one signal has far higher frequencies present than the other and also possesses a substantial mean (zero frequency) component. If we now took a second portion of each random signal of the same duration T_0 and again evaluated the spectra, we would be bound to find them different from the first pair because of detailed differences in the waveforms. But the longer the duration T_0, the more confident would we be that each waveform portion was typical of its parent signal, and therefore that spectra calculated for different portions of a particular random signal would show broad similarities. In other words, we expect that some useful average measure of spectral components can be found, even if the spectrum of any finite portion of a waveform can never be expected to match it perfectly. The average measure most widely adopted is the so-called power spectrum, or its associated time-domain function, the autocorrelation function.

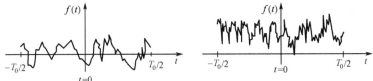

Figure 4.7 Portions of two random signals

4.4.2 The Autocorrelation Function

The autocorrelation function (ACF) of a signal waveform is an average measure of its time-domain properties, and is therefore likely to be especially relevant when the signal is a random one. Furthermore, we shall see that the ACF is not only an interesting and valuable function in its own right, but that it also provides the key to a random signal's spectrum. Formally, the ACF is defined as

$$r_{xx}(\tau) = \lim_{T_0 \to \infty} \frac{1}{T_0} \int_{-T_0/2}^{T_0/2} f(t) f(t + \tau) \mathrm{d}t \qquad (4\text{-}22)$$

It is therefore equal to the average product of the signal $f(t)$ and a time-shifted version of itself, and is a function of the imposed time-shift, τ. The above expression applies to the case of a continuous signal of infinite duration. If used for a signal of limited duration such as an isolated pulse, the average product over a very long interval T would tend to zero at all values of τ. In such case it is therefore normal to use a modified version of the ACF, generally called the 'finite ACF' which is defined as

$$r'_{xx}(\tau) = \int_{-\infty}^{\infty} f_1(t) f_1(t+\tau) dt \tag{4-23}$$

In the case of a sampled-data signal the product of the signal and its shifted version only has non-zero values when the shift is equal to a multiple of the sampling interval T, and the ACF is therefore defined as

$$r_{xx}(k) = \lim_{T_0 \to \infty} \frac{1}{(2N+1)} \sum_{m=-N}^{N} x_m x_{m+k} \tag{4-24}$$

where x_m and x_{m+k} represent two sample values separated by kT seconds, and the summation has $(2N+1)$ terms with the integer parameter m taken between $-N$ and $+N$. In the case of a signal of limited duration it is once again appropriate to use a 'finite' version of the above expression which takes either the average or just the sum of a number of terms corresponding to the available length of signal. [2] It should be noted that the precise definition of the autocorrelation function for finite and infinite signals tends to vary somewhat from text to text; the important thing to remember, however, is that they are all measures of the average product of a signal and its time-shifted version. (the subscript 'xx' or '11' is widely used for the ACF to denote that a signal is being multiplied by a delayed version of itself. This distinguishes it from the closely related cross-correlation function, which we shall use in the next chapter to describe the effects of time-shifts imposed between two different signals, and which is generally given the subscript '12' or 'xy'.)

Like the average measures described in the previous section, the ACF may be applied to deterministic as well as to random signals, and some of its principal properties are most conveniently illustrated in this way. Suppose for example we have a signal $f(t)$ composed of two cosinusoidal waves of different frequency and phase angle.

Thus

$$f(t) = A_1 \cos(\omega_1 t + \theta_1) + A_2 \cos(\omega_2 t + \theta_2) \tag{4-25}$$

The ACF is then given by

$$r_{11}(\tau) = \frac{A_1^2}{2} \cos\omega_1 \tau + \frac{A_2^2}{2} \cos\omega_2 \tau \tag{4-26}$$

Therefore each of the frequency components in the signal $f(t)$ gives rise to a term in the autocorrelation function having the same period in the time-shift variable τ as the original component has in the time variable t, and an amplitude equal to half the squared valued of the original. This result is illustrated by Figure 4.8. the phase shifts θ_1 and θ_2 of the component waves of $f(t)$ do not figure in the ACF, and hence all information about relative phase has been lost. Although the above ACF has been obtained for a function $f(t)$ consisting of just two component frequencies, the orthogonal properties of sine and cosine functions mean that it may be extended to the case of a signal containing any number of components. Regardless of its phase, each and every component gives rise to a simple cosine term in the ACF.

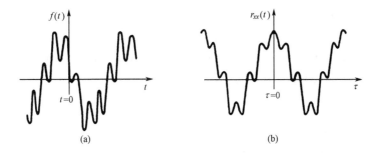

Figure 4.8 (a) A signal with two discrete frequency components, and (b) its autocorrelation function

Since any ACF is composed of cosine it must be an even function of τ. In order words, the averaged product of a signal and its time-shifted version is the same whether the shift is a forward or backward one, a conclusion which is indeed quite simple to demonstrate. When $\tau=0$, all the cosine function are at their peak positive value and hence reinforce one another to give the largest possible value of the ACF. Whether this peak value is ever attained again at other value of τ depends upon whether or not the various components in the signal are harmonically related, but in any event it can never be exceeded. The peak value is given by

$$r_{xx}(\tau)\mid_{\tau=0} = r_{xx}(0) = \lim_{T_0 \to \infty} \frac{1}{T_0} \int_{-T_0/2}^{T_0/2} [f(t)^2] dt \tag{4-27}$$

which is simply the mean square value, or average power, of the signal. Thus $r_{xx}(0)$ is equal to the second moment of the signal which, as we have seen, may also be derived from its probability density function. On the other hand, the curve traced out by the ACF as τ varies gives information about the time-domain structure of a signal which is not contained in the probability density function. It should be noted that in the case of a signal of finite duration, for which we use the 'finite' version of the ACF, the value of the ACF relevant to $\tau=0$ is given by

$$r_{xx}(0) = \int_{-\infty}^{\infty} [f(t)^2] dt \tag{4-28}$$

which equals the total energy in the signal waveform rather than its power averaged over a very long time interval.

4.4.3 The Power Spectral Density Function

We have seen how a term such as $A_1 \cos(\omega_1 t + \theta_1)$ in a signal waveform contributes a term $(A_1^2/2) \cos \omega_1 \tau$ to its autocorrelation function. It has been shown in section 4.4.2 that the mean square value, or average power, of any wave of sinusoidal form having an amplitude A_1 is equal to $(A_1^2/2)$. Therefore the amplitudes of the various cosine terms in the ACF merely indicate the average power of the corresponding spectral terms in the signal itself.

Just as a signal waveform may be described in terms of its frequency spectrum, so an autocorrelation function (which is a function of the time-shift variable τ) has a counterpart in the frequency domain.[3] As the above argument shows, this counterpart will have a number of spectral lines representing the power in the various components, and it is therefore given the name 'power spectrum'. The relationship between the frequency components of a typical periodic signal and those of its ACF are illustrated in Figure 4.9. In the case of an aperiodic signal with a continuous frequency spectrum, the frequency-domain counterpart of its ACF is also continuous and is known as the 'power spectral density'.

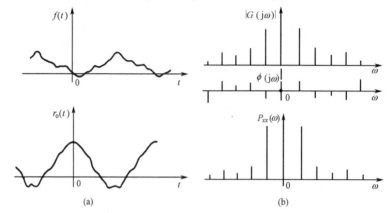

Figure 4.9 (a) A signal and its autocorrelation function, and (b) their corresponding frequency-domain descriptions $G(j\omega)$ and $P_{xx}(\omega)$ • $G(j\omega)$ involves both magnitude and phase terms, whereas $P_{xx}(\omega)$ is purely real

Separate provision must again be made for signals of limited duration, whose average power, measured over a very longtime interval, tends to zero.[4] In such case we refer to the 'energy spectrum' rather than the power spectrum. Just as the power spectrum is the frequency domain counterpart of the normal version of the ACF, so the energy spectrum is the equivalent of the 'finite' version of the ACF, and describes the distribution of signal energy in the frequency domain. The discussion which follows (both here and in subsequent sections) will be largely in terms of the power spectrum of a signal which continuous for ever, but it may be assumed that the energy spectrum of a time-limited signal displays essentially similar properties. Both power and energy spectra will be denoted by the symbol $P_{xx}(\omega)$.

We therefore see that the ACF and power spectrum, or power spectral density are equivalent measures in the time and frequency domains. In other words, they are related to one another by the Fourier transform. This fact is formally expressed by the so-called Wiener-Khinchin relations.

$$P_{xx}(\tau) = \int_{-\infty}^{\infty} r_{xx}(\tau) e^{-j\omega\tau} d\tau \qquad (4-29)$$

and

$$r_{xx}(\tau) = \frac{1}{2\pi} \int_{-\infty}^{\infty} P_{xx}(\omega) e^{j\omega\tau} d\tau \qquad (4-30)$$

where $P_{xx}(\omega)$ is the power spectral density. These equations are identical in form to the Fourier transform pair which relate a signal $f(t)$ to its spectrum $G(j\omega)$, expect that we are now working in terms of the delay variable τ rather than the time variable t.

Words and Expressions

axiom [ˈæksiəm] n. 公理
Bayes' rule 贝叶斯准则
binomial distribution 二项分布
correlation function 相关函数
cosinoidal [kəusiˈnɔidəl] adj. 余弦的
cumulative [ˈkjuːmjulətiv] adj. 累积的
deterministic [diˌtəːmiˈnistik] adj. 确定性的
elaborate [iˈlæbərit] adj. 精心制作的；vt. 详细描述
ergodic [əːˈgɔdik] adj. 遍历的
fluctuation [ˌflʌktjuˈeiʃən] n. 起伏，波动
ill-defined 不清楚的，欠明了的
moment [ˈməumənt] n. 矩
Morse code 莫尔斯码
neurophysiology [ˌnjuərəuˌfiziˈɔlədʒi] n. 神经生理学
normal distribution 正态分布
poisson distribution 泊松分布
probability [ˌprɔbəˈbiliti] n. 概率
pseudo-random 伪随机的
quotient [ˈkwəuʃənt] n. 商
rectangular [rekˈtæŋgjulə] n. 矩形
standard deviation 标准偏差
summarize [ˈsʌməraiz] vt. 相加，概括
symbol [ˈsimbəl] n. 符号，记号
tantamount [ˈtæntəmaunt] vt. 等价
time-shifted version 时移形式
tone [təun] n. 音调
vanishingly [ˈvæniʃiŋli] adv. 趋于零地，难以觉察地

Notes

1. Signal in which the times of occurrence of some definite even or transition between states are random arise in many diverse fields of study, such as queuing theory, nuclear particle physics and neurophysiology, as well as in electronic communications.
   某种确定事件发生的时间或在两种状态之间跃变的时间是随机信号，这种随机信号存在于许多不同的研究领域中，例如，在排队论、核粒子物理学和神经心理学领域中，以及在电子通信领域中。

2. In the case of a signal of limited duration it is once again appropriate to use a 'finite' version of the above expression which takes either the average or just the sum of a number of terms corresponding to the available length of signal.

   在信号持续时间有限的情况下，应采用上述表达式的"有限"形式，即对信号有效长度相对应的一些项取平均值或单纯地求和。

3. Just as a signal waveform may be described in terms of its frequency spectrum, so an autocorrelation function (which is a function of the time-shift variable τ) has a counterpart in the frequency domain.

   正如一个信号波形可以根据它的频谱来描述一样，一个自相关函数（关于时移变量 τ 的函数）在频域中有其相对应的部分。

4. Separate provision must again be made for signals of limited duration, whose average power, measured over a very longtime interval, tends to zero.

   对于在一个非常长的时间区间上平均功率趋近于零的持续时间有限的信号，必须对其做出单独的规定。

Exercises

Translate the following passages into English or Chinese.

1. 随机信号是不能用精确的数学关系式描述的信号，但它的变动服从统计规律，可以用概率统计特性来描述。对随机信号按时间历程所做的各次长时间观测记录称为样本函数。

2. 随机信号的时域和频域特性可以用统计方法进行研究，其幅值特性用信号的均值、均方值和概率密度函数表示。信号的时域特性也可以用自相关函数描述，频域特性用功率谱密度函数描述。

3. The definition of a random process may be compared with that of a random variable. A random variable maps events into constants, whereas a random process maps events into functions of the parameter t.

4. In summary, if a process is ergodic, all time and ensemble averages are interchangeable. The time average cannot be a function of time since the time parameter has been averaged out. Furthermore, the ergodic process must be stationary, since otherwise the ensemble averages (such as moments) would be a function of time. However, not all stationary processes are ergodic.

Reading Material

The Processing of Random Signal

When a random waveform passes through a linear signal processor its time-averaged properties as measured by its amplitude distribution, autocorrelation function, or power spectrum are usually modified. Whether the random waveform represents a useful signal or an unwanted interference, it is clearly desirable to quantify the effects which a linear signal processing operation has upon it.

We have seen that a linear system is normally characterized either by its frequency response (or transfer function), or by its impulse response, and that these functions are equivalent descriptions of the system in frequency and time domain respectively. Not surprisingly, therefore, it is a relatively straightforward matter to discuss the effect which a linear system has on the frequency and the time domain measures of a random signal. On the other hand, the effects of signal processing on such properties of a random waveform as its amplitude distribution or central moments are not simply related to its time and frequency domain structure.

Let us begin by considering the effect of the system on the spectral properties of the random signal. As we have seen, the power spectral density defines the average power of the various frequency components in a signal, but ignores their relative phase. Hence the power spectral density of the output from a linear system cannot depend upon the phase response of the system, but only upon its magnitude response. Suppose, for example, we apply a sinusoidal wave $A\sin \omega t$ to the input of a linear system and obtain an output $AB\sin(\omega t+\phi)$. The average input power is $A^2/2$ and the average output power is $(AB)^2/2$. Therefore a system having a response magnitude B at a frequency ω modifies the power of a component at that frequency by a factor B^2. More generally, if the input signal contains many frequency components and has a power spectral density $P_{xx}(\omega)$, then the output power spectral density still be

$$P_{yy}(\omega) = P_{xx}(\omega) \mid H(j\omega) \mid^2 = P_{xx}(\omega)H(j\omega)H^*(j\omega)$$

where $H(j\omega)$ is the frequency response of the system, and the asterisk denotes the complex conjugate. Each input component is multiplied by the square of the response magnitude at the relevant frequency.

We note that the output power spectrum from a linear system is found by multiplying the input power spectrum by the system function $\mid H(j\omega) \mid^2$; since the power spectrum and autocorrelation function of a signal are related as Fourier transform pair, it therefore follows that the ACF of the output signal from a linear system may be found by convoluting that of its input with the time function corresponding to $\mid H(j\omega)\mid^2$. This latter time function is in fact equal to the ACF of the system's impulse response.

Although our major interest in this lesson is to describe how signals are modified by linear processing, there is another side to the same coin: knowledge of the properties of input and output signals allows us to infer those of an unknown linear system. For example, if we know the input signal spectrum $G_1(j\omega)$ and the output signal spectrum $G_2(j\omega)$ of a system, we may express its frequency response quite simply as

$$H(j\omega) = G_2(j\omega)/G_1(j\omega)$$

Similarly if a random signal with a power spectrum $P_{xx}(\omega)$ is applied to a linear system and we measure the output power spectrum $P_{yy}(\omega)$, we are in a position to define the system's response magnitude $|H(j\omega)|$, since

$$|H(j\omega)|^2 = P_{yy}(\omega)/P_{xx}(\omega)$$

Of course, any practical estimates of $P_{xx}(\omega)$ and $P_{yy}(\omega)$ based on finite portions of random input and output power spectra fails to reveal any information about the way in which a system modifies the phases of various frequency components applied to its input, and it is interesting to consider whether any other type of comparison of random input and output signals might be used to define the system's response in phase as well as magnitude. We discussed signal comparison using the cross spectral density and cross-correlation functions, and saw that these measures reflect not only magnitudes but also relative phases of common frequency components in two signals $f_1(t)$ and $f_2(t)$. If we now consider $f_1(t)$ and $f_2(t)$ represent random input and output waveforms of a linear system, it is clear that any measure which indicates relative phases of their various common components must indicate phase changes introduced by the system itself. This is the clue to a complete definition of a linear system by examination of the properties of its random input and output. In order to examine this question rather more carefully, recalling that correlation and convolution are similar time-domain operations apart from reversal of one of the time functions, we may write down the following list of equivalent operations.

(1) cross-correlation of $f_1(t)$ and $f_2(t)$.

(2) convolution of $f_1(-t)$ and $f_2(t)$.

(3) multiplication of $G_1(-j\omega)$ and $G_2(j\omega)$.

But $G_1(-j\omega) = G_1^*(j\omega)$ when $f_1(t)$ is a real time function; furthermore, $G_2(j\omega) = G_1(j\omega) \cdot H(j\omega)$, so that

$$G_1(-j\omega)G_2(j\omega) = G_1^*(j\omega)G_1(j\omega)H(j\omega) = |G_1(j\omega)|^2 H(j\omega)$$

Now $|G_1(j\omega)|^2$ is the power spectral density of the signal, which is the Fourier transform of its autocorrelation function $r_{xx}(\tau)$; therefore operation (3) above is equivalent to

(4) multiplication of $|G_1(j\omega)|^2$ and $H(j\omega)$.

And hence to

(5) convolution of $r_{xx}(\tau)$ and $I(t)$.

This last result is most interesting: we started by assuming cross-correlation of the input and output waveforms of the system, and result (5) shows that this is equivalent to convoluting the impulse response of the linear system with the ACF of its input. Formally the relationship may be stated by the convolution integral

$$r_{xx}(\tau) = \int_{-\infty}^{\infty} r_{xx}(\tau-1)I(t)\,dt$$

If we know the input ACF and evaluate the cross-correlation function relating the system input $f_1(t)$ and its output $f_2(t)$, we may therefore evaluate the impulse response of the system, the above formula becomes particularly useful when the input $f_1(t)$ is a very wideband random signal, which has an ACF approximating a Dirac pulse at $r=0$. In this case we have

$$r_{xx}(\tau) = \int \delta(\tau-1)I(t)\,dt$$

which, by the shifting property of Dirac pulse, reduces directly to

$$r_{xx}(\tau) = I(\tau)$$

In other words, the input-output cross-correlation function has the same shape as the impulse response of the linear system. It is at first rather hard to believe this result, since we are apparently deriving complete and deterministic information about a system by comparing its random input and output waveforms. The essential point, however, is that out estimate of $I(t)$ will involve sampling errors and willingly be accurate if we cross-correlate very long portions of the input and output random signals.

This technique of system identification by input-output cross-correlation is of considerable practical interest. Since the cross-correlation function takes the same form as the system's impulse response only when the input random signal is wideband (in practice this means that the input power spectrum must be constant or 'flat' over the full range of frequencies significantly transmitted by the system), the value of wideband random or pseudo-random signals for system identification is emphasized. It should also be noted that a random disturbance is quite often present at the input (and hence output) of a linear system anyway; normally such 'noise' is merely a nuisance, but it can sometimes allow the above technique to be put to good use. The characteristic of electronic circuits, chemical process plants, and physiological systems have all been explored by input-output cross-correlation.

Although knowledge of the spectral properties of a linear system allows us to definite its effect on the power spectrum (or autocorrelation function) of a random signal, and same is unfortunately not true of the signal's amplitude distribution, the relationship between input and output signal amplitude distributions is not generally a simple one; neither is it unique. To take an example, suppose that two random binary signals with the same amplitude distribution from alternative inputs to a typical linear system: the amplitude distribution of the output wave form is quite different in the two cases. Indeed a given input amplitude distribution can give rise to any number of output distributions, and the precise relationship between the two may not easily be described. There are in fact only two points of contract between the description of a random signal by its amplitude distribution and by its spectrum-namely the mean value, and the average power. The mean, or average value of a signal described by an amplitude probability density function $p(y)$ is $\bar{y} = \int_{-\infty}^{\infty} p(y)y\,dy$. The mean value of a random output signal from a linear system may, of course, be found by multiplying the mean of the input by the value of the system's response relevant zero frequency. Secondly, the average power or mean square value of a signal is given by

$$\overline{y^2} = \int_{-\infty}^{\infty} p(y) y^2 \, dy$$

and is also equal to the value of its autocorrelation function at $\tau=0$. Further, since

$$r_{xx}(\tau) = \frac{1}{2\pi} \int_{-\infty}^{\infty} P_{xx}(\omega) e^{j\omega\tau} \, d\omega$$

then

$$r_{xx}(0) = \frac{1}{2\pi} \int_{-\infty}^{\infty} P_{xx}(\omega) \, d\omega$$

which is equal to $(1/2\pi)$ times the area under the signal's power spectral density characteristic. To summarize, if we know the spectral properties of a random input signal, we may estimate the changes to its mean and mean square value as it passes through a linear system having a known frequency response. This gives us information about the first and second moments of its amplitude distribution at the output although only a complete set of moments would allow us to define the form of this distribution precisely.

There is one important type of amplitude distribution which is preserved by linear processing the normal, or gaussian distribution. When a random signal with a normal amplitude distribution forms the input to a linear system, the output signal is also normal although its mean and variance are generally different from those of the input (and so, of course, are its spectral properties). The reason for this important result may be understood by considering once again the graphical interpretation of the convolution integral. Suppose we have a normally distributed sampled-data signal forming the input to a linear system response $I(t)$ beneath the input signal, cross-multiply, and sum all the terms. Therefore each output is formed from a weighted set of input samples. But the sum of a number of normally distributed variables is itself normal; hence the output signal will be normal in this case. Indeed, this arrangement suggests that, since the sum of a large number of random variables having any form of amplitude distribution tends to be normal (this was Gauss original result), the amplitude distribution of a random output signal from a linear system will tend to be normal, even when the input is not.

Although these arguments are simpler to appreciate in the case of random sampled data signals, they apply equally well to continuous ones.

Lesson 5 Static Performance

5.1 The Ideal Measuring System

The ideal measuring system is one where the output signal has a linear relationship with the measurand, where no errors are introduced by effects such as static friction, and where the output is a faithful reproduction of the input no matter how the input varies. This is, of course, a theoretical case and serves only as a comparison for actual results obtained from a measurement. Failure of a measuring system or instrument to match up to the perfect case is usually specified in terms of errors, where error is defined as the difference between the indicated value and the 'true value'.

The term 'true value' is taken here to mean the value obtained from an instrument or measuring system deemed by experts [1] to be acceptably accurate for the purposed to which the results are being put. Thus, in calibrating a pressure gauge against a dead-weight tester,[1] the readings from the latter would be taken as 'true values'.

It is convenient to examine measuring and control system performance in two ways:

(1) when steady or constant input signals are applied, comparison of the steady output with the ideal case gives the static performance of the system;

(2) when changing input signals are applied, comparison with the ideal case gives the dynamic performance of the system.

5.2 Sensitivity

Static sensitivity is defined as the ratio of the change in output to the corresponding change in input under static or steady-state conditions.

$$\text{Static sensitivity} \quad k = \frac{\Delta \theta_o}{\Delta \theta_i} \tag{5-1}$$

where $\Delta \theta_o$ is the change in output, and $\Delta \theta_i$ is the corresponding change in input.

Sensitivity may have a wide variety of units, depending on the instrument or measuring system being considered. The platinum resistance thermometer, for example, gives a change of resistance with increase of temperature and therefore its sensitivity would have units of $\Omega/^\circ C$.

Figure 5.1(a) shows a linear relationship between output and input, and sensitivity therefore equals the slope of the calibration graph. In the case of the non-linear input/output relationship shown in Figure 5.1(b), the sensitivity will vary according to the value of the output.

Manufacturers of recording and display equipment tend to quote values of sensitivity which are the inverse of those given by the above definition; for example, an oscilloscope sensitivity would be quoted in V/cm rather than cm/V which would be expected from the definition.

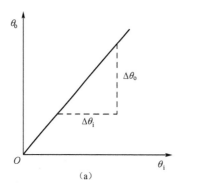
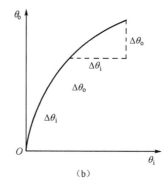

Figure 5.1 Static sensitivity

If elements of a system having static sensitivities of K_1, K_2, K_3, ···, etc. are connected in series or cascade as illustrated in Figure 5.2, then the overall system sensitivity K is given by

$$K = K_1 \cdot K_2 \cdot K_3 \cdot \cdots \tag{5-2}$$

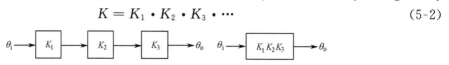

Figure 5.2 Overall system sensitivity

Provided that there is no alteration in the value of K_1, K_2, K_3, ···, etc. due to loading effects.

It is common to find system elements that have inputs and outputs of the same form—for example the voltage amplifier—and in this case the term gain, or more specifically voltage gain, is used rather than sensitivity. In mechanical system using lever systems it is more usual to use the term magnification to describe the increase in displacement, the terms sensitivity, gain, and magnification all mean the same.

Example 5.1 A measuring system consists of a transducer, an amplifier, and a recorder, with individual sensitivities as follows:

Transducer sensitivity   0.2mV/°C
Amplifier gain   2.0V/mV
Recorder sensitivity   5.0 mm/V

Determine the overall system sensitivity, using equation (5-2)

$$K = K_1 \cdot K_2 \cdot K_3 = 0.2\text{mV/°C} \times 2.0\text{V/mV} \times 5.0 \text{ mm/V} = 2.0\text{mm/°C}$$

5.3 Accuracy and Precision

Since all measurement involves error, the question 'is the system accurate?' is meaningless, since it can always be answered in the negative. What is more important is the answer to the question 'how accurate is the system?'

Accuracy is normally stated in terms of the errors introduced, where

$$\text{percentage error} = \frac{\text{indicated value} - \text{true value}}{\text{true value}} \times 100\%$$

However, it is common practice to express the error as a percentage of the measuring range of the equipment,

$$\text{percentage error} = \frac{\text{indicated value} - \text{true value}}{\text{maximum scale value}} \times 100\% \qquad (5\text{-}3)$$

For example, if a 0 to 1 bar pressure gauge is accurate to within ±5% of full-scale deflection, then the maximum error will be ±0.05. Note that if the gauge is used at the lower end of its range then the ±0.05bar error will result in a larger percentage error from the 'true value'.

Example 5.2 A 0 to 10 bar pressure gauge was found to have error of ±0.15 bar when calibrated by the manufacturer. Calculate (1) the percentage error of the gauge and (2) the possible error as a percentage of the indicated value when a reading of 2.0 bars was obtained in a test.

(1) Using equation (5-3),

$$\text{percentage error} = \frac{0.15 \text{ bar}}{10 \text{ bar}} \times 100\% = \pm 1.5\%$$

(2) Possible error= ±1.5 bar,

$$\text{error at 2.0 bars} = \pm \frac{0.15 \text{ bar}}{2.0 \text{ bar}} \times 100\% = \pm 7.5\%$$

The gauge is therefore more unreliable at the lower end of its range, and an alternative gauge with a more suitable range should be used.

'Precision' is a term that is sometimes confused with accuracy, but a precise measurement may not be an accurate measurement device is subjected to the same input on a number of occasions and the indicated results lie closely together, then the instrument is said to be of high precision. The term used to specify the closeness of results is the reproducibility of the instrument.

If a good-quality voltmeter[3] is used to measure a constant voltage on a number of different occasions and (with the limits of accuracy of the instrument) all the readings are the same, the precise readings are said to have been obtained. Suppose, however, that when putting the voltmeter away after the test, it is noticed that the pointer is offset and not reading zero. All the reading obtained would be precise but not accurate. Figure 5.3 illustrates diagrammatically the difference between the two terms.

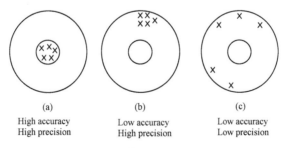

(a) High accuracy High precision

(b) Low accuracy High precision

(c) Low accuracy Low precision

×=Result, center circle represents true value.

Figure 5.3 Illustration of difference between accuracy and precision

5.4 Possible and Probable Errors

Consider a measurement that involves the use of three devices with maximum possible errors of $\pm a\%, \pm b\%$, and $\pm c\%$ respectively. It is unlikely that all three devices will have their maximum errors at the same time; therefore a more practical way of expressing the overall system error is to take the square root of the sum of the squares of the individual errors, i.e.

$$\text{root-sum-square error of overall system} = \pm \sqrt{(a^2 + b^2 + c^2)} \qquad (5\text{-}4)$$

Example 5.3 For a general measuring system where the errors in the transducer, signal conditioner, and recorder are $\pm 2\%$, $\pm 3\%$, and $\pm 4\%$ respectively, calculate the maximum possible system error and the probable or root-sum-square error.

$$\text{maximum possible error} = \pm(2+3+4)\% = \pm 9\%$$

Using equation (5-4),

$$\text{root-sum-square error} = \pm \sqrt{(2^2 + 3^2 + 4^2)}\% = \pm \sqrt{29}\% = \pm 5.4\%$$

Thus the error is possibly as large as $\pm 9\%$ but probably not larger than $\pm 5.4\%$.

5.5 Other Static-Performance Terms

Reproducibility A general term applied to the ability of a measuring system of instrument to display the same reading for a given input applied on a number of occasions.

Repeatability The reproducibility when a constant input is applied repeatedly at short intervals of time under fixed conditions of use.

Stability The reproducibility when a constant input is applied over long periods of time compared with the time of taking a reading conditions of use.

Constancy The reproducibility when a constant input is presented continuously and the conditions of test are allowed to vary within specified limits, due to some external effect such as a temperature variation.

Range The total range of values which an instrument or measuring system is capable of measuring.

Span The range of input signals corresponding to the designed working range of the output signal.

Tolerance The maximum error.

Linearity The maximum deviation from a linear relationship between input and output i.e. from a constant sensitivity-expressed as a percentage of full scale.

Resolution The smallest change of input to an instrument which can be detected with certainty, expressed as a percentage of full scale.

Dead-band The largest change of input to which the system does not respond due to friction or backlash effects, expressed as a percentage of full scale.

Hystersis The maximum difference between reading for the same input when approached from opposite directions-i.e when increasing and decreasing the input-expressed as a percentage of full scale.

Words and Expressions

accuracy ['ækjurəsi] *n.* 准确度
alternative [ɔːl'təːnətiv] *adj.* 替换的，交替的，（供选择的）比较方案
cascade [kæs'keid] *n.* ; *adj.* ; *vt.* 串联，串级
constancy ['kɔnstənsi] *n.* 不变，恒定
corresponding to　符合于，相应于
dead-band [dedbænd] 死区
deem [diːm] *vt.* 认为，认定
diagram ['daiəgræm] *adj.* 图表的，简图的，用图表示的
diagrammatically [ˌdaiəgrə'mætikəli] *adv.* 利用图表地，用图解法地
faithful ['feiθful] *adj.* 可靠的，正确的
hysteresis [ˌhistə'riːsis] *n.* 滞后
hysteresis over　滞后，误差
linearity [ˌlini'æriti] *n.* 线性度
magnification [ˌmægnifi'keiʃən] *n.* 放大，放大倍数
measurand ['meʒərənd] *n.* 被测量
occasion [ə'keiʒən] *n.* 时机，机会，场合
offset ['ɔːfset] *n.* ; *vt.* 偏移
oscilloscope [ə'siləskəup] *n.* 示波器
performance [pə'fɔːməns] *n.* 特性
precision [pri'siʒən] *n.* 精确度
put … away　拿开，送走
reproducibility [riːprəˌdjuːsə'biliti] *n.* 可再现性
resolution [ˌrezə'luːʃn] *n.* 分辨率
static friction　静态摩擦
sensitivity [ˌsensi'tiviti] *n.* 灵敏度
span [spæn] *n.* 范围
tolerance ['tɔlərəns] *n.* 公差

Notes

1. measuring system deemed by expert　由专家认定的测量系统
2. dead-weight tester　净重测量仪
3. a good-quality voltmeter　高质量的电压表

Exercises

Translate the following passages into English or Chinese.

1. 当被测量不随时间变化或变化缓慢时，输出 y 与 x 之间的关系称为静态特性；当被测量随时间迅速变化时，输出 y 与 x 之间的关系称为动态特性。
2. 测量是为了准确了解被测物理量，但是人们通过测量是永远测不到被测物理量的真实值的，只能观测到经过测试系统各个环节对被测物理量传输后的输出量。研究系统的特性就是为了能使系统尽可能准确、真实地反映被测物理量。
3. A random error is due to a large number of independent small effects that cannot be identified or controlled, it is a statistical quantity. As such, it will vary for each replication of the observations. If a large number of readings is observed for the same quantity, the scatter of the data about a mean value can be evaluated. The scatter generally follows a guassian distribution about a mean value, which is assumed to be the true value.
4. Accuracy is the deviation of the output from the calibration input or the true value. If the accuracy of a voltmeter is 2% full scale as described in the preceding section, the maximum deviation is ±2units for all readings.

Reading Material

Noncontact Temperature Measurement

Any object at any temperature above absolute zero radiates energy. This radiation varies both in intensity and in spectral distribution with temperature. Hence, temperature may be deduced by measuring either the intensity or the spectrum of the radiation.

The total energy density radiating from an ideal 'blackbody' (more on that later) is given by the Stefan-boltzmann law, $E=\sigma T^4$, where E is energy density in W/cm^2, σ is the Stefan-boltzmann constant (5.6697×10^{-12} $W/cm^2 K^4$) and T is the absolute temperature (K). In other words, the total radiated energy is proportional to the fourth power of the absolute temperature.

All objects, particularly ideal blackbody objects, also absorb incident radiation. Given time to equilibrate, and presuming they are insulated from the heating or cooling effects of surrounding air or other materials, they will eventually reach a point where they absorb and radiate energy at equal rates. One consequence of this is that if an object (a temperature sensor, for example) is an ideal blackbody, is perfectly insulated, and is flooded on its entire surface with radiation from a radiating source, it will eventually reach an equilibrium sources and blackbody calibration sources are available, the temperature of the sensor is a measure of the temperature of the radiating object.

An infrared radiation thermometer may be created in a manner similar to that in Figure 5.4 the radiated energy from the hot (or cold) object is focused on a temperature sensor, whose temperature then is indicative of the intensity of the radiation falling upon it. The sensor should be small and low mass for reasonable response time. Thermistors offer high sensitivity for low temperature measurements while thermocouples provide the operating range necessary for high levels of radiated energy. In some designs, the sensor is insulated from ambient conditions by placing it in a vacuum. The sensor's output is amplified, linearized, and fed to an output indicator or recorder.

The optics are apt to be a bit different than shown in diagram. In most applications, particularly at lower temperatures, much of the radiation will be far infrared, which is not passed well by most glasses. It may be preferable to use a reflective concave mirror to focus the incoming energy, rather than a lens. There may also be a red or infrared filter over the inlet to keep down interference due to stray ambient light. For higher temperature use it may be necessary to reduce the total incoming energy using a gray filter, shutter, or other obstruction. The Stefan-boltzmann law, and the proper operation of these thermometers, presumes that the radiation is coming from a perfect blackbody radiator. To oversimplify and it is not our intention here to which does not reflect any radiation which may fall upon it. All incident energy is absorbed. A non-blackbody object which reflects external radiation will also reflect internally generated radiation, lowering the amount of energy radiated at any given temperature.

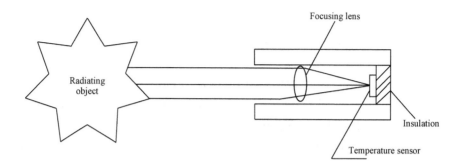

Figure 5.4 Noncontact infrared temperature measurement

Any surface has a reflectivity and an emissivity. Reflectivity, r, is simply the ratio of reflected energy to incident energy; a perfect reflector has a reflectivity of one; a blackbody, zero. Emissivity, ε, turns out to be simply $\varepsilon = 1 -$ reflectivity. A perfect blackbody has emitted by an object at a given temperature is proportional to its emissivity; a reflectivity object has emissivity (we expect more heat from a rough, black radiator than from a smooth, polished one).

All this has a serious impact on radiation thermometry. An infrared radiation thermometer calibrated against a blackbody radiator will read seriously low when aimed at a reflective object. Most commercial radiation thermometers include a control allowing the user to dial in the emissivity of the object being measured, plus a table of typical emissivity values. Mist organic and nonmetallic materials have emissivity values. Most organic and nonmetallic materials have emissivities between 0.85 and 0.95, while metals range roughly between 0.1 and 0.5 (interestingly, both white and black paints have similar emissivities—between 0.9—at temperatures up to 100℃).

Variations in emissivitiy can cause serious errors, especially with metal surfaces. Highly polished surfaces have lower emissivity still farther. As an oxidation or coating of the surfaces raises emissivity still farther. As an example, the emissivity of stainless steel at 800℃ is when polished, 0.5 when rough machined, 0.7 when rough machined and lightly oxidized and 0.8 to 0.9 when heavily oxidized. If at all possible, the surface to be measured should be painted, oxidized, or otherwise made black and nonreflective. Liquid metals, a frequent application for infrared thermometry, are not as variable in their emissivity, but may be affected by layers of slag on their surface. It is a good ideal to calibrated the infrared reading by making a contact temperature measurement or in the case of liquid metal, by plunging in a thermocouple as described in the previous section.

Also affecting the readings are atmospheric attenuation. Water vapor strongly attenuates certain infrared wavelengths while dust smoke, and particulate matter will attenuate the radiation between the source and the sensor. Such problems are apt to be most troublesome in industrial applications.

The dependence of the measurement upon emissivity can be reduce by the use of two-color pyrometry. As was mentioned at the start of this section, both the intensity and the spectral distribution of the radiation vary with temperature. The radiant intensity at any wavelength, λ, is given by

$$J = \frac{C_1 \varepsilon \lambda^{-5}}{\exp(C_2/\lambda T) - 1}$$

where J is the radiant energy, ε is the emissivity, λ is the wavelength, and T is the absolute temperature (K). On the assumption that emissivity is not a function of wavelength (this assumption is not entirely true) the ratio of intensities at two wavelengths becomes

$$\frac{J_1}{J_2} = \frac{\lambda_1^{-5}/[\exp(C_2/\lambda_1 T) - 1]}{\lambda_2^{-5}/[\exp(C_2/\lambda_2 T) - 1]}$$

which may be simplified to

$$\frac{J_1}{J_2} = \text{const}_1 \exp\left(\frac{\text{const}_2}{T}\right)$$

where

$$\text{const}_1 = (\lambda_2/\lambda_1)^5$$
$$\text{const}_2 = C_2(1/\lambda_2 - 1/\lambda_1)$$

Lesson 6 Dynamic Performance

Many industrial processes require the measurement of parameters which remain constant or change every slowly, for example a constant pressure or temperature in a chemical process. In cases like this, the static performance of the measuring system is of prime concern. However, with the increasing of automation, a greater emphasis is being placed on whether a device can respond adequately to changing signals. If a transducer responds sluggishly to a sudden change of input parameter, then automatic control of that parameter may become difficult if not[1] impossible. As a further example, consider a vibration-measuring system, where the parameter is by its very nature [2] a changing quantity. If the system could not respond to the frequencies of the vibration then the results would be totally useless.

The dynamic perform of both measuring and control systems is extremely important and is specified by responses to certain standard test inputs. There are:

(1) The step input, which takes the form of an abrupt change from one steady value to another. This indicate how well the system can cope with the change and results in the transient response.

(2) The ramp input, which varies linearly with time and gives the ramp response, indicating the steady-state error in following the input.

(3) The sine-wave input, which gives the frequency response or harmonic response of the system. This shows how the system can respond to inputs of a cyclic nature as the frequency f Hz or ω rad/s (where $\omega=2\pi f$) varies.

These three inputs illustrated in Figure 6.1 and may be encountered in front section. However, in this chapter only the transient and frequency response will be considered, since these will be needed to interpret measuring-system specifications.

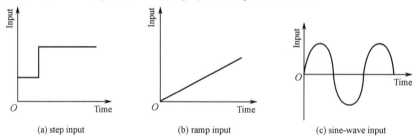

(a) step input   (b) ramp input   (c) sine-wave input

Figure 6.1 Different forms of dynamic test inputs

All systems will respond to some extent fail to follow exactly a changing input, and a measure of how well a system will respond is indicated by its dynamic specifications. These are expressed as step or transient parameters or frequency-response parameters, depending on the type of input applied. Many systems, although different in nature, produce identical forms of response, and this is due to the fact that the system dynamics are similar, the dynamic or differential equations are of the same form.

6.1 Zero-order Systems[3]

The ideal measuring system mentioned in section 5.1 is one whose output is proportional to the input no matter how the input varies, the mathematical equation relation them is of the form,

$$\theta_o = K\theta_i \tag{6-1}$$

where K is the sensitivity of the system.

This is the equation of a zero-order system, since there are no differential coefficients present. Alternatively, equation (6-1) can be expressed in terms of the ration θ_o/θ_i to give

$$\frac{\theta_o}{\theta_i} = K \tag{6-2}$$

and this can represented by the block diagram shown in Figure 6.2.

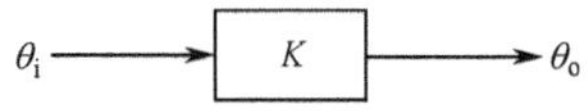

$$\theta_i \longrightarrow \boxed{K} \longrightarrow \theta_o$$

Figure 6.2 Block-Diagram Representation of a zero-order system

In practice the measuring system which approaches the ideal zero-order system is the potentiometer, which gives an output voltage proportional to the displacement of the wiper.

6.2 First-order Systems

A first-order system is one whose input/output dynamics are represented by a first-order differential equation of the form,

$$a\frac{d\theta_o}{dt} + b\theta_o = c\theta_i$$

where θ_o is the output variable, θ_i is the input variable, and a, b, and c are constants.

This equation can be rewritten to obtain a unity coefficient of θ_o, giving

$$\frac{a}{b} \cdot \frac{d\theta_o}{dt} + \theta_o = \frac{c}{b} \cdot \theta_i \tag{6-3}$$

and this can be expressed in a standard form as

$$\tau \frac{d\theta_o}{dt} + \theta_o = K\theta_i \tag{6-4}$$

and K is the static sensitivity, with the units of the units of the ratio θ_o/θ_i.

Comparison of equation (6-3) and (6-4) shows that

$$\tau = \frac{a}{b} \quad \text{and} \quad K = \frac{c}{b}$$

equation (6-4) can be written in terms of the D operator, where

$$D \equiv \frac{d}{dt} \quad \text{and} \quad D^2 \equiv \frac{d^2}{dt^2} \quad \text{etc.}$$

hence $\tau D\theta_o + \theta_o = K\theta_i$, i.e. $(\tau D+1)\theta_o = K\theta_i$,

$$\frac{\theta_o}{\theta_i} = \frac{K}{1+\tau D} \tag{6-5}$$

the ratio θ_o/θ_i expressed in terms of the D operator is known as the transfer operator of the system.

For the first-order systems, equation (6-5) represents the standard form of the transfer operator and can be represented by the block diagram in Figure 6.3, note that equation (6-3), (6-4) and (6-5) are all alternative ways of expressing the same differential equation.

$$\theta_i \longrightarrow \boxed{\frac{K}{1+\tau D}} \longrightarrow \theta_o$$

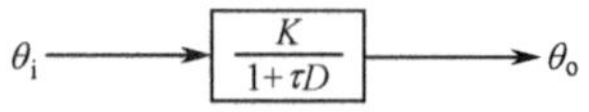

Figure 6.3 Block-diagram Representation of a first-order system

Example of first-order systems include:

(1) the mercury-in-glass thermometer, where the heat conduction through the glass bulb to the mercury is described by a first-order differential equation.

(2) the build-up of air pressure in a restrictor/bellows system.

(3) a series resistance-capacitance network.

Example 6.1 The differential equation describing a mercury-in-glass thermometer is

$$4\frac{d\theta_o}{dt} + 2\theta_o = 2 \times 10^{-3}\theta_i$$

where θ_o is the height of the mercury column in meters and θ_i is the input temperature in ℃. determine the time constant and the static sensitivity of the thermometer.

For the standard form shown in equation (6-4), the θ_o term must have unity coefficient and therefore dividing all terms by 2 gives

$$2\frac{d\theta_o}{dt} + \theta_o = 1 \times 10^{-3}\theta_i$$

comparing this with equation (6-4), i.e.

$$\tau\frac{d\theta}{dt} + \theta_o = K\theta_i$$

It can be seen that

$$\tau = 2\text{s}$$

and $K = 10^{-3}\text{m/℃}$ or 1mm/℃.

The standard form indicated in equation (6-4) and (6-5) is very convenient, because a first-order system always produces a standard response to either a step or a sin-wave input.

6.2.1 Step response

Figure 6.4(a) shows the exponential rise to the final value which is characteristic of the first-order system. The dynamic error is the difference between the ideal and actual responses, and comparison of the two shows that this error decrease with time. The step response is shown in more detail in Figure 6.4(b) and from this the definition of time constant is obtained:

Time constant is the time taken to reach the final value if the initial rate had been maintained, or the time taken to reach 63.2% of the step change.

It is important to note the initial slope at time $t=0$, as this distinguishes it from a second-

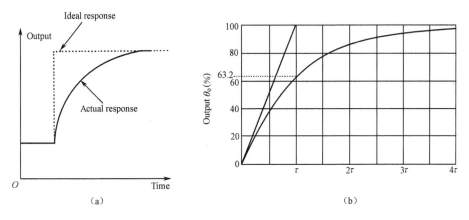

Figure 6.4 The step response of a first-order system

order step response, which has zero initial slope.

6.2.2 Frequency response

This response is obtained by applying sine waves of a known amplitude at the input and examining how the output responds as the frequency of the sine wave are varied, Figure 6.5 illustrates the inability of the system to follow the input faithfully, and it can be seen that the output lags behind the input. As the frequency is increased, the output falls further behind and perhaps more importantly-decreases in amplitude.

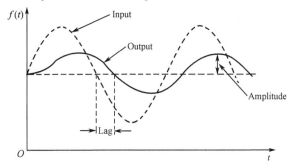

Figure 6.5 Output response to a sin-wave input

The ratio of output amplitude to input amplitude is called the amplitude ratio and should equal a constant irrespective of the input frequency, i. e. $\theta_o = K\theta_i$ at all times. Figure 6.6 illustrates the standard frequency-response curve for a first-order system, showing the variation of amplitude ratio with frequency. It is valid for $K=1$ or if the amplitude ratio is regarded as the ratio of actual output amplitude to the ideal output amplitude. The frequency axis is in generalized $\omega\tau$ form and enables the curve to be used for different values of time constant. To obtain the ω values in rad/s for a particular time constant τ, the horizontal-axis units would have to be multiplied by the factor $1/\tau$.

Example 6.2 If a first-order measuring system has a time constant of 0.01s, determine the approximate range of input signal frequencies that the system error could follow to within 10%.

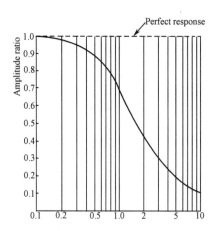

Figure 6.6 Frequency response of a first-order system

From Figure 6.6 the amplitude ratio is greater than 0.9, i.e. there is less than 10% error, up to $\omega\tau \approx 0.5$,

$$\therefore \omega\tau = 0.5 \text{ define the upper frequency limit}$$

i.e. $\omega = \dfrac{0.5}{0.01} \text{rad/s} = 50 \text{rad/s}$

Therefore the required frequency range is from 0 to $(50/2\pi)$ Hz. i.e. from 0 to 8 Hz.

Example 6.3 A first-order system has a time constant of 6 ms determine the frequency corresponding to the condition $\omega\tau = 1$ and calculate the approximate percentage error at this frequency,

$$\omega = \frac{1}{\tau} = \frac{1}{6 \times 10^{-3}} \text{rad/s} = 166.7 \text{rad/s}$$

$$\therefore \text{ frequency } \quad f = \frac{166.7}{2\pi} \text{Hz} = 26.5 \text{Hz}$$

from Figure 6.6, the amplitude ratio ≈ 0.7 at $\omega\tau = 1$,

$$\therefore \% \text{ error} = 30\%$$

6.3 Second-order Systems

A second-order system is one whose input/output relationship is described by

$$a \frac{d^2\theta_o}{dt^2} + b \frac{d\theta_o}{dt} + c\theta_o = e\theta_i \tag{6-6}$$

where a, b, c and e are constants.

This can be rearranged to give a unity coefficient of the $\dfrac{d^2\theta_o}{dt^2}$ term, i.e.

$$\frac{d^2\theta_o}{dt^2} + \frac{b}{a} \frac{d\theta_o}{dt} + \frac{c}{a}\theta_o = \frac{e}{a}\theta_i$$

and this can be written in a standard form as

$$\frac{d^2\theta_o}{dt^2} + 2\xi\omega_n \frac{d\theta_o}{dt} + \omega_n^2 \theta_o = K'\theta_i \tag{6-7}$$

where ω_n is the undamped natural frequency in rad/s, ξ ('zeta') is the damping ratio, and K' is a constant which would equal ω_n^2 if θ_i were equal θ_o under static conditions.

In terms of the D operator, equation (6-7) becomes

$$(D^2 + 2\xi\omega_n D + \omega_n^2)\theta_o = K'\theta_i$$

giving the transfer operator,

$$\frac{\theta_o}{\theta_i} = \frac{K'}{D^2 + 2\xi\omega_n D + \omega_n^2} \tag{6-8}$$

alternatively, diving through by ω_n^2 gives

$$\frac{\theta_o}{\theta_i} = \frac{K}{(1/\omega_n^2)D^2 + (2\xi/\omega_n)D + 1} \tag{6-9}$$

where K is the static sensitivity.

This form of the transfer operator can by represented by the block diagram shown in Figure 6.7.

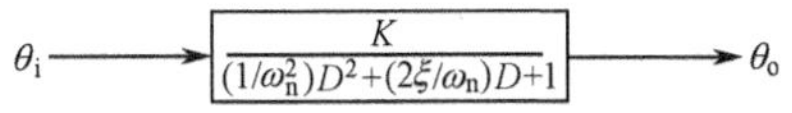

Figure 6.7 Block-diagram representation of a second-order system

ω_n is a measure of the speed of response of the second-order system, a high ω_n would mean that the system would more rapidly to sudden changes. ξ is a measure of the damping present in a system and is equal to the ratio of actual damping to critical damping, its value determines the forms of the step and frequency responses, as follows.

(1) when $\xi < 1$ this system is here said to be underdamped and results is in oscillations occurring in the step response and (for values of $\xi < 0.707$) resonance effects in the frequency response. 'resonance' here means an output signal greater in amplitude than the idea output.

(2) when $\xi = 1$ this is the critically damped condition; no oscillations or overshoots appear in the step response and there is no resonance in the frequency response. This is the point of change-over from an underdamped condition to an overdamped condition.

(3) when $\xi > 1$ here the system is overdamped and responds in a sluggish manner, again with no overshoot in the response and no resonance in the frequency response.

The most common example of a second-order system is a mass-spring system with damping. A large number of devices and mechanisms are of this type, u. v. galvanometer,[4] the piezo-electric transducer, and the open-control system on x-y plotters.

6.3.1 Step response

The step response for second-order systems does not have one unique form but may have one of an infinite number depending on the value of ξ.

Figure 6.8 shows the step responses for a number of ξ values, and these illustrate the fact that as the damping in the system reduces so the overshoot and oscillation increase. A generalized time scale in terms of $\omega_n t$ is used to enable the same curves to apply to very fast systems with high ω_n values and to slow systems with low values of ω_n. A useful curve showing the relationship between the first overshoot and the damping ration is shown in Figure 6.9, and this can be used to estimate ξ from an underdamped step response. The term 'percentage overshoot' is used to express the magnitude of the first overshoot as a percentage of the final or steady-state value.

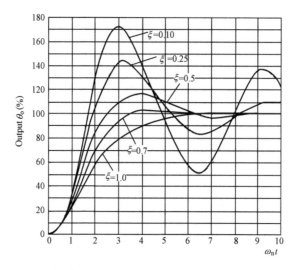

Figure 6.8   Step response of a second-order system

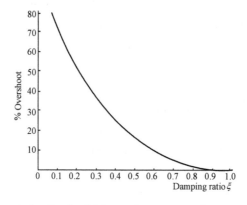

Figure 6.9   Graph of % overshoot against damping ratio ξ

Example 6.4   A mass-spring-damper system has a first overshoot of approximately 40% of its final value when subjected to a step input force. Estimate the values of the damping ratio ξ and ω_n if the time taken to reach the first overshoot is 0.8s from the application of the step.

From Figure 6.9, $\xi=0.28$ results in a 40% overshoot. If an approximation of $\xi=0.25$ is made then Figure 6.8 can be used to determine ω_n,

units at the first overshoot $\qquad \omega_n t = 3.2$

$$\therefore \omega_n = \frac{3.2 \text{rad}}{0.8 \text{s}} = 4 \text{ rad/s}$$

The technique outlined in example 6.4 provides a reasonably quick and convenient method of estimating the value of ξ from the percentage overshoot of the step response, provided the standard curve are available.

As far as the best step response is concerned, [5] it can be seen that lower ξ values exhibit a faster response but increased overshoot, while higher values result in very sluggish responses but no overshoot. The optimum condition is therefore a compromise between an acceptable speed of response and amount of overshoot, and a value $\xi \approx 0.7$ is usually stated for most second-order measurement systems.

6.3.2 Frequency response

Figure 6.10 illustrates a typical set of frequency-response curves for a second-order system. Examination shows that the amplitude-ratio axis is correct for $K=1$, and a generalized-frequency axis is plotted to allow use of the curves for any second-order system. Resonance effects are observed for lightly damped cased as the frequency of the input approaches the natural frequency of the system. This abnormally high output response occurs only for damping ratios of less than 0.7. As the input frequency is increased well beyond the natural frequency, then the amplitude ratio falls as the system fails to respond to the higher rates of change.

The ideal frequency response for either a measuring system or an automatic-control system would be one which had an amplitude ratio of unity for all frequencies. The nearest response to this is for a ξ value of between 0.6 and 0.7, which has a constant amplitude ratio within $\pm 3\%$ of unity for a range of frequencies up to $\pm 60\%$ of the undamped natural frequency f_n.

In both the step and the frequency responses the optimum value of ξ lies between 0.6 and 0.7, with 0.64 often quoted, and manufacturers of measuring system must ensure that the correct amount of damping is present. The word 'optimum' is used here to indicate the best response to step and sine-wave signals, but there are devices which are specifically designed with overdamped so as not to follow varying signals but to give average values. In these cases a higher value of ξ would be optimum.

Example 6.5 In a frequency-response test on a second-order system, resonance occurred at a frequency of 216Hz, giving a maximum amplitude ratio of 1.36. Estimate the values of ξ and ω_n for the system.

Examination of Figure 6.10 indicates that $\xi=0.4$ gives an amplitude ratio of 1.4,

$$\therefore \xi \approx 0.4$$

this resonance occurs at $\omega/\omega_n=0.8$,
but $\omega=2\pi f=2\pi \times 216\text{Hz}=1357\text{rad/s}$,

$$\omega_n = \frac{\omega}{0.8} = \frac{1357\text{rad/s}}{0.8} = 1696\text{rad/s}$$

Note: resonance occurs at frequencies which are lower than the undamped natural frequency except in the one case where the damping is zero.

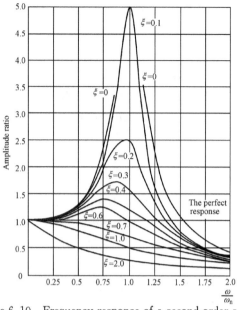

Figure 6.10　Frequency response of a second-order system

6.4　Step-response Specification

Three terms are used to specify a system's step response time, rise time, and settling time and these are defined as follows.

Response time (t_{res})　the time taken for the system output to rise from 0 to the first crossover point of 100% of the final steady-state value. Applicable only to underdamped systems.

Rise time (t_r)　the time taken for the system output to rise from 10% to 90% of its final steady-state value.

Settling time (t_s)　the time taken for the system output to reach and remain within a certain percentage tolerance band of the final steady-state value. Typical values would be 2% and 5% settling times.

These parameters are illustrated in Figure 6.11, which shows a step response containing oscillations. In the figure, t_s refers to the 5% settling, which is taken to reach and remain within the tolerance band of 95% to 105% of the final value.

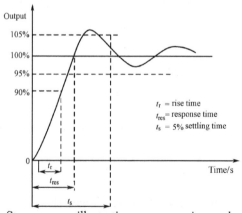

Figure 6.11　Step response illustrating response, rise, and settling times

6.5 Frequency-response Specification

There are numerous ways of specifying the frequency response of measuring systems, instruments, and control systems and this can lead to some confusion. A distinction can be made between devices which are designed to work over a range or band of frequency-called AC devices-and those which can respond to DC, i. e. down to zero frequency, called DC devices.

6.5.1 AC devices

For devices of this type, the gain or amplitude ratio is usually constant over a given frequency range, but at low or high frequencies the gain falls off, as illustrated in Figure 6.12. the term used to specify the frequency range is bandwidth, which is equal to (f_2-f_1) Hz.

Bandwidth is the range of frequencies between which the gain or amplitude ratio is constant to within -3dB (this corresponds to a reduction in gain).

Thus an AC amplifier on an oscilloscope will have a typical specification of a bandwidth of 8Hz to 10Hz, which means that at 8Hz and 10Hz the trace size for a constant-amplitude input wave will be 70% of that for a mid frequency, say 1kHz. Therefore at these frequencies the value read from the oscilloscope screen will be 30% lower than the actual values.

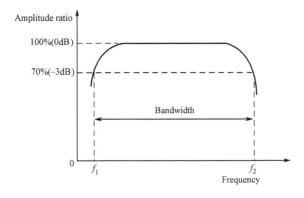

Figure 6.12 Typical frequency response of an AC device

6.5.2 DC devices

In this context,[6] DC device does not mean a device which will respond only to steady or DC, signals but on which will respond to both AC and DC signals. A typical frequency response for devices of this is shown in Figure 6.13 and might apply, for example, to a pen-recorder. Here the operating range may be expressed as the upper frequency at which the gain or amplitude ratio falls outside a certain tolerance band. A value of $\pm 3\%$ is illustrated in Figure 6.13, but other common value are ± 5, -3dB as mentioned in or ± 1dB which is equivalent to $\pm 10\%$.

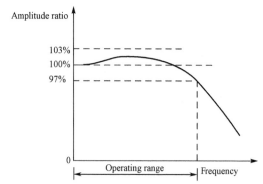

Figure 6.13 Typical frequency response of a DC device

Words and Expressions

abrupt [ə'brʌpt] *adj.* 突然的，意外的
adequately ['ædikwitli] *adv.* 足够地，充分地，相当地，满足要求地
bulb [bʌlb] *n.* 球，球管（头），（温度计）水银球
comparison [kəm'pærisn] *n.* 比较，对照
compromise ['kɔmprəmaiz] *n.*；*vt.* 妥协，兼顾，综合
concern [kən'sə:n] *vt.*；*n.* 与……有关，涉及，关心，担心，挂念
confusion [kən'fju:ʒn] *n.* 混乱，混淆
cope with *vt.* 解决
context ['kɔntekst] *n.* 上下文，前后关系
encounter [in'kauntə] *vt.* 遇到
exponential [ˌekspəu'nenʃəl] *adj.* 指数的
generalized ['dʒenərəlaizd] *adj.* 一般的，普通的，扩大的
interpret [in'tə:prit] *vt.* 说明，解释
irrespective of 考虑，注意，关心
lag [læg] *vt.* 落后，滞后
lag behind 不管，不顾
outline ['autlain] *n.*；*vt.* 外形，轮廓，大纲，画轮廓，概述
overshoot ['əuvəʃu:t] *n.* 过调量，超调量
ramp [ræmp] *n.* 斜坡
resonance ['rezənəns] *n.* 共振
restrictor [ris'triktə] *n.* 节流阀，限流阀，限制器
step input 阶跃响应
trace [treis] *n.* 踪迹，图形
transient ['trænziənt] *adj.* 瞬态的
wiper ['waipə] *n.* 滑动针

Notes

1. if not   甚至
2. by its (very) nature   就其本性而言
3. zero-order system   零阶系统
4. u. v. galvanometer   紫外线镜反射式电流计
5. as far as ... is concerned   就……而论
   as far as the best step response is concerned   就最好的阶跃响应而论
6. in this context   在这方面，由于这个原因，在这个意义上

Exercises

Translate the following passages into English or Chinese.

1. 输出信号稳态值与响应曲线在垂直方向的差值称为系统的动态误差。它与时间 t 有关，当 $t \to \infty$ 时，动态误差趋于零。显然，在相同时刻，输出量与输入量之间的差异也越小，所以应尽可能采用时间常数 τ 值小的测试系统。
2. 实际的测试系统往往难以做到完全符合不失真测试条件，被测信号也不可能包含所有的频率分量。根据测试精度的要求，被测信号的频带宽度应处于测试的带宽。
3. Uncertainty is generally stated as a number, indicating the tolerance from the true value of the measurand. The tolerance is only estimated. It represents the confidence level of the investigator in the results, since the true value of the measurement is unknown.
4. The purpose of the sensor is to obtain dimensional information from the workpiece. It is like a transducer in many instances because it converts one energy form to another. This other energy form is always an electrical signal, since we are considering sensors which provide an electrical signal to be used as feedback to the process or machine control.

Reading Material

Eddy Current

The eddy current sensor is similar in concept and performance to the capacitive sensor. It does have some differences, though, that may make it more suitable in some applications. The major difference between the two sensors is the principle of operation.

The eddy current sensor utilizes an electromagnetic field as opposed to the capacitive electric field. It is the electromagnetic field that makes this noncontacting sensor much less sensitive to the effect of excited by a high-frequency alternating-current source. The resultant alternating magnetic field emanating from the coil generates eddy currents in the near surface of the material being inspected. These currents, in turn, create their own magnetic field, which couples to the coil, superimposing a current on the driving current. Demodulating circuitry detects this current signal, which can be calibrated to derive a distance is possible since the field strengths are a function of the distance from the sensor to the target.

Progressing required for the analog output of this type of sensor is identical to that for the capacitive sensor. Resolution is limited by the A/D converter. Spot size, range, and standoff distance are comparable to those available in a capacitive device. A conductive work piece is required to support the induced currents.

Concerning spot size, the eddy current probe has limitations in terms of how small a coil, with enough turns to generate a sufficient magnetic field, can be wound. Since fringe fields have an effect on the output, the eddy current probe is also limited to constant-geometry situations where these errors can be determined and calibrated out. Since this device depends on eddy currents, which are a near-surface phenomenon, there are other variables peculiar to this sensor that must understood.

The strength of the induced magnetic field is a function of the condition of the material. Because different materials have different resistivities, the sensor must be calibrated for a specific material. In addition, other material characteristics such as porosity and density affect the output. Near-surface conditions including defects such as cracks or inclusions alter the output and may be indistinguishable from changes in the distance being measured. Indeed, the primary use of this type of sensor is for surface and near-surface material defects, to which it is very sensitive.

Ultrasound

Sound waves can be utilize by ranging to obtain dimensional information. The configuration of the sensor involves a transmitter of sound energy and receiver. In many cases the transmitter and receiver are in the same unit. Distance information is obtained by measuring the transit time required for the echo the return. Since the speed of sound in the medium is known, distance can be determined directly from time-of-flight information.

Ultrasonic energy, which is beyond the audible frequency range, is normally utilized since the wavelength of audible sound energy is relatively long compared with the resolution required for most dimensional measurements. For high-resolution applications, with measurement resolutions finer than 0.1 in., a liquid couplant (typically water) is required because the sound energy at the higher frequencies is highly attenuated in air. In an inprocess measurement application, water-or oil-based cutting fluids can serve as a coupling medium. The workpiece under measurement need not be immersed as a continuous stream of liquid can adequately convey the sound.

The transmitter/receiver for an ultrasonic sensor is typically a piezoelectric crystal. The crystal itself is mechanically damped in order to attenuate oscillation and avoid masking the weak incoming echoes. The pulser is a fast, high-voltage switch that drives the transducer with a short rise time pulse that is converted into a mechanical pressure wave. The longitudinal sound wave is conveyed by the medium to the target, where it is reflected back to the transducer. The same crystal or an identical one converts the echo to an electrical impulse, which is amplified by a tuned amplifier.

A timer is generally triggered by the pulse unit and disabled by receipt of the incoming echo. The value held by the timer is the time of flight of the sound wave, that is, the time it takes the sound is constant for homogenous materials at a fixed temperature, the distance to the target can be determined by multiplying the speed by the time and dividing the result by two to get a single path length. Digital counter/timers are presently utilized with ultrasonic sensors and provide direct digital information for transmission over serial or parallel data links.

The spot size of the noncontact ultrasonic sensor is determined by the size of the wave front of the sound energy. This size id in turn governed by the configuration of the transducer, which may have a lens incorporated to it to permit a focused spot. This spot size may range from approximately 0.05 to 1.0 contacting transducers discussed here, tends to average the information received from the target.

Other different form sensors, however, the returning sound waves reflected from varying distances to the target integrated over the spot size cause a dispersion of the item signal which may result in trouble establishing an exact time reading, especially in cases where a constant thresholding circuit encounters varying amplitude signals. This problem can be overcome, however, when the geometry of the workpiece is known or when the sensor is kept normal to the workpiece. Under these circumstances, accuracies for this type of sensor of 0.0001 to 0.001 in. can be achieved.

Standoff distances in the range of a fraction of an inch to tens of feet are possible with ultrasonic transducers, again, as previously inferred, with a ling standoff distance and a lower-frequency sound wave, resolution miss be compromised. In a machining application, accuracies mentioned in the previous paragraph may be realized with standoff distances in the range of a fraction of an inch to several inches. Like the eddy current probe, the ultrasonic sensor has the added benefit of being able to detect material defects. Unlike the eddy current probe, however, this sensor's sound energy can penetrate more deeply into the material and, by utilizing clever time discrimination of the resultant echoes, surface measurements may be distinguished from subsurface material defects, allowing the sensor to be used for both dimensional mensuration and material integrity of the workpiece.

Lesson 7 Basic Knowledge of Transducers and Resistance Transducers

A transducer is a device which converts the quantity being measured into an optical, mechanical, or-more commonly-electrical signal. The energy-conversion process that takes place is referred to as transduction.[1]

Transducers are classified according to the transduction principle involved and the form of the measurand. Thus a resistance transducer for measuring displacement is classified as a resistance displacement transducer. Other classification examples are pressure bellows, force diaphragm, pressure flapper-nozzle, and so on.

7.1 Transducer Elements

Although there are exceptions, most transducers consist of a sensing element and a conversion or control element, as shown in the two-block diagram of Figure 7.1.

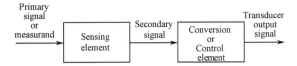

Figure 7.1 Two-block-diagram representation of a typical transducer

For example, diaphragms, bellows, strain tubes and rings, bourdon tubes, and cantilevers are sensing elements which respond to changes in pressure or force and convert these physical quantities into a displacement. This displacement may then be used to change an electrical parameter such as voltage, resistance, capacitance, or inductance. Such combination of mechanical and electrical elements form electromechanical transducing devices or transducers. Similar combinations can be made for other energy input such as thermal. Photo, magnetic and chemical, giving thermoelectric, photoelectric, electromagnetic, and electrochemical transducers respectively.

7.2 Transducer Sensitivity

The relationship between the measurand and the transducer output signal is usually obtained by calibration tests and is referred to as the transducer sensitivity K_1, i.e.

$$K_1 = \frac{\text{output-signal increment}}{\text{measurand increment}}$$

In practice, the transducer sensitivity is usually known, and, by measuring the output signal, the input quantity is determined from

$$\text{input} = \frac{\text{output-signal increment}}{K_1}$$

The following example, in which a spring shown in Figure 7.2 deflects 0.05m when subjected to a force of 10kN, find the input force for an output displacement of 0.075m.

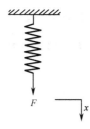

Figure 7.2 Loaded spring

$$\text{Sensitivity} \quad K_1 = \frac{x}{F} = \frac{0.05}{10\text{kN}}$$

∴ input force requtred for 0.075m deflection $= 0.075\text{m} \times \frac{10\text{kN}}{0.05\text{m}} = 15\text{kN}$

7.3 Characteristics of an Ideal Transducer

The high transducer should exhibit the following characteristics.

(1) High fidelity-the transducer output waveform shape be a faithful reproduction of the measurand; there should be minimum distortion.

(2) There should be minimum interference with the quantity being measured; the presence of the transducer should not alter the measurand in any way.

(3) Size. The transducer must be capable of being placed exactly where it is needed.

(4) There should be a linear relationship between the measurand and the transducer signal.

(5) The transducer should have minimum sensitivity to external effects, pressure transducers, for example, are often subjected to external effects such vibration and temperature.

(6) The natural frequency of the transducer should be well separated from the frequency and harmonics of the measurand.

7.4 Electrical Transducers

Electrical transducers exhibit many of the ideal characteristics. In addition they offer high sensitivity as well as promoting the possible of remote indication or measurement.

Electrical transducers can be divided into two distinct groups.

(1) Variable-control-parameter types, which include:

a) resistance;

b) capacitance;

c) inductance;

d) mutual-inductance types.

These transducers all rely on an external excitation voltage for their operation.

(2) Self-generating types, which include:

a) electromagnetic;

b) thermoelectric;

c) photoemissive;

d) piezo-electric types.

These all themselves produce an output voltage in response to the measurand input and their effects are reversible. For example, a piezo-electric transducer normally produces an output voltage in response to the deformation of a crystalline material; however, if an alternating voltage is applied across the material, the transducer exhibits the reversible effect by deforming or vibrating at the frequency of the alternating voltage.

7.5 Resistance Transducers

Resistance transducers may be divided into two groups as follows.

(1) Those which experience a large resistance change, measured by using potential-divider methods. Potentiometers are in this group.

(2) Those which experience a small resistance change, measured by bridge-circuit methods. Examples of this group include strain gauges and resistance thermometers.

7.5.1 Potentiometers

A linear wire-wound potentiometer consists of a number of turns of resistance wire wound around a non-conducting former, together with a wiping contact which travels over the barwires. The construction principles are shown in Figure 7.3(a) and (b) which indicate that the wiper displacement can be rotary, translational, or a combination of both to give a helical-type motion. The excitation voltage may be either AC or DC and the output voltage is proportional to the input motion, provided the measuring device has a resistance which is much greater than the potentiometer resistance.

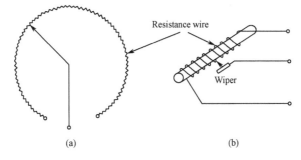

Figure 7.3 Construction principles of resistance potentiometers

Such potentiometers suffer from the linked problem of resolution and electrical noise. Resolution is defined as the smallest detectable change in input and is dependent on the cross-sectional area of the windings and the area of the sliding contact. The output voltage is thus a serials of steps as the contact moves from one wire to next, as shown in Figure 7.4(a).

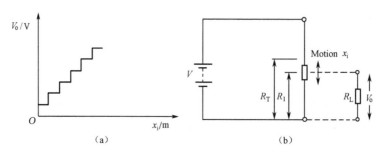

Figure 7.4 Resolution effects and circuit diagram of resistance potentiometer

Electrical noise (i. e. unwanted signals) may be generated by variation in contact resistance, by mechanical wear due to contact friction, and by contact vibration transmitted from the sensing element. In addition, the motion being measured may experience significant mechanical loading by the inertia and friction of the moving parts of the potentiometer. The wear on the contacting surface limits the life of a potentiometer to a finite number of full strokes or rotations, usually referred to in the manufacturer's specification as the 'number of cycles of life expectancy', a typical value being 20×10^6 cycles.

The output voltage V_o of the unload potentiometer circuit shown in Figure 7.4(b) is determined as follows.

Let resistance $\quad R_1 = \dfrac{x_i}{x_T} R_T$

where x_i = input displacement (m), x_T = maximum possible displacement (m), R_T = total resistance of the potentiometer (Ω). Then

output voltage $\quad V_o = V \dfrac{R_1}{R_1 + (R_T - R_1)} = V \dfrac{R_1}{R_T} = V \dfrac{x_i}{x_T} \cdot \dfrac{R_T}{R_T} = V \dfrac{x_i}{x_T}$ \hfill (7-1)

This shows that there is a straight-line relationship between output voltage and input displacement for the unloaded potentiometer.

It would seen that high sensitivity could be achieved simply by increasing the excitation voltage V. however, the maximum value of V is determined by the maximum power dissipation P of the fine wires of the potentiometer winding and is given by

$$V = \sqrt{PR_T} \quad (7\text{-}2)$$

Example 7.1 A potentiometer resistance transducer has a total winding resistance of 10kΩ and a maximum displacement range of 4mm. If the maximum power dissipation is not to exceed 40mW, determine the output voltage of the device when the input displacement is 1.2mm, assuming the maximum permissible excitation voltage is used.

Using equation (7-2), excitation voltage $V = \sqrt{PR_T} = \sqrt{0.04\text{W} \times 10000\Omega} = 20\text{V}$, from equation (7-1),

$$V_o = V \dfrac{x_i}{x_T} = 20\text{V} \times \dfrac{1.2\text{mm}}{4\text{mm}} = 6\text{V}$$

Loading potentiometer

When the potentiometer is loaded by placing across its terminals a measuring device such as a meter, having a resistance R_L, a current flows into the meter, this has loading effect on the potentiometer and causes the output/input graph to depart from the linear relationship as shown in Figure 7.5.

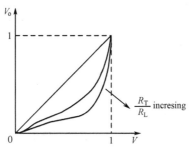

Figure 7.5 Characteristic of a loaded potentiometer

An analysis of the circuit in the loaded condition gives

$$V_o = V\left[\frac{x_T}{x_i} + \frac{R_T}{R_L}\left(1 - \frac{x_i}{x_T}\right)\right]^{-1} \tag{7-3}$$

which is far from linear, and the non-linearity increases as the ratio R_T/R_L increases.

Example 7.2 Calculate the error, at 50% full-scale travel of the wiper, of a resistance potentiometer when loaded with a meter having resistance equal to twice the potentiometer resistance.

Using equation (7-1) and (7-3), unload $V_o = \dfrac{V}{2} = 0.5\text{V}$. Using equation (7-3), loaded

$$V_o = V\left[\frac{1}{2 + 1/2(1 - 0.5)}\right] = \frac{V}{2.25} = 0.44\text{V}.$$

Hence

$$\text{error} = \frac{0.44\text{V} - 0.5\text{V}}{0.5\text{V}} \times 100\% = -1.2\%$$

(note the negative sign, which shows that the reading is too low.)

7.5.2 Resistance Strain Gauges

Resistance strain gauges are transducers which exhibit a change in electrical resistance in response to mechanical strain. They may be of the bonded or unbonded variety.

1. Bonded strain gauges

Using an adhesive, these gauges are bonded, or cemented, directly on to the surface of the body or structure which is being examined.

Examples of bonded gauges are:

a) fine wire gauges cemented to paper backing;

b) photo-etched grids of conducting foil on an epoxy-resin backing;

c) a single semiconductor filament mounted on an epoxy-resin backing with copper or nickel leads.

Resistance gauges can be made up as single elements to measuring strain in one direction only, or a combination of elements such as rosettes will permit simultaneous measurements in more than one direction.

2. Unbonded Strain Gauges

A typical unbonded-strain-gauge arrangement is shown in Figure 7.6, which shows fine resistance wires stretched around supports in such a way that the deflection of the cantilever spring system changes the tension in the wires and thus alters the resistance of wire. Such an arrangement may be found in commercially available force, load, or pressure transducers.

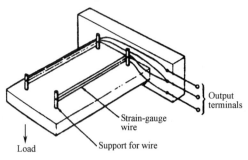

Figure 7.6 Unbonded strain gauges

7.5.3 Resistance Temperature Transducers

The materials for these can be divided into two main groups.

(1) Metals such as platinum, copper, tungsten, and nickel which exhibit and increases in resistance as the temperature rises; i.e. they have a positive temperature coefficient of resistance.

(2) Semiconductors, such as thermistors which use oxides of manganese, cobalt, chromium, or nickel. These exhibit large non-linear resistance changes with temperature variation and normally have a negative temperature coefficient of resistance.

1. Metal resistance temperature transducers

These depend, for many practical purpose and within a narrow temperature range, upon the relationship

$$R_1 = R_0[1+\alpha(\theta_1 - \theta_0)] \tag{7-4}$$

where α coefficient of resistance in $°C^{-1}$, and R_0 resistance in ohms at the reference temperature $\theta_0 = 0°C$ at the reference temperature range $°C$.

IPTS (the International Practical Temperature Scale)[2] is based on the platinum resistance thermometer, which covers the temperature range from $-259.35°C$ to $630.5°C$.

Typical characteristic curves for a platinum resistance thermometer are shown in Figure 7.7.

Example 7.3 If the resistance of a platinum resistance thermometer is 100Ω at $0°C$, calculate the resistance at $60°C$ if $\alpha = 0.00392°C^{-1}$.

Using equation (7-4),

$$R_1 = R_0[1+\alpha(\theta_1 - \theta_0)] = 100\Omega \times [1+0.00392 \times 60] = 123.5\Omega$$

2. Thermistor (semiconductor) resistance temperature transducers

Thermistors are temperature-sensitive resistors which exhibit large non-liner resistance changes with temperature variation. In general, they have a negative temperature coefficient as illustrated in Figure 7.8.

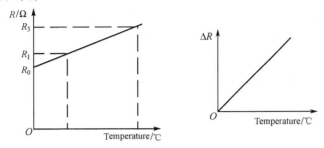

Figure 7.7 Characteristics of a platinum resistance thermometer

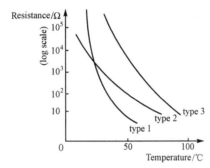

Figure 7.8 Characteristics of thermistors

For small temperature increments the variation in resistance is reasonably linear; but, if large temperature changes are experienced, special linearizing techniques are used in the measuring circuits to produce a linear relationship of resistance against temperature.

Thermistors are normally made in the form of semiconductor discs enclosed in glass vitreous enamel. Since they can be made as small as 1mm, quite rapid response times are possible.

Example 7.4 Use the characteristic curve for the type-1 thermistor shown in Figure 7.8 to determine the temperature measured when the meter in the circuit shown in Figure 7.9 reads half full scale.

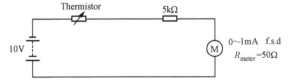

Figure 7.9 Circuit for Example 7.5

Total resistance $R = \dfrac{V}{I} = \dfrac{10\text{V}}{0.5 \times 10^{-3}\text{A}} = 20\text{k}\Omega$

∴ thermistor resistance $= 20\text{k}\Omega - 5\text{k}\Omega = 15\text{k}\Omega$ (neglecting meter resistance)
Hence, from the characteristic temperature ≈ 20 ℃.

7.5.4 Photoconductive cells

The photoconductive cell, Figure 7.10 uses a light-sensitive semiconductor material. The resistance between the metal electrodes decrease as the intensity of the light striking the semiconductor increases. Common semiconductor materials used for photoconductive cells are cadmium sulphide, lead sulphide, and copper-doped germanium.

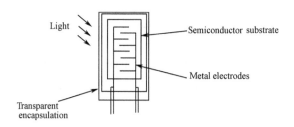

Figure 7.10 Photoconductive cell

The useful range of frequencies is determined by the material used. Cadmium sulphide is mainly suitable for visible light, whereas lead sulphide has its peak response in the infra-red region and is, therefore, most suitable for flame-failure detection and temperature measurement.

7.5.5 Photoemissive cells (variable conduction or inverse resistance)

When light strikes the cathode of the photoemissive cell shown in Figure 7.11 electrons are given sufficient energy to arrive the cathode. The positive anode attracts these electrons, producing a current which flows through resistor R_L and resulting in an output voltage v_o.

Photoelectrically generated voltage $\qquad v_o = I_p R_L \qquad$ (7-5)

where I_p = photoelectric current (A), and photoelectric current $I_p = K_t \Phi$.
where K_t = sensitivity (μA/lm), and Φ = illumination input (lumen).

Although the output voltage does give a good indication of the magnitude of illumination, the cells are more often used for counting or control purpose, where the light striking the cathode can be interrupted.

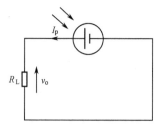

Figure 7.11 Photoemissive cell

Example 7.5 A photoemissive cell is connected in series with a 5kΩ resistor. If the cell has a sensitivity of 30μA/lm, calculate the input illumination when the output voltage is 2V.

Using equation (7-5),

$$v_o = I_p R_L = K_T \Phi R_L$$

$\therefore$ illumination $\Phi = \dfrac{v_o}{K_t R_L} = \dfrac{2\text{V}}{3 \times 10^{-6} \text{A/lm} \times 5 \times 10^3 \Omega} = 13.3 \text{lm}$

Words and Expressions

adhesive [əd'hi:siv] *adj.* 黏结的

bellows ['beləuz] *n.* 真空，膜盒，皱纹管

bond [bɔnd] *n.*；*vt.* 黏结

cadmium ['kædmiəm] *n.* 镉

capacitance [kə'pæsitəns] *n.* 电容

cantilever ['kænti‚li:və] *n.* 悬臂

cement [si'ment] *n.* 黏结剂

chromium ['krəumjəm] *n.* 铬

cobalt [kə'bɔ:lt] *n.* 钴

coefficient [‚kəui'fiʃnt] *n.* 系数

commercially [kə'mə:ʃəli] *adj.* 大宗的

crystalline ['kristəlain] *adj.* 结晶的，晶状的

deflection [di'flekʃən] *n.* 偏移

deformation [‚di:fɔ:'meiʃən] *n.* 变形

diaphragm ['daiəfræm] *n.* 膜（片盒），薄膜，振动膜

dissipation [‚disi'peiʃən] *n.* 消耗

distinct [dis'tiŋkt] *adj.* 不同的，有差别的

distortion [dis'tɔ:ʃən] *n.* 变形，扭曲

dope [dəup] *n.*；*vt.* 掺杂物，掺入，掺杂

electromagnetic [i‚lektrəumæg'neitik] *adj.* 电磁的

enamel [i'næml] *n.* 搪塞

encapsulation [in‚kæpsju'leiʃən] *n.* 密封，封装

enclose [in'kləuz] *vt.*；*n.* 包装，封入

epoxy [i'pɔksi] *n.*；*adj.* 环氧的

epoxy resin 环氧树脂

etch [etʃ] *vt.*；*n.* 蚀刻，酸洗

fidelity [fi'deliti] *n.* 保真

filament ['filəmənt] *n.* 丝，游丝，细丝线

flame-failure [fleim'feiljə] *n.* 火焰，故障

flapper ['flæpə] *n.* 舌阀

foil [fɔil] *n.* 箔，薄片

former ['fɔ:mə] *n.* 框架，成型设备

friction ['frikʃən] *n.* 摩擦

gauge [geidʒ] *n.* 表，仪器，计

germanium [dʒəːˈmeiniəm] n. 锗

harmonics [hɑːˈmɔniks] n. 谐波

illumination [iˌljuːmiˈneiʃən] n. 照度

inductance [inˈdʌktəns] n. 电感，感应

inertia [iˈnəːʃə] n. 惯性

infrared [ˌinfrəˈred] adj. 红外的

inverse [ˈinˈvəːs] adj. 反向的

life expectancy 概率寿命

linearize [ˈliniəˌraiz] vt. 线性化

lumen [ˈljuːmin] n. 流明（光通量单位）

natural frequency 固有频率

nickel [ˈnikl] n. 镍

mutual-inductance [ˈmjuːtjuəl inˈdʌktəns] n. 互感

paper backing 纸基

photoconductive [ˌfəutəukənˈdʌktiv] adj. 光导电的

photoconductive cell 光电导管

photoelectric [ˌfəutəuiˈlektrik] adj. 光发射的

photoelectric effect 光电效应

piezo-electric [paiˌiːzəuiˈlektrik] adj. 压电的

platinum [ˈplætinəm] n. pt. 铂

potential [pəˈtenʃ(ə)l] adj.; n. 位的，电位的，电势

potential divider 分压器

potentiometer [pəˌtenʃiˈɔmitə] n. 电位计，电位器

pressure flapper-nozzle 压力舌阀

promote [prəˈməut] vt. 促进，发展，发扬，助长

resistance [riˈzistəns] n. 电阻

resistance thermometer 电阻温度计

resolution [ˌrezəˈluːʃən] n. 分辨率

reversible [riˈvəːsəbl] adj. 可逆的，双向的

rosette [rəuˈzet] n. 三相应变计

significant [sigˈnifikənt] adj. 有意义的，重要的

simultaneous [ˌsiməlˈteinjəs] adj. 同时的

strain [strein] n. 应变

substrate [ˈsʌbstreit] n. 基片，感光胶片

tension [ˈtenʃən] n.; vt. 张力，拉伸

thermistor [θəːˈmistə] n. 热敏电阻

thermoelectric [ˌθəːməuiˈlektrik] adj. 热电的

transduce [trænzˈdjuːs] vt. 转换，传感

transparent [trænsˈpɛərənt] adj. 透明的

vitreous ['vitriəs] *adj.* 玻璃状态的
wipe [waip] *vt.*；*n.* 摩擦，接触
wire-wound ['waiə¡wu:nd] *adj.* 绕线的

Notes

1. be referred to as…　　被称为……
   例如，The energy-conversion process that takes place is referred to as transduction. 所发生的能量转换过程称为转换。
2. IPTS　the International Practice Temperature Scale 国际实用温标。

Exercises

Translate the following passages into English or Chinese.
1. 电阻应变式测力传感器是目前使用最广泛的一种测力传感器。它不仅灵敏度高，而且可用于测量力的瞬时值，另外，还可以通过合理的应变片位置及其接桥方式，来消除被测切削分力的相互干扰，从而使这种传感器的结构简化。
2. 半导体应变片的突出优点是：灵敏度高，可测微小应变。其不足之处是：电阻温度系数大，对环境温度的变化敏感；测量大应变时，灵敏度的非线性严重。
3. Resolution and accuracy of the sensor are determined by the quality requirements of the part itself. Resolution is the smallest increment of distance that the sensor can resolve. In cases where the machine tool encoders are used to read location, the resolution may be determined by the encoders themselves.
4. As discussed earlier, sensors can be broken down into four divisions: contact, noncontact, direction and indirect. No one sensor can be described as being the best overall, independent of application, since in-process sensing is heavily application dependent.

Reading Material

Laser

Several measurement methodologies, all noncontact, are possible using light sources. The utilization of laser light sources has made these methods increasingly sensitive due to the intensity, monochromaticity, and directionality of the laser beam. This is happening while the cost of laser light source is decreasing. The two primary approaches to be discussed in this section are the laser triangulator and the shadow technique.

A traditional device improved by the use of a laser, the interferometer, can be used as an extremely accurate measurement sensor; however, it is not really suitable for on-machine use because of its high-cost requirement for stability and a secular reflecting surface. In an interferometer configuration, the laser's light beam is divided with a beam splitter into a measurement beam and a reference beam. The measurement beam is shone on the object. The reflected return beam is compared with the reference beam through a phase detector. Since in-phase light waves add constructively, the interferometer acts in principle like an optical encoder using the monochromatic light beam as an encoder scale. This enables it to determine relative motion to and from the object under measurement by counting light 'fringes' caused by the alternating bands from the constructive and destructive interference. Multifrequency and Doppler techniques, which do not count such fringes, permit absolute measurements.

Using the interferometer approach, practical resolutions to 0.000001 in. are achievable with a standoff distance of a fraction of an inch and no theoretical upper limit. In reality, intensity of ht laser, refraction of the light, and changes in the speed of light over long distances through the medium (the latter due to thermally induced changes in density in the primary medium, air) limit n the standoff to several feet. For reasons discussed here, mainly lack of ruggedness and sensitivity, the laser interferometer has not found widespread use as an in-process measurement sensor. However, because of its extreme accuracy and resolution, the interferometer is being utilized to set up and calibrate machine tools and coordinate measuring machines. Improvements in this technique, such as those made by Hewlett-packard to reduce the sensitivity to atmospheric changes, are making interferometer usage more widespread.

The two laser-based techniques that have been applied successfully as sensors operate on different principles, although both use the laser as a light source. The laser shadow gauge senses the blockage of a beam of scanned light, whereas the laser triangulator measures a spatial shift in reflected or scattered light to provide a measurement. Both type of sensors have a high degree of accuracy with reasonable standoff distances and small spot sizes.

Because of the configuration of this type of sensor, only features which are strictly convex can be measured. Other features may be shadowed by the part geometry itself, preventing measurement. Accuracies down to 0.00001 in. are achievable with this type of noninterferometric system. The field of view of the sensor is limited to the size of the lenses utilized, which for commercial units may range from 2 to 6. If larger objects of variable size must be measured, the sensors may be adapted to a positioning system. Where objects of fixed size larger than the lenses are encountered, a configuration of multiple sensors or single sensors with beamsplitters and mirrors may be constructed.

The standoff distance of the shadow gauge is relatively large compared with that of other noncontacting sensors. This distance can be in the range of 10 in. for the accuracy previously stated. Although theoretically very large, the standoff distance is limited by the accuracy of the lenses used to refract the light into parallel rays over the field of view. Moreover, at large distances conditions that affect the interferometer, such as atmospheric disturbances, will also affect this type of system.

An advantage of this approach is that it is not affected by surface condition or reflectivity as many light-based systems are. In a machining application, it can be readily used in a cylindrical grinding or turning operation. Although it is bulkier than other sensors, through the use of mirrors it is possible to direct the beam immediately behind the cutting tool, minimizing delays in obtaining measurements data. Since clear and translucent cutting fluids can refract the light at the workpiece edge, an air stream or other means must be used to remove cutting fluids for highly accurate measurements. These types of systems are relatively insensitive to the presence of cutting fluids and chips (unless they completely and continuously block the light path) since clever averaging and false-signal refection techniques have been implemented by the manufacturers.

The system can provide a measurement as fast as the mirror can scan the beam. The effective rate may be several hundred measurements a second. This rate, however, is reduced depending on the measurement environment and the accuracy required, which determines the degree of signal refection required. The effective spot size is the width of the laser beam, which might typically be 0.030in. For an HeNe laser used in this application. The sensor will tend to average a measurement, as opposed to a micrometer used for an outer-diameter measurement, which would read the peak. Also, imperfect alignment of the facets of each reflecting surface in the rotating mirroring provides a dither of the beam, which tends to average the measurement.

Since the laser shadow gauge is normally supplied as a system and not only a sensor, the cost is generally higher than that of most sensors. However, the units are easy to align, provide a high degree of accuracy, are relatively impervious to machining conditions (assuming that cutting fluids are not flooding the measurement area), and provide a direct digital output which can normally be interfaced to a computer or machine tool control via a serial interface.

The laser triangulation technique, as the name implies, uses a laser light source and the geometric principle of triangulation to make a distance measurement. The laser is commonly oriented normal to the surface and uses scattered light as opposed to direct reflection.

Because this sensor is light based, the light path must be unobscured and lenses kept clean when used in an machining application. The triangulator has been demonstrated in process control applications and is supplied commercially. There is a trend among makers of coordinate measuring machines to equip the machines with this type of sensor, which will allow the machines to scan surfaces and measure point at a much higher rate.

Lesson 8 Capacitance, Inductance Transducers and Some Others

8.1 Capacitive Transducers

The capacitance of a parallel-plate capacitor is given by

$$C = \varepsilon_0 \varepsilon_r \frac{A}{d} \tag{8-1}$$

where $\varepsilon_0 =$ the permittivity of free space $= 8.854 \times 10^{-12}$ F/m

$\varepsilon_r =$ relative permittivity of the material between the plates

$A =$ overlapping or effective area between plates(m^2)

and $d =$ distance between plates (m)

The capacitance can thus be made to vary by changing either the relative permittivity ε_r, the effective area A, or the distance separating the plates d. Some examples of capacitive transducers are shown in Figure 8.1.

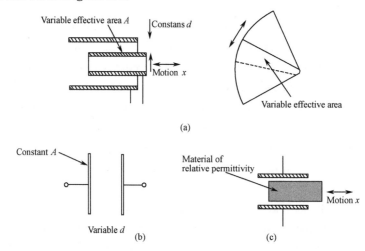

Figure 8.1 Examples of capacitive transducers

The characteristic curves shown in Figure 8.2 indicate that variations of area A and relative permittivity ε_r give a linear relationship only over a small range of spacings.

By differentiating equation (8-1), we can find the sensitivity in farads/s, i.e.

$$\frac{dC}{dd} = -\frac{\varepsilon_0 \varepsilon_r A}{d^2} \tag{8-2}$$

Thus the sensitivity is high for small values of d. Unlike the potentiometer, the variable-distance capacitive transducer has an infinite resolution, which making it most suitable for measuring small increments of displacement or quantities which may be changed to produce a displacement.

Example 8.1 A parallel-plate air-spaced capacitor has an effective plate area of $6.4 \times 10^{-4} \text{m}^2$, and the distance between the plates is 1mm, if the relative permittivity for air is 1.000 6, calculate the displacement sensitivity of the device.

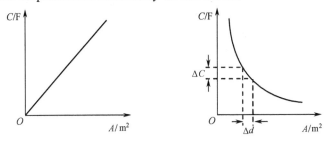

Figure 8.2 Characteristics of capacitive transducers

Using equation (8-1),

$$C = \varepsilon_0 \varepsilon_r \frac{A}{d}$$

differentiating

$$\frac{dC}{dd} = -\frac{\varepsilon_0 \varepsilon_r A}{d^2} = \frac{-8.854 \times 10^{-12} \text{F/m} \times 1.000\,6 \times 6.4 \times 10^{-4} \text{m}^2}{(1 \times 10^{-3} \text{m})^2} = -56.6 \times 10^{-10} \text{F/m}$$

The minus sign indicates a reduction in the capacitance value for increasing d.

8.2 Inductive Transducers

The inductance of a coil wound around a magnetic circuit is given by

$$L = \frac{\mu_0 \mu_r N^2 A}{l} \tag{8-3}$$

where μ_0 = permeability of free space

μ_r = relative permeability

N = number of turns on coil

l = length of magnetic circuit(m)

and A = cross-sectional area of magnetic circuit (m^2)

This can be rewritten as

$$L = \frac{N^2}{S} \tag{8-4}$$

where S is the magnetic reluctance of the inductive circuit.

The inductance can thus be made to vary by changing the reluctance of the inductive circuit. Some examples of variable-reluctance transducers are shown in Figure 8.3(a) to (c).

A typical characteristic curve for an inductive transducer is shown in Figure 8.4.

Example 8.2 Determine the sensitivity of a single-coil inductive transducer for a) variations in relative permeability μ_r; b) variations in length of magnetic circuit.

a) differentiating equation (8-3) with respect to μ_r,

$$\frac{dL}{d\mu_r} = \frac{\mu_0 N^2 A}{l}$$

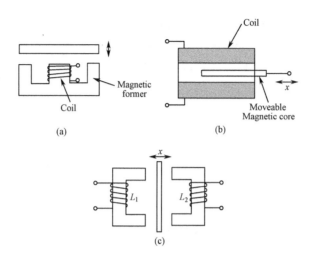

Figure 8.3 Examples of variable-reluctance inductive transducers

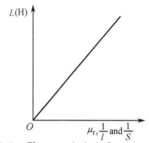

Figure 8.4 Characteristic inductive transducer

b) differentiating equation (8-3) with respect to l,

$$\frac{dL}{dl} = -\frac{\mu_0 \mu_r N^2 A}{l^2}$$

Measuring techniques used with capacitive and inductive transducers:

(1) AC excited bridges using differential capacitors inductors.

(2) AC potentiometer circuits for dynamic measurements.

(3) DC circuits to give a voltage proportional to velocity for a capacitor.

(4) Frequency-modulation methods, where the change of C or L varies the frequency of an oscillation circuit.

Important features of capacitive and inductive transducers are as follows:

(1) resolution infinite;

(2) accuracy $\pm 0.1\%$ of full scale is quoted;

(3) displacement ranges from 25×10^{-6} m to 10^{-3} m;

(4) rise time less than $50 \mu s$ possible.

Typical measurands are displacement, pressure, vibration, sound, and liquid level.

8.3 Linear Variable-differential Transformer(l. v. d. t.)

A typical differential transformer, as illustrated in Figure 8.5, has a primary coil, two secondary coils, and a movable magnetic core.

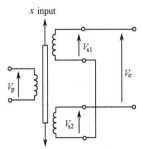

Figure 8.5 Details of an l. v. d. t.

A high-frequency excitation voltage V_p is applied to the primary winding and, due to transformer action, voltage V_{s1} and V_{s2} are induced in the secondary coils. The amplitudes of these secondary voltages are dependent on the degree of electromagnetic coupling between the primary and secondary coils and hence on the core displacement x.

Since the secondary coils are connected in serials opposition, the displacement x of the core which produces an increase in V_{s1} will produce a corresponding decrease in V_{s2}. Ideally the voltages V_{s1} and V_{s2} should be 180° out of phase with each other, so that at the central position there is zero output voltage. However, the voltages generally are not exactly 180° out of phase and there is a small null output voltages as illustrated in Figure 8.6.

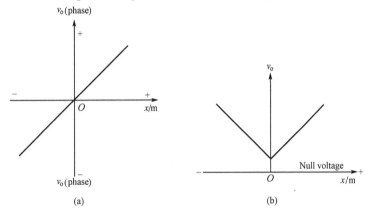

Figure 8.6 Output characteristics of an l. v. d. t.

Some important characteristics and features of the l. v. d. t are as follows:

a) infinite resolution;

b) linearity better than 0.5%;

c) excitation frequency 50Hz to 20kHz;

d) null voltage less than 1% of full-scale output voltage;

e) maximum displacement frequency 10% of the excitation frequency;

f) displacement ranges available from 2×10^{-4}m to 0.5m;

g) no wear of moving parts;

h) amplitude-modulated output, i. e. the output voltage is constant frequency waveform with an amplitude depending on the displacement input.

Typical measurands are any quantities which can be transduced into a displacement, e. g. pressure, acceleration, vibration, force, and liquid level.

8.4 Piezo-electric Transducers[1]

When a force is applied across the faces of certain crystal materials electrical charges of opposite polarity appear on the faces due to the piezo-electric effect ('piezo' comes from the Greek for 'to press'). Piezo-electric transducers are made from natural crystals such as quartz and Rochelle salt, synthetic crystals such as lithium sulphate, or polarized ceramics such as barium titanate. Since these materials generate an output charge proportional to applied force, they are most suitable for measuring force-derived variables such as pressure, load, and acceleration as well as force itself.

Piezo-electric materials are good electrical insulators; therefore, with their connecting plates, they can be considered as parallel-plate capacitors as shown in Figure 8.7(a). When a force is applied, the capacitor simply 'charges up' due to the piezo-electric effect, as illustrated by the equivalent electric circuit shown in Figure 8.7(b). Unfortunately, any measuring instrument electrically connected across the capacitor C will tend to discharge it; hence the inducer's steady-state response is poor. This can be overcome by using measuring amplifiers with very high input impedances (10^{12} to 10^{14} Ω being typical) known as charge amplifiers, but these make the measuring system increasingly expensive.

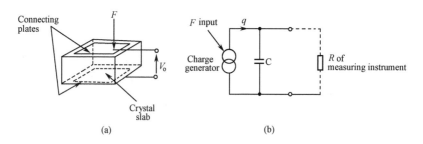

Figure 8.7   Piezo-electric transducer

Example 8.3   A piezo-electric pressure transducer has a sensitivity of 80 pC/bar. If it has a capacitance of 1nF, determine its output voltage when the input pressure is 1.4bar.

$$\text{Charge} \quad q = \text{sensitivity} \cdot \text{pressure} = 80\,\frac{\text{pC}}{\text{bar}} \times 1.4\text{bar} = 80 \times 1.4\text{pC}$$

$$\text{Output voltage} \quad V = \frac{q}{C} = \frac{112 \times 10^{-12}\text{C}}{1 \times 10^{-9}\text{F}} = 112\text{mV}$$

8.5 Electromagnetic Transducers

These employ the well-known generator principle of a coil moving in a magnetic field. The output voltage of the electromagnetic transducer is given as follows.

1. For a coil with changing flux linkages

$$\text{Output voltage} \quad v_o = -N\frac{d\Phi}{dt}$$

where N=number of turns on coil, and $\frac{d\Phi}{dt}$=rate at which flux changes (Wb/s).

2. For the single conductor moving in a magnetic field

$$\text{Output voltage} \quad v_o = Blv$$

where B= flux density (T), l= length of conductor (m), and v= velocity of conductor perpendicular to flux direction (m/s).

Both relationships are used in commercially available velocity transducers, the construction principles of which are illustrated in Figure 8.8(a) to (c).

Some important features of the electromagnetic transducer are as follow:

(1) output voltage is proportional to the velocity of input motion.

(2) usually they have a large mass, hence they tend to have low natural frequencies.

(3) high power outputs are available.

(4) limited low-frequency response-ranges from 10Hz to 1kHz are quoted in manufacture's literature.

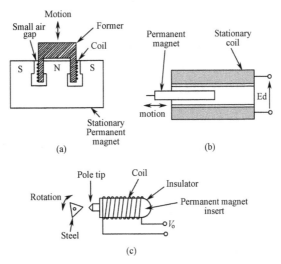

Figure 8.8 Electromagnetic transducers

8.6 Thermoelectric Transducers

When two dissimilar metals or alloys are joined together at each end to form a thermocouple as shown in Figure 8.9 and the ends are at different temperatures, an e.m.f. will be developed causing a current to flow around the circuit. The magnitude of the e.m.f. depends on the temperature difference between the two junctions and on the materials used. This thermoelectric effect is known as the Seebeck effect and is widely used in temperature-measurement and control systems.

The main problems with thermocouples are corrosion, oxidation, or general contamination by the atmosphere of their location. These problems can be overcome by the selection of a protective sheath which does not react with atmosphere or fluid.

Although they do give a direct output voltage, this is generally small-in the order of

Figure 8.9 Thermocouple circuit

millivolts-and often requires amplification.

Advantages of thermocouples include:
(1) temperature at localized points can be determined, because of the small size of the thermocouple.
(2) they are robust, with a wide operating range from $-250\,^\circ\text{C}$ to $2600\,^\circ\text{C}$.

8.7 Photoelectric Cells (self-generating)

The photoelectric or photovoltaic cell makes use of the photovoltaic effect, which is the production of an e.m.f. by radiant energy-usually light-incident on the junction of two dissimilar materials. The construction of a typical cell is illustrated in Figure 8.10(a), which shows a sandwich layer of metal, semiconductor material, and a transparent layer. Light traveling through the transparent layer generates a voltage which is a logarithmic function of light intensity, the device is highly sensitive; has a good frequency response; and, because of its logarithmic relationship of voltage against light, is very suitable for sensing over a wide range of light intensities, the characteristic of device is shown in Figure 8.10(b).

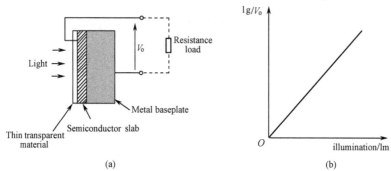

Figure 8.10 Photoelectric cell

8.8 Mechanical Transducers and Sensing Elements

Many transducing systems consist of two different types of transducer operating in series, or cascade. In the electrical-transducer section it was assumed that the input to the transducer was provided by the sensing element. The sensing element itself is often a mechanical transducer which converts the measured into a displacement or force which is then used to change some electrical parameter.

Some of the more common mechanical transducers are shown in the following examples.

8.8.1 Force-to-displacement Transducers

1. Spring

The spring shown in Figure 7.2 is the simplest form of mechanical transducer.
For equilibrium, we have
$$F = \lambda x$$
where λ = spring stiffness (N/m),
$$x = \frac{F}{\lambda} \quad \text{or} \quad \lambda = \frac{F}{x}$$
but sensitivity $K_1 = \frac{x}{F}$, so sensitivity $= \frac{1}{\lambda}$, i.e. the stiffer the spring, the smaller the sensitivity[2].

2. Cantilever

When the cantilever shown in Figure 8.11 is loaded, it experiences a deflection y. The relationship between the force F and deflection is given by

deflection y = constant • force

$y = kF$

where the constant k depends on the material and dimensions of the cantilever.

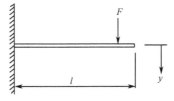

Figure 8.11 Cantilever

8.8.2 Pressure-to-displacement Transducer

1. Diaphragms

Pressure can be measured using a steel diaphragm as shown in Figure 8.12. The displacement of the diaphragm is proportional to the pressure difference ($P_1 - P_2$) if the displacement is less than one third of the diaphragm thickness t. The relationship between pressure differential ($P_1 - P_2$) and diaphragm displacement is thus given by

deflection = constant • pressure differential

$$x = k(P_1 - P_2)$$

where k depends on the material and dimensions of the diaphragm.

The diaphragm, usually a thin flat plate of spring steel, may be used with an electrical transducer to produce a small transducer with high sensitivity.

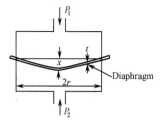

Figure 8.12 Diaphragm

2. Bourdon tubes

This type of transducer, illustrated in Figure 8.13, is used in many commercially available pressure gauges. The main feature of bourdon tubes is their large deflection.

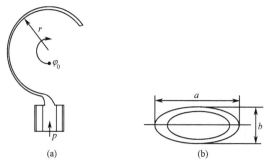

Figure 8.13 Bourdon tubes

Provided the major axis a of the cross-section is considerably larger than the minor axis b, the following relationship holds between the input pressure p and the tube-tip deflection φ_0

$$\varphi_0 = \text{constant} \cdot p$$

3. Bellows

This is basically a pneumatic spring, as illustrated in Figure 8.14, and is in general used in pneumatic instruments.

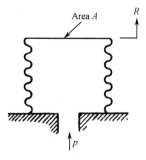

Figure 8.14 Bellows

Equating the forces acting on the bellows, for equilibrium

$$pA = \lambda x, \quad x = \frac{A}{\lambda} p$$

where A = cross-sectional area of bellows (m^2), p = input pressure (N/m^2), λ = bellows stiffness (N/m).

8.8.3 Displacement-to-pressure transducers

A typical arrangement of a flapper-nozzle system is shown in Figure 8.15(a).

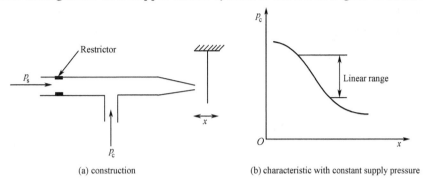

(a) construction (b) characteristic with constant supply pressure

Figure 8.15 Flapper-nozzle system

For a constant supply pressure p_s, movement of the flapper allows a variable bleed-off of air which varies the control pressure p_c. The transducer characteristic, illustrated in Figure 8.15(b), is non-linear but has a narrow linear region which is usually extended by employing feedback when the transducers are used in pneumatic instruments. The flow rates used are very small. And a pneumatic amplifier must be employed to boost the controlled pressure.

Words and Expressions

barium ['bɛəriəm] n. 钡
bleed of vt. 放血，流血
boost [buːst] vt.; n. 推出，助推，帮助，提高，加速
contamination [kɔn:tæmi'neiʃən] n. 污染，混杂
coupling ['kʌpliŋ] n. 耦合
differentiate [ˌdifə'renʃieit] vt. 微分，差动
equilibrium [ˌiːkwi'libriəm] n. 平衡，均衡
flux [flʌks] n. 磁通
impedance [im'piːdəns] n. 阻抗
incident light   入射光
infinite ['infinit] adj. 不定的
lithium ['liθiəm] n. 锂
linkage ['liŋkidʒ] n. 连接，耦合，磁链
logarithmic [ˌlɔgə'riθmik, -iðm-] adj. 对数的
overlapping ['əuvə'læpiŋ] n. 覆盖着，重叠
permeability [ˌpəːmiə'biliti] n. 磁导率
permittivity [ˌpəːmi'tiviti] n. 电容率，介电系数
polarize ['pəuləraiz] vt. 使极化

quote [kwəut] vt. 引用
reluctance [ri'lʌktəns] n. 磁阻
robust [rəu'bʌst] adj. 坚固的，耐用的，加强的
Seebeck effect　塞贝克效应
sheath [ʃi:θ] n. 鞘，套，壳，涂料，包层
stiffness ['stifnis] n. 刚度
sulphate ['sʌlfeit] n.；vt. 硫酸盐，用硫酸处理
synthetic [sin'θetik] adj. 综合的
titanate ['taitəneit] n. 钛酸盐

Notes

1. piezo- [词头] 压（力，电）
   piezo-electric adj. 压电的
2. The stiffer the spring, the smaller the sensitivity.
   弹簧的刚度越大，其灵敏度越小。

Exercises

Translate the following passages into English or Chinese.
1. 电感式传感器的应用十分广泛，凡是与位移有关的物理量都可通过它转换成电量输出。它常用来测量振动、厚度、应变、压力、加速度等各种物理量。
2. 根据电容式传感器的原理，利用极板间距和极板遮盖面积的变化，可以测量直线位移或角位移。通过弹性膜片可以测量压力，通过弹性梁可以测量振动和加速度等。
3. A simple air sensor consists of a regulated source of air, an adjustable restriction, and a nozzle with an orifice, with an object, the workpiece, blocking the nozzle, the pressure across the restriction is zero.
4. Available sensors have a variety of characteristics that must be understood before application. The major characteristics, physical size, resolution, accuracy, type of signal output, signal processing requirements, effective spot size are outlined herein.

Reading Material

Video Camera

Perhaps the newest, most flexible noncontacting sensor, and the one with the most potential for future, is the video camera-based measurement sensor. These sensors are normally categorized under the title of machine vision. The technology categorized is based on a video camera (vidicon) and a computer-based image analyzer. The technology of the components is fairly mature, dating back to television and the digital computer. Ironically, machine vision has been used in an industrial setting only within the past five years. Its use is primarily due to the availability of stable solid-state vidicons, faster, less expensive minicomputers, and image processing algorithms. A significant advantage of this approach is that an entire scene can be captured in an instant, allowing a multitude of measurements to be made without moving the sensor. The main disadvantage stems from the volume of data that must be processed by the image analyzer, resulting in either a slow response or a requirement for an extremely fast processor. Although a camera can acquire only a two-dimensional scene, structured lighting approaches combined with triangulation can be implemented to allow the video camera-based sensor to collect measurements in three dimensions.

The component parts of a machine vision sensor are simply a vidicon, interface, and computer. In practice, a configuration as shown in Figure 8.16 would be implemented. The video camera views the workpiece under measurement through an optical system, normally a magnifying lens. The sensing element is a solid-state photodiode array, although, in principle, a conventional vidicon tube can be utilized. The sold-state sensor has many advantages, including lower cost, ruggedness, less sensitivity to blooming, wider dynamic light range, greater image stability, smaller size, less weigh, and ease of interfacing. the spacing of the photodiodes is typically on the order of 0.001 to 0.005 in., which would limit the resolution of the sensor. For this reason, magnifying optics are usually used to achieve resolutions in the range of 0.0001 to 0.01 in., which would normally be required in manufacturing environment.

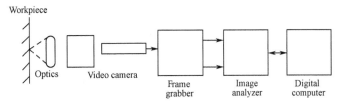

Figure 8.16 Block diagram of video camera-based measurement sensor

The camera is connected to a frame grabber, which is a specialized, high-speed memory designed to collect and store single or multiple images, called frames, from the camera. For each frame the grabber stores, it must have a capacity of at least the same number of photodiodes as the camera has in its array. The individual diodes are called pixels, for picture elements. The frame grabber, in turn, must accommodate the same number of pixels. Since the photodiode itself merely provides a light level as an analog signal, this must be converted to a digital value for the memory and the digital computer. This conversion is normally done by means of an analog-to-digital converter. The a/d converter resides in either the camera or the frame grabber. The resolution of the converter determines the number of levels of light that can be resolved. This range of levels is normally termed the gray scale and the typical resolution is 8 bits.

The frame grabber is connected to an image analyzer, which is another piece of specialized circuitry designed to unburden the digital computer from handling all the data collected from the camera. Repetitively, high-speed processing of the data can be performed by the image analyzer breaking the picture down into features for further analysis, including measurement by a less costly computer of moderate speed. Operations that the image analyzer might handle include averaging, convolution, edge detection, thresholding, and windowing. Operations for image analysis originally developed at SRI and GM spurred use of the video camera for inspection purposes.

To utilize the camera as a measurement sensor, features in the field of view of the camera must be located. This is normally done through edge detection. Edge detection, as the name implies, is used to locate the edge (light-to-dark transition) of an object so that it can be used as a measurement point itself or so that may be used with other points to describe a feature such as a line, circle, arc, or ellipse. For objects that cannot be contained within one field of view, points obtained in one frame can be used to refer other views. Edge detection is a differentiation process usually done by examining a small region of pixels for their light-to-dark transition in the vicinity of the edge. Since differentiation is inherently a noisy process, normally it is combined with some averaging, statistical, or convolution techniques to 'clean up' the image. After edge detection, the resultant image can be subtracted from a threshold to establish a clean edge.

After points on the edges are found, they can be fit-for example, by using a least squares approach-into lines that describe the object. A least squares approach has the advantage that it is an averaging process which aids in removing random variation that might affect feature location. The features can then be passed up to the computer, where measurements can be made by subtraction of absolute feature locations of preselected points. Where objects are too large for the field of view of the camera and lens combination, the computer can combine images into data that describe the entire object. Features of the object might also be directly compared with entities from a computer-aided design database, for example.

Because of limitations in the size of photodiode arrays and a reduction in field of view due to the magnifying lens required to get proper resolution, in many cases a positioning system is also required for a video camera sensor. When a large object is measured, the camera is moved in the vicinity of the points to be measured. In some cases a third axis of motion may be required to focus the camera. Again, when magnifying optics are used, the depth of field of the system, like the field of view, is reduced. When few points must be measured, the image can be windowed to improve the response time of a measurement, windowing reduces the area of interest in the scene to the area where the point to be measured is expected to be found, thus reducing processing time greatly.

Accuracies in the range of 0.0001 in. can be achieved with a video camera-based sensor. Standoff distances may range from 0.05 to several inches, with the former number typical of a microscope objective lens. Other optical solutions are available in the form of macro focusing telescope lenses where standoff distances on the order of feet may be obtain with magnification. Limitations due to spot size associated with other sensors do not exit, since the camera can view a whole scene at once. As with other light-based sensors, limitations due to dirty lenses and objects blocking the lens such as coolant and chips apply to this camera-based sensor.

An extension of this sensor which permits determination of three-dimensional information involves the use of structured light scenes. In the general case, when mapping a three-dimensional scene onto a plane, a ray of light emanating from a point on a close object may be mapped onto the same pixel in the camera's photodiode array as a ray emanating from a point displaced from the first on a distant object. This confusion must be resolved in order to obtain a distance measurement. To do this, another type of structured light application can be used to make measurements in three dimensions from an unknown scene. In this application a shuttered, structured lighting system has been developed to control the illumination of the being measured, thereby permitting a one-for-one mapping from the three-dimensional scene to the two-dimensional image. To accelerate the data processing, a binary sot routine was utilized to analyze structured light patterns from the shutter.

The video camera-based measurement sensor is extremely flexible for in-process measurements. Its accuracy is well within the range for all but the most demanding applications and is certainly comparable to that of other contacting and noncontacting sensors. The camera can be configured with a variety of lenses and light sources to accommodate different field-of-view and standoff requirements. The primary disadvantages of this type of measurement sensor at this time are its cost, complexity, and response time for complicated scenes, however, this will change as the technology matures and less expensive high-speed computers become available along with specialized very high-speed integrated circuit chips to perform the image analysis functions.

Lesson 9 Analog Instruments

Basic to all scientific investigation, electronic circuit design, circuit evaluation, and electronic servicing are accurate measurements, selective analysis, and mathematical formation. First and foremost is measurement, defined as the process by which physical, analog quantities are converted to numbers that are used to analyze and formulate conclusions about the world around us.

Converting real-life quantities into numbers that we can deal with is done using test instruments. Test instruments come in a wide range of types and descriptions, each designed with a specific measurement to perform.

The most basic, and most venerable, of all test instruments is the deflection pointer, or analog meter. There was a time when the analog meter was the only source of reliable electric measurements. Recent advances in digital technology, however, have permitted the design of instruments that eliminate the measuring error introduced by the mechanical components of an analog meter. [1] While it can be argued that deflection meters are not as accurate as modern digital laboratory test instruments, it is doubtful that the overall results obtained by a more precise test instrument would necessarily be more meaningful than those attainable from an inexpensive meter. [2] Consequently, analog meters have secured their place in the laboratory and field as a reliable source of measurements.

9.1 Meter Basics

All deflection instruments require three forces for proper operation: a deflecting force, a controlling force, and a damping force. The deflecting force is commonly generated by a magnetic field produced by a current flowing through a coil. This field interacts with a controlling magnetic field to produce a repelling force. When a mechanical pointer is associated with one of the magnetic forces, movement is noted.

The earliest deflection device was Oersted's experimental measuring apparatus which have been called a galvanometer. Essentially, it was a compass needle placed below a wire in which current was to be measured. With no current flowing, compass needle and wire were aligned in the north-south direction. When current passed through the wire, the compass needle was deflected toward the east-west direction by an amount proportional to the current flowing in the wire. In this device, the magnetic field of the earth is the controlling force, and the magnetic field generated by current flowing through the wire is the deflecting force. Damping, in the form of friction within the needle pivot, prevents the pointer from oscillating (or, as it is sometimes called, seeking).

In later years, Lord Kelvin improved the sensitivity of the device by wrapping turns of the wire around the body of the compass, thus intensifying the strength of the deflecting force at lower current levels.

Most meters today are of the D'Arsonval design. The D'Arsonval movement consists of a moving coil or pivoted coil suspended in the force field of a permanent magnet. The magnet is much larger than the coil and shaped so (Figure 9.1 and Figure 9.2). The torque is calculated by multiplying the force times the distance to the point of suspension, and is equal to

$$T = NBILW$$

where N = number of turns, B = magnetic field strength, I = current, L = vertical length of coil, and $W/2$ is the distance to the point of suspension.

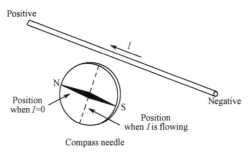

Figure 9.1 Oersted's galvanometer

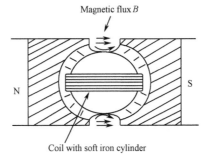

Figure 9.2 Torquing lines of force. Note that the magnetic flux lines of B are radially directed by the soft iron cylinder.

This torque will cause the coil to rotate until an equilibrium position is reached at an angle ϕ with its original position. In this position, the torque exerted by the suspension is exactly equal to that due to the current and the field. The angle of rotation is equal to

$$\phi = \frac{NBLW}{k} I$$

where k is a constant determined by the thickness, the width, and the spring quality of the suspending strip or restoring spring (Figure 9.3).

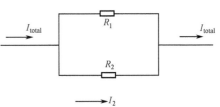

Figure 9.3 Current divider rule

9.2 Ammeters

The suspension D'Arsonval-type meter can detect and measure currents as small as 1 microampere, or 1×10^{-6} amp. To measure larger currents, the meter must be modified to increase its range. This is easily done using an external current shunt across the meter movement. A current shunt is nothing more than a resistor placed in parallel with the meter, thus forming a current divider.

The current divider rule states: The ratio of the currents in any two parallel branches is inversely proportional to the ratio of the resistances of the branches, or $\dfrac{I_1}{I_2}=\dfrac{R_2}{R_1}$. If the meter movement, with its internal resistance of R_m, is substituted for R_1, and the shunt resistance R_s is substituted for R_2, the equation reads

$$\frac{I_m}{I_s}=\frac{R_s}{R_m}$$

If $I_s = I_{total} - I_m$, then $\dfrac{I_m}{I_{total}-I_m}=\dfrac{R_s}{R_m}$ and $R_s=\dfrac{I_m R_m}{I_{total}-I_m}$.

This last equation can be used to calculate the value of a shunt resistor to increase the current range of the meter to display I_{total}. To allow a 1-mA meter with an internal resistance of 100Ω to measure 100 mA of current(Figure 9.4), for example, the shunt resistance is

$$R_s = \frac{0.001 \times 100}{0.1-0.001} = \frac{0.1}{0.099} = 1.01\Omega$$

Using this principle, any microammeter, milliammeter, or ammeter may measure a larger current of any magnitude by using suitable shunt resistors to change the range. In fact, this is how commercial instruments measure a wide range of currents using a single meter movements. A schematic diagram of a seven-range commercial ammeter is shown in Figure 9.5. Notice that a rotary switch is used to insert the various shunt resistors across the meter for the range selected.

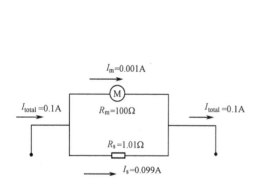

Figure 9.4 Current values in meter shunt

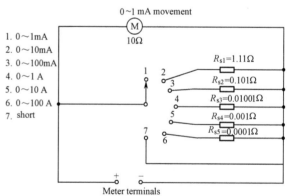

Figure 9.5 Simpson 373 ammeter

9.3 Current Measuring Errors

An ammeter is always connected in series with the circuit in which current is to be measured. To avoid affecting the current level in the circuit during measurement, the ammeter must have a resistance much lower than the circuit resistance. Ideally, the ammeter would have zero resistance, but such a meter is not possible. The effect of meter loading is demonstrated in the following example (Figure 9.6).

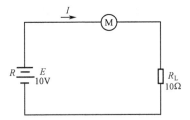

Figure 9.6 Current measuring circuit

Consider the circuit below, in which an ammeter is connected to measure the current in a 10-ohm load supplied from a 10-volt source. Ohm's law says that under ideal conditions the current flow through the circuit is

$$I = \frac{E}{R} = \frac{10}{10} = 1A$$

If the resistance of the ammeter is 1Ω, however, the meter will not register 1A, but will indicate a current value of 0.91A. The discrepancy between the calculated and measured values is made apparent when the circuit is redrawn to include the resistance of the ammeter, as indicated in Figure 9.7.

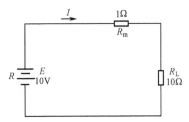

Figure 9.7 Equivalent current measuring circuit

Using Ohm's law, we can now conclude that the ammeter is telling the truth. With $I = 10V/11\Omega$, the actual current flow is only 0.909A. The amount of error introduced by the insertion of the measuring device is equal to

$$\text{error} = \frac{\text{ideal value} - \text{measured value}}{\text{ideal value}} \times 100\%$$

$$= \frac{1A - 0.909A}{1A} \times 100\%$$

$$= 9.1\%$$

If the resistance of the meter was reduced to 0.1Ω, then the measured value would be $I = 10V/10.1\Omega = 0.99A$, and the amount of error created by the meter would be approximately 1%.

9.4 DC Voltmeters

The basic microammeter movement may also be adapted to measure voltage. Because the deflection of the instrument is proportional to the current of the pointer and also proportional to the voltage across the coil. Therefore, the scale of the meter can be calibrated to indicate voltage rather than current.

The coil resistance of most microammeters is normally quite small, on the order of 1000Ω or less, thus the coil voltage is also usually very small. Full-scale voltage deflection of a 50-μA meter with a 2000-Ω coil, for example,

$$E = IR = 5 \times 10^{-5} \times 2 \times 10^{3} = 1 \times 10^{-1} (V)$$

or 0.1V. Unfortunately, it is not very often that one needs to measure voltage values of 0.1V or less.

The range of a voltmeter can easily be increased by connecting a resistance in series with the instrument, as illustrated in Figure 9.8. The resistor limits the current through the meter so that the instrument can measure higher voltages than it can when using its own internal resistance. These resistors, whether mounted in or out of the meter case, are called multipliers. The value of the multiplier resistor can be determined by the current for full-scale deflection and by the range of the voltage to be measured. Since the current through the meter is directly proportional to the applied voltage, Ohm's law can be used to calculate its value accordingly

$$R_s = \frac{E}{I_m} - R_m$$

where R_m = resistance of the meter coil
R_s = multiplier resistor
I_m = full-scale deflection current of the meter

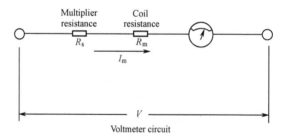

Figure 9.8 Basic voltmeter circuit

Consider a microammeter with an internal resistance R_m of 100Ω and a full-scale current value of 100μA. For this meter, to indicate the range of voltages from 0 to 1V, the value of the multiplier resistor is

$$R_s = \frac{1}{1 \times 10^{-4}} - 100 = 10000 - 100 = 9.9 (k\Omega)$$

When a 9.9kΩ resistor is placed in series with this meter movement and the scale is calibrated in volts, the result is a 0 to 1V voltmeter.

As with the ammeter, multirange voltmeters can be constructed using one meter movement and a number of multiplier resistors and a rotary switch. Two circuit configurations are possible. In Figure 9.9(a), individual multiplier resistors are selected by the rotary switch for insertion in series with the meter movement. In Figure 9.9(b), the multiplier resistors are wired in series, and the junction between each resistor is connected to one of the switch terminals.

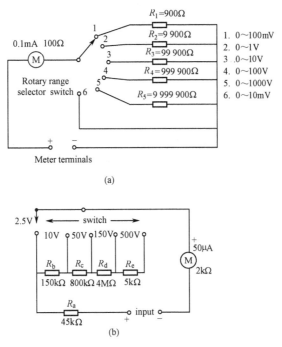

Figure 9.9 Multirange voltmeter circuits

Of the two circuits, configuration 9.9(b) is the least expensive to construct because only one resistor (R_a) is a nonstandard value. The nonstandard value is required to compensate for the internal resistance of the meter. Once the meter's internal resistance has been brought up to standard value, all other resistor values will fall within the scope of standard (precision) resistor devices, leading to lower construction costs. Configuration 9.9(a), on the other hand, requires each resistor to be a nonstandard value because each is individually affected by the meter's internal resistance.

9.5 Voltage Measuring Errors

In contrast to the ammeter, voltmeters are always connected in parallel with the circuit under test. Because the voltmeter requires a finite amount of current (however small) to operate, the circuit under test is affected. This is known as voltmeter loading effect. Such a loading effect tends to lower the voltage across the source, which in turn affects the measurement.

The effect can be described using the example in Figure 9.10. In this example, a voltmeter is used to monitor the voltage across one resistor of a series of pair. In Figure 9.10, the potential across points A and B is 1V without the voltmeter connected. This is obvious because resistance values are equal, thus dividing the source voltage exactly in half.

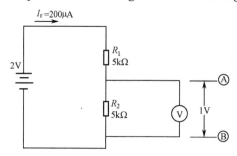

Figure 9.10 Series-resistance voltage divider

When the voltmeter is applied, however, the equivalent configuration of the circuit changes. The voltmeter can be represented by a defined resistance in parallel with R_2, as illustrated in Figure 9.11.

Assuming the voltmeter has an effective resistance of 5000Ω, that being the sum of R_s and R_m, the equivalent resistive load R_L between points A and B can be calculated as being:

$$R_L = \frac{R_2(R_s + R_m)}{R_2 + (R_s + R_m)} = \frac{5000 \times 5000}{5000 + 5000} = 2500(\Omega)$$

Application of Ohm's law, using the new circuit values, yields:

$$I = \frac{2V}{7500\Omega} = 2.66 \times 10^{-4} A$$

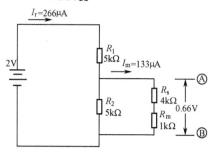

Figure 9.11 Voltmeter loading effect

Calculation for the voltage across A and B yields:

$$V_{A-B} = (2.66 \times 10^{-4}) \times 2500 = 0.665 V$$

As we can see, the voltmeter will not indicate the correct voltage of the circuit under test, but instead modifies the values to give a false measurement. Instead of indicating the true value of 1V, the voltmeter reads 0.66V.

The voltmeter loading factor is called the sensitivity, and is expressed in ohms per volt, where:

$$\text{sensivity} = \frac{R_s + R_m}{\text{full-scale-voltage-value}}$$

The higher the sensitivity of the voltmeter, the less current will be drawn and the less circuit loading. In the example above, the sensitivity of the voltmeter involved is

$$\text{sensivity} = \frac{5000}{1} = 5000(\Omega/\text{V})$$

The ideal voltmeter has infinite ohm-per-volt sensitivity in that the circuit parameters are not affected by its presence. Laboratory and service quality analog instruments commonly have a sensitivity of 20 000Ω/V.

One way to reduce the effects of voltmeter in Figure 9.9 is a full-scale deflection of 10V instead of 10V instead of 1V, its internal resistance would be

$$R_{in} = (\text{full-scale voltage value})(\text{sensitivity}) = 10 \times 5000 = 50(\text{k}\Omega)$$

This is ten times more resistance than the previous meter setting, and represents a lighter load that has less effect on the circuit under test. When this value is substituded for the values in Figure 9.11, the resultant meter reading is 0.95V, an error of only 5 percent.

Unfortunately, analog meters are notoriously inaccurate at the lower end of their scale. Because of frictional forces within the bearings of the coil mounting, a certain amount of torque must be applied before pointer deflection is achieved. As the current through the coil increases, the repelling forces are greater, and the effects of bearing friction are less noticeable. Consequently, readings that fall on the lower one-third of a mechanical meter's scale are less accurate than those taken at half scale or full deflection.

9.6 Ohmmeter and Resistance Measurements

Analog meters can also be used to measure resistance. Resistance, as defined by Ohm's law, is the constant of proportionality between the current and the voltage of the current. Consequently, if we know the amount of current flowing through a device and the voltage drop across it, both quantities of which can be measured by a meter, we can calculate its resistance using Ohm's law. Mathematically, the quantity is expressed as

$$R = \frac{E}{I}$$

The ammeter-voltmeter method of measuring resistance is an indirect method inasmuch as it is a calculated value. Neither meter reading is a direct indication of resistance value.

By setting either voltage or current constant and monitoring the variable value, we can scale the meter so that it does the resistance calculation for us. There are two types of analog ohmmeters in popular use: the series ohmmeter and the shunt ohmmeter.

9.7 Series Ohmmeter

The series ohmmeter provides a constant voltage source and derives its resistance measurement from the variable current value. Figure 9.12 illustrates the circuit of the basic series ohmmeter.

A battery (E) is connected in series with a variable resistor (R_1), a fixed meter resistance (R_m), and a 0 to 1 mA meter movement. The two test points P_1 and P_2 represent test probes that are connected across the unknown resistance (R_x) to be measured. Since the meter reads 1 mA at full scale, the value of R_1 and R_m must be such that

$$\frac{E}{R_1+R_m} = 1\text{mA}$$

So that full-scale deflection is achieved when the test probes are shorted together. This condition indicates an R_x value of zero ohms, and the meter scale can be marked accordingly.

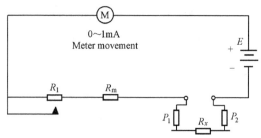

Figure 9.12 Series ohmmeter

When a resistance of any value greater than zero is inserted between P_1 and P_2, the meter current decreases proportionally. The meter current indicated by the instrument is equal to

$$I_m = \frac{E}{R_1+R_x+R_m}$$

An R_x value equal to R_1+R_m, for example, would yield a meter reading of

$$I_m = \frac{E}{R_1+R_m+(R_1+R_m)}$$
$$= \frac{E}{2R_1+2R_m}$$
$$2I_m = \frac{E}{R_1+R_m}$$

In other words, the meter would indicate half scale, and this point could be marked with the appropriate resistance value. If the total internal meter resistance was 100Ω, for example, then this point would be marked with a value of 100. By proceeding with this type of calibration using a wide range of resistance values, one is able to construct a meter scale reflective of the unknown resistance. Such a meter scale is illustrated in Figure 9.13.

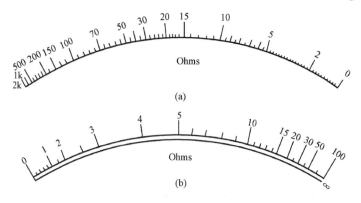

Figure 9.13 Series (a) and shunt (b) ohmmeter scales

Notice that the scale is nonlinear and tends to be overcrowded near the high-resistance end of the scale. As the value of R_x increases, the internal resistance of the instrument becomes less of a factor and has less influence on the current and meter reading. As R_x approaches infinity, the instrument's current rapidly decreases to near zero, virtually swaping the meter's internal resistance by its shear magnitude, thus creating the crowding conditions noted on the scale.

9.8 Shunt Ohmmeter

When very low values of resistance are to be measured, the shunt ohmmeter is superior to the series ohmmeter just described. The shunt ohmmeter relies on the internal resistance of the meter movement (R_m) to derive its resistance measurements. The circuit of a shunt ohmmeter is shown in Figure 9.14.

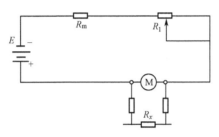

Figure 9.14 Shunt ohmmeter

In operation, the current in the ohmmeter is first adjusted by R_1 to provide full-scale deflection on the meter with the test probes separated. This reading represents an unknown resistance R_x of infinity, and the scale is marked accordingly.

When an unknown resistance is connected between the probes, R_x acts like a current shunt and passes a portion of the current around the meter, decreasing the meter reading proportionally. When $R_x = R_m$, the amount of current shunted by the unknown resistor is equal to the amount of current recorded by the meter, and the meter pointer indicates half scale. If the internal meter resistance is 100 ohms, then R_x is equal to 100 Ω. This value is marked on the scale, and the calibration process continues. When finished, the scale looks like the drawing in Figure 9.13(b).

Notice that the shunt ohmmeter scale is calibrated from left to right, as opposed to the series-type ohmmeter, which is calibrated from right to left. An inherent disadvantage of this circuit is the fact that the battery is always in the circuit, so that unless the instrument is turned off when not in use, the battery runs down very rapidly. [3]

9.9 Ohmmeter Accuracy

Inaccuracies in the measurement of resistance using analog meters can occur in a number of ways. The nonlinearity of the scale puts an immediate stress on the precision of the measurement made near the infinity region. A small displacement in needle position within this region can translate into a resistance difference of up to a magnitude.

Another problem occurs in the way that the factory rates the accuracy of the meter itself. The factory-rated accuracy of a meter is based on a full-scale DC voltage indication. The amount of error is determined by an area of indecision brought about by mechanical limitations inherent to the meter, described as the arc of error. Frictional losses in the bearings of the coil pivot or taut-band suspension create a hysteresis effect that the meter must overcome before the pointer comes to final rest. Depending on the direction of pointer movement up or down scale, the needle can fall to either side of the arc of error, or anywhere in between.

Suppose that the rated accuracy of the meter is specified as $\pm 1\%$ of full-scale voltage indication. This means that the voltage reading at the center of the scale (one-half full deflection) is $\pm 1\%$ the full-scale value, or $\pm 2\%$ of the indicated voltage.

Because of the nonlinearity of the ohmmeter scale, the arc of error can cause a relatively greater error in resistance measurements than it does in voltage or current measurements. The relationship between the arc of error for DC voltages and the accuracy of the ohmmeter scale is demonstrated in Figure 9.15. While a series-type scale is illustrated, the same reasoning applies to a shunt-type scale.

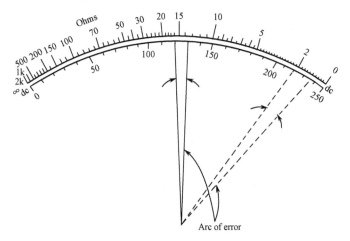

Figure 9.15 Arc of error measurement

When one considers the arc of error in light of scale nonlinearity, it is evident that the greatest accuracy can be achieved when the meter gives an indication as close to half scale as possible. This occurs when R_x is equal in value to the effective internal resistance that exists across points P_1 and P_2, whether the instrument be a series or shunt type.

9.10 Volt-Ohm-Milliammeters

Seldom are analog ohmmeters found as stand-alone instruments. An ohmmeter is normally part of a volt-ohm-milliammeter (VOM) or multifunction meter. The VOM is a multifunction instrument capable of operating as an AC or DC voltmeter, an AC or DC ammeter, or a multirange ohmmeter. A function switch is provided to permit selection of any function, and a range switch facilitates range changing. Additional functions, such as decibel measurements, are available with the use of special adapters or probes.

In essence, the VOM is many separate instruments in one, with all using the same meter movement as a readout. The circuitry for each function is basically the same meter, ammeter, or ohmmeter circuit described above. The precision resistors used in the instrument normally serve in as many functions of the circuitry is simply arranged so that only two selector switches are required to configure the instrument: function and range.

To use the VOM, one simply chooses the function and range desired, and proceeds to measure accordingly. A typical VOM multimeter is shown in Figure 9.16.

Figure 9.16 Simpson Model 260 (courtesy of Simpson Electric)

The function switch facilitates selection of $+$ DC volts and current, $-$ DC volts and current, or AC volts and current. The range switch, located in the center of the instrument below the meter, selects the scale for: voltage (2.5V through 500V), current (1mA through 500mA), and resistance (1Ω through 100kΩ). Additional functions are available by using the labeled probe jack located on the front of the VOM.

The terminals marked COMMON ($-$) and ($+$) are those normally employed for all voltage, current, and resistance measurements. The $+10$A and -10A terminals are used to measure DC current levels ranging from 0~10A when the range switch is in the 10A position. These jacks permit direct access to the meter movement and use a meter shunt for proper ranging. Direct access to the meter movement with no current shunt is through the $+50$ AMPS/250 mV jack, with the range switch set to 50V and COMMON used as the return line. Maximum input voltage in this mode is 250kV.

To measure voltages, both AC and DC, greater than 500 volts, the test probe is removed from the $+$ input terminal and plugged into the 1000V terminal. The range switch is set to the 500V position, effectively doubling the range of the instrument. For voltages less than 2.5V but greater than 250kV, one can plug into the $+1$V terminal and twist the range switch to the 2.5V position while reading from a 0~10 scale.

Over the years, the VOM has become more than a versatile instrument, with a wide range of features. For some it is an institution. Consider the plight of the field engineer of technician, whose job would be made very inconvenient, if not impossible, if he had to carry around an arsenal of instruments for service.

9.11 Electronic Voltmeters

Voltmeters constructed of moving-coil meters and multiplier resistors have some important limitations. As illustrated above, the most severe is an inherently low internal resistance, which leads to significant loading of many circuits under test and subsequent measurement errors. The instrument is also incapable of measuring very low voltage levels. These limitations can be overcome using an instrument that can amplify and buffer the input signal from the mechanical meter movement. Examples of such devices are the electronic voltmeter and the electronic multimeter.

The electronic multimeter was originally called the vacuum tube voltmeter (VTVM) because it was designed using vacuum tubes. With the advent of the transistor and later the integrated circuit, vacuum tubes were dropped in favor of more stable semiconductor devices and the instrument became known simply as the electronic voltmeter (Figure 9.17).

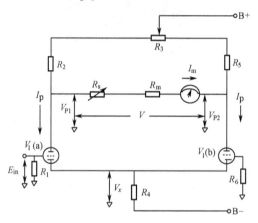

Figure 9.17 VTVM schematic

9.12 Transistorized Electronic Voltmeters

The concept of the transistorized electronic voltmeters is basically one of placing a metering device within a differential amplifier circuit. The differential amplifier provides both amplification and isolation of the input signal from the meter movement. Refer to Figure 9.18 for the following analysis.

Transistor Q_1 and Q_2, together with R_2, R_5, and R_4, constitute a differential, emitter-coupled amplifier. The amplifier gets its name from the fact that when the voltage at the base of Q_2 is zero and a signal voltage is applied to the base of Q_1, the difference between the two base voltages is amplified and reflected in the collector voltages.

Differential amplifiers require both a positive and a negative power supply to work properly. The common junction of the two power sources is the virtual ground. With the bases of both Q_1 and Q_2 held at ground (zero volts), the voltage drop across emitter resistor R_4 is

$$V_{R4} = 0 - V_{BE} - (-V_{EE})$$

or approximately V_{EE} less 0.7V. The current through R_4 is

$$I_{E1} + I_{E2} = \frac{V_{R4}}{R_4}$$

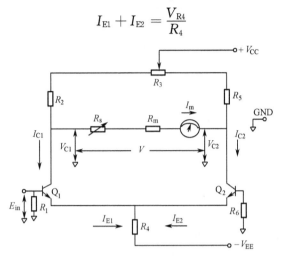

Figure 9.18 Differential amplifier stage

When $E_{in}=0$, then $I_{E1}=I_{E2}$. If the transistors are sufficiently matched in characteristics, then $I_{C1}=I_{C2}$, and $V_{R2}=V_{R5}$.

This condition forces the voltage at the collector of Q_1 equal to the voltage at the collector of Q_2. If a meter or other voltage monitoring device is now placed between the collectors of the two transistors, the meter would register no difference in voltage, and remain at rest.

Because the voltage drop across R_4 must be maintained to satisfy the V_{BE} requirements of transistor Q_2, a positive voltage applied to the base of Q_1 causes the current through Q_1 to increase and the current through Q_2 to decrease. This in turn produces a larger voltage drop across R_2 and a smaller voltage drop across R_5, with respectively lower and higher voltages at the collectors of Q_1 and Q_2.

The consequence of this is a differential voltage between the collectors, which the meter sees as a positive potential at the collector of Q_2 and a negative potential at the collector of Q_1. The meter voltage is directly proportional to the input voltage E_{in}, resulting in a meter reading that is representative of the signal voltage at E_{in}.

A voltage divider made of a resistor ladder limits the voltage to the input of the differential amplifier and essentially sets the instrument's range. The voltage divider, which consists of resistors R_a and R_e, accurately divides the voltage to be measured before it is applied to the input transistor, as illustrated in Figure 9.19. This arrangement limits the input to the input transistor to less than 1V when the selector switch is properly set for the voltage under test.

This attenuator configuration has a constant input resistance to the circuit under test, regardless of the voltage range settings. The term sensitivity no longer has meaning with an electronic voltmeter. The term generally applied to electronic voltmeters is input impedance.

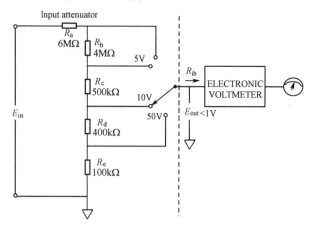

Figure 9.19　Input divider network

When using the attenuator described above with a bipolar transistor, however, the transistor input resistance has an effect on the instrument's input impedance. Because the base resistance is in parallel with the divider, some error is introduced into the attenuator in terms of amplifier loading. Assuming that R_4 in figure 9.18 is 10kΩ, the resistance seen at the base of Q_1 is approximately 20kΩ to 40kΩ, depending on the characteristics of the transistor. When this parallel load is placed across the divider resistors, it lowers the total resistance by an amount that can be determined by simple parallel resistor laws. For the sake of argument, assume that the base input resistance is 40kΩ. When the selector switch is in the 50-volt position, this 40kΩ is in parallel with R_e, leading to an effective resistance of

$$R_{e'} = \frac{R_{ib}R_e}{R_{ib} + R_e}$$
$$= \frac{40\text{k}\Omega \times 100\text{k}\Omega}{40\text{k}\Omega + 100\text{k}\Omega}$$
$$= 28.6\text{k}\Omega$$

This is a far cry from the intended 100k value for R_e, and is totally unacceptable. To resolve the problem, the input impedance of the differential amplifier must be increased.

9.13　FET Voltmeters

A device capable of the task is the field-effect transistor, or FET. When an FET is connected at the input of the differential amplifier, the input resistance of the electronic voltmeter is significantly increased. Because the FET gate current is typically less than 50 nA, it has an equivalent input impedance of 2×10^{10}. MOSFET devices exhibit even larger input impedances. The resulting circuit is illustrated in Figure 9.20. In this configuration, the source terminal voltage changes by the same amount as the voltage at the FET gate terminal, and is called a source follower.

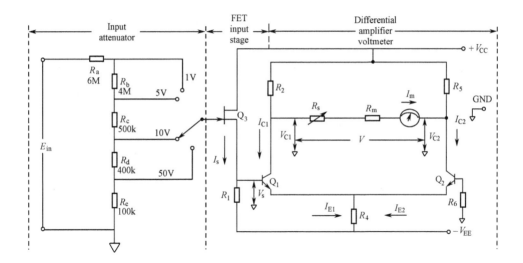

Figure 9.20 FET electronic voltmeter

Although a definite value of h_{FE} has been assumed in Figure 9.18 for transistors Q_1 and Q_2, h_{FE} can vary enough from one transistor to another to create an imbalance within the differential stage, even when the transistors are closely matched. For this reason, an electrical zero control (h_{FE}) is provided to balance the currents flowing through Q_1 and Q_2 for proper meter operation. With the test probes of the instrument shorted together, this control is adjusted so that the meter registers zeros volts.

9.14 Operational Amplifier Electronic Voltmeters

Recent advances in integrated circuits have yielded operational amplifiers with characteristics stable enough for use in electronic voltmeters. The input resistance of IC op amps, as they are called, is very high. Those with bipolar transistor input stages have typical input currents on the order of 0.2A, while FET input op amp is ideally zero ohms, and is quite capable of supplying the current necessary to drive a deflection instrument.

Operational amplifier electronic voltmeters come in three circuit configurations, depending on application. They are voltage follower, noninverting, and voltage-to-current converters.

The op-amp voltage follower is analogous to a single transistor emitter follower circuit, whose sole purpose is to change a high-impedance input onto a low-impedance output with no amplification. Referring to Figure 9.21, the gain of the op-amp is set at unity by tying the output of the op-amp to its inverting input. By setting the gain of the op-amp at unity, the voltage follower accurately reproduces the input signal at the noninverting input at its does not suffer from a base-to-emitter voltage drop as it mimics the input signal. The output voltage is always equal to the input voltage.

The meter is connected to the output of the op-amp through a network of series resistances. Resistance R_m is the internal meter resistance and is an inherent characteristic of the meter itself. R_s is a series current-limiting resistor used to calibrate the meter reading. The attenuator circuit feeds the noninverting input.

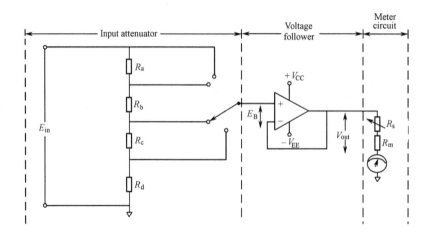

Figure 9.21 Voltage follower op amp voltmeter

When the instrument is called on to measure voltages that are smaller than the meter can display, the op-amp circuit can assume the role of both buffer and amplifier, using a noninverting amplifier circuit similar to the one in Figure 9.22. [4]

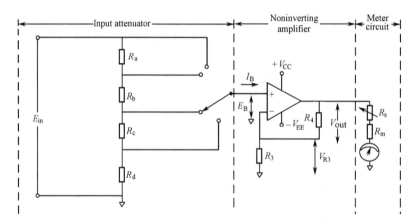

Figure 9.22 Noninerting op-amp voltmeter

A noninverting amplifier is very easily designed. A feedback network, consisting of R_4 and R_3, returns current from the op-amp output to the inverting input, and thus sets the amplification factor. The gain of the circuit is calculated using the equation

$$A_v = \frac{R_3 + R_4}{R_3}$$

The potential divider current I is selected very much larger than the input bias current (I_B) of the op-amp so that I_B has no significant effect upon the feedback voltage. The total resistance of the network ($R_3 + R_4$) is calculated as

$$R_3 + R_4 = \frac{V_{out}}{I}$$

The voltage across R_3 is always equal to E_{in}, i.e.

$$R_3 = \frac{E_{in}}{I}$$

Apart from any error due to the deflection instrument or inaccuracies in the resistors, there is a small error introduced by the internal gain of the operational amplifier. In both circuits, however, the error is negligible compared to the normal deflection meter error.

The voltage-to-current converter is a variation on the noninverting amplifier configuration of Figure 9.22. In the voltage-to-current converter, a meter is substituted in place of resistor R_4, and R_3 is made partially adjustable. As with the noninverting configuration, the voltage drop across R_3 is always equal to input voltage E_{in}. If the input signal increases or decreases, V_{R3} follows it precisely. So, the current through the meter (I) is equal to

$$I = \frac{E_{in}}{V_{R3}}$$

In other words, the meter reading is a direct function of the input voltage. Part of R_3 is made adjustable for instrument calibration. Calibration is performed by inputting a voltage to the noninverting input of the op amp and adjusting R_3 until the meter pointer indicates the applied voltage. For proper circuit operation, the voltage developed across $R_m + R_3$ must always be less than the V_{CC} supply voltage.

9.15 Electronic Current Measurements

The electronic multimeter is basically a voltage measuring device. Consequently, when using an electronic voltmeter to measure current, the current must first be converted to a voltage. The network in Figure 9.23 is used for current measurement.

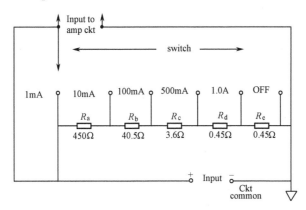

Figure 9.23 Electronic current measurement

This circuit represents a selectable resistance shunt, across which is generated a voltage. When the voltage is measured and displayed on an analog meter with the proper scale markings, it is representative of a current value according to ohm's law, i.e.

$$I = \frac{E_R}{V_m}$$

where E_R is the voltage across the current shunt and V_m is the meter reading.

For each range starting at 1A, an additional resistor is added in series until a total of 450Ω is inserted into the circuit under test. Note that the resistance values are selected so that the shunt voltage never exceeds 0.455 V at each range setting. The lower the insertion loss of the shunt resistor, the less error the instrument introduces to the circuit under test and the more accurate the measurement.

Unfortunately, the current function of most electronic multimeters inserts a resistance value into the circuit under test that is considerably greater than that presented by a laboratory current shunt, which typically displays a 50 mV voltage drop at the same current value. While the induced error caused by the multimeter is unacceptable for serious laboratory studies, the convenience of an all-in-one instrument for field service and bench work far out-weighs the disadvantage in these applications.

9.16 Resistance Measurements

The electronic voltmeter can also be used to measure resistance by providing a fixed current through an unknown resistance, then measuring the voltage drop across the unknown resistance and correlating the voltage to a resistance value. All that is required is a battery, several standard resistors, and a suitable switching arrangement. There are three methods normally used by electronic voltmeters to measure resistance.

1. Frequency Response

One very important characteristic that affects the AC measuring process is frequency. The various types of instruments used have internal resistance, capacitance, and inductance. Because the latter two parameters relate to lumped impedance in a frequency-sensitive manner, the frequency response and bandwidth of an instrument become major influences in the measurement of an AC signal.

As in the case of DC measurements, the AC meter is inserted into the circuit under test - voltmeters in parallel and ammeters in series. The net effect of the meter's effect is detailed in Figure 9.24.

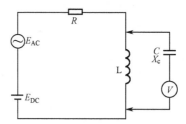

Figure 9.24 AC meter loading effect

As the frequency varies, certain elements of each component come into play. At very low frequencies, the capacitive reactance of the input-blocking capacitor must be such that the input impedance of the meter is 100 times greater than the reactance of the capacitor. The instrument capacitor should be isolated enough by the test probe that its effect on the waveform is minimized (see page 161 for further explaination).

Because of the proliferation of oscilloscopes as general-purpose measuring instruments, we have become accustomed to viewing on the screen a nonsinusoidal AC waveform that appears to be comprised only of what we see occurring at a certain repetitive rate. From Fourier analysis, however, we know this is far from the truth. The harmonic content of many waveforms—particularly those, like square waves, with sharp corners—extend far beyond the fundamental frequency and are intricate parts of the waveform energy spectrum. As a rule of thumb, the usable passband of the test analog test instrument should be at least ten times the repetition rate of any nonsinusoidal waveform under test.

2. AC Electronic Voltmeters

The measurement of AC voltages presents some unique problems not present in DC measurements. The problems stem from the fact that AC voltage has additional parameters, besides magnitude, that affect the measurement. Let's begin with an analysis of the most common AC waveform, the sine wave.

3. Sine Wave Analysis

Figure 9.25 illustrates the parameters of a sinusoidal waveform, or sine wave. The voltage of the waveform goes from zero to maximum in the negative direction and back to zero. This compete alternation is called one cycle. The time duration for one cycle is called the period (T). The number of periodic cycle repetitions in a one-second interval is called the frequency.

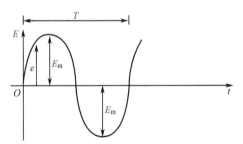

Figure 9.25 Sine wave parameters

The instantaneous value of the voltage of the sine wave is given by the mathematical expression:

$$e = E_m \sin(\omega t)$$

where e = instantaneous value (volts), E_m = maximum amplitude (volts, peak value), ω = radian frequency (radians per second), t = time (second).

The following expressions are used to further define the properties of a sine wave:

Peak-to-peak value(E_{p-p}): The peak-to-peak voltage is twice the maximum (E_m) value. $E_{p-p} = 2E_m$.

Average value (E_{avg}): The average value of any voltage is defined algebraically as the total area under the voltage curve divided by the period in radians. In the case of the sine wave, the average value is zero because the positive portion is exactly equal to the negative portion; hence, the total area is zero.

Root-mean-square (E_{rms}): The root-mean-square, or effective value, of the AC is defined as that value of AC voltage that has the same heating effect as a DC voltage. For example, it requires 14.14 volts of DC. Therefore, the E_{rms} value can be expressed as

$$\frac{E_m}{E_{rms}} = \frac{14.14}{10} = 1.414 = \sqrt{2}$$

therefore

$$E_m = \sqrt{2} E_{rms}$$

$$E_{rms} = \frac{E_m}{\sqrt{2}} = 0.77 E_m$$

Of the three defined values, the Erms is by far the most important quantity, since it is the only quantity that permits a direct and accurate comparison between the effects of DC and AC signals, regardless of the wave shape.

Words and Expressions

apparatus [ˌæpəˈreitəs] n. 仪器
attainable [əˈteinəbl] adj. 可获得的
attenuator [əˈtenjueitə(r)] n. 衰减器
bandwidth [ˈbændwidθ] n. 带宽
battery [ˈbætəri] n. 电池
be inversely proportional to 与……成反比
be proportional to 与……成正比
calibrate [ˈkælibreit] vt. 修正，校正
capacitor [kəˈpæsitə] n. 电容
cathode [ˈkæθəud] n. 阴极
collector [kəˈlektə] n. 集电极
conduct current 导通电流
decibel [ˈdesibel] n. 分贝
deflection pointer n. 偏转指针计
discrepancy [disˈkrepənsi] n. 差异
emitter [iˈmitə] n. 射极
equivalent [iˈkwivələnt] adj. 等效的
feedback [ˈfiːdbæk] n. 反馈
friction [ˈfrikʃən] n. 摩擦力
gain [ɡein] n. 增益
hysteresis [ˌhistəˈriːsis] n. 滞后作用
impedance [imˈpiːdəns] n. 阻抗
in light of 根据
intricate [ˈintrikit] adj. 复杂的，错综的，难以理解的
instantaneous [ˌinstənˈteinjəs] adj. 瞬间的

investigation [inˌvestiˈgeiʃən] n. 研究
mimic [ˈmimik] vt. 模仿
nonsinusoidal [ˈnɔnsainəˈsɔidəl] adj. 非正弦的
noticeable [ˈnəutisəbl] adj. 显著的
nonlinearity [ˌnɔnliniˈæriti] n. 非线性
operational amplifier 运算放大器
oscillate [ˈɔsileit] vi. 振荡
pivot [ˈpivət] n. 中轴，枢纽
plight [plait] n. 困境，情况，状态
potential [pəˈtenʃ(ə)l] n. 电势
proliferation [prəuˌlifəˈreiʃən] n. 增殖，分芽繁殖
reading [ˈriːdiŋ] n. 读数
readout [ˈriːdaut] n. 读出
rectification [ˌrektifiˈkeiʃən] n. 整流
rectifier diode 整流二极管
schematic [skiˈmætik] adj. 示意性的
semiconductor [ˈsemikənˈdʌktə] n. 半导体
sensitivity [ˈsensiˈtiviti] n. 灵敏度
suspend [səsˈpend] vi. 悬浮
terminal [ˈtəːminl] n. 终端
test probe 探针
torque [tɔːk] n. 力矩
transistor [trænˈsistə] n. 晶体管
transistorize [trænˈsistəraiz] vt. 装上晶体管
venerable [ˈvenərəbl] adj. 古老的

Notes

1. Recent advances in digital technology, however, have permitted the design of instruments that eliminate the measuring error introduced by the mechanical components of an analog meter.
近年来，随着数字技术的发展，使得能设计出消除由模拟表的机械元件而产生误差的仪器成为可能。

2. While it can be argued that deflection meters are not as accurate as modern digital laboratory test instruments, it is doubtful that the overall results obtained by a more precise test instrument would necessarily be more meaningful than those attainable from an inexpensive meter.
然而，有争议的是动圈表不如现代数字试验测试仪测量精确。值得怀疑的一点是，由比较精确的测试仪获得的结果是否一定比那些由廉价仪器获得的结果更有意义呢？

3. An inherent disadvantage of this circuit is the fact that the battery is always in the circuit, so that unless the instrument is turned off when not in use, the battery runs down very rapidly.

   该电路的天生缺陷在于：电池总是接在电路上，只有在分流欧姆表不同时，电池才不消耗，而其他时候它的电池消耗特别快。

4. When the instrument is called on to measure voltages that are smaller than the meter can display, the op-amp circuit can assume the role of both buffer and amplifier, using a noninvering amplifier circuit similar to the one in Figure 9.22.

   当仪表被用来测量比仪表本身所能显示的更小电压时，则运算放大器电路既是缓冲器也是放大器，其应用电路类似于图 9.22 所示的同相放大电路。

Exercises

Translate the following passages into English or Chinese.

1. 电子测量分两种：一种是对诸如电压、电容或场强这些电量所进行的测量，另一种是对其他一些量（如压力、温度或流速）用电方法所进行的测量。
2. 一只理想的安培表具有非常低的等效电阻，无论它存在于何种电路时，它对该电路性质的影响都会尽可能小。
3. All electronic voltmeters use a DC meter movement for their readout. Consequently, the AC signal input must be converted from AC to DC before it can be displayed. The conversion must be such that the true rms value of the waveform is retained. There are a number of ways to do the conversion.
4. The equivalent resistance of a set of inter-connected resistors is the value of the single resistor that can be substituted for the entire set without affecting the current that flows in the rest of any circuit of which it is a part.

Reading Material
Rectifier Meters

1. Half-wave Rectifier Meters

Basically, the conversion of AC to DC is done using rectifier diode is a one-way electronic value that essentially shears off the negative peak of the waveform when used alone, thus converting the wave shape to a pulsating DC waveform, as illustrated in Figure 9.26. The process is called half-wave rectification because only half the original sine wave exists after it takes place.

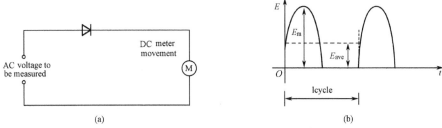

Figure 9.26 Half-wave rectifier

The integrated circuit operational amplifier is an ideal candidate for AC to DC conversions using diode rectification. Connecting a rectifier in series with the meter circuit of an op-amp voltage-follower voltmeter, as illustrated in Figure 9.27, converts the instrument into a half-wave rectifier AC voltmeter.

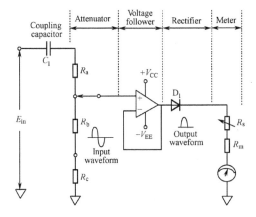

Figure 9.27 Half-wave rectifier voltmeter

In this design, the AC signal is input to the attenuator stage through coupling capacitor C_1. The coupling capacitor is provided at the input of the AC voltmeter to block unwanted DC voltages from affecting the measurement. As with the DC electronic voltmeter, the attenuator stage provides the instrument with a wide selection of input voltage ranges using a simple switching arrangement. The attenuated waveform is input directly to the noninverting input of the op amp. The output of the amplifier is returned to the inverting input, thus setting the gain of the amplifier at unity.

A rectifier may also be connected to the input of the op amp, where it converts the AC to DC before it is impressed across the meter(Figure 9.28). By placing the rectifier before the op amp, the amplifier deals only with DC values, and the bandwidth of the amplifier is not a critical factor in instrument design. This design is not as desirable as that of Figure 9.27 because of the nonlinearity properties inherent to solid-state rectifier diodes. Backward diodes, which are variations of tunnel diodes, are an exception and work well with this configuration for measuring low-value, high-frequency AC voltages.

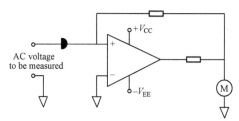

Figure 9.28 Backward diode voltmeter places the rectifier ahead of the op amp

Analog meters are average-voltage-responding instruments. The average value of the half-wave-rectified output is given by

$$E_{avg} = \frac{E_m}{\pi} = 0.318 E_m$$

since $\quad E_m = 1.41 E_{rms}$

then $\quad E_{avg} = 0.318(1.41) E_{rms} = 0.45 E_{rms}$

or $\quad E_{rms} = \dfrac{E_{avg}}{0.45}$

$$E_{avg} = 2.22 E_{rms}$$

The coefficient 2.22 is called the form factor, which is the ratio of $E_{rms}/E_{avg} = 2.22$. This means that when a meter reads an E_{rms} value of 100V, it is actually measuring

$$\frac{E_{rms}}{2.22} = 45 \text{ V}$$

The form factor allows us to calibrate the meter scale in E_{rms} while actually measuring average values.

Unfortunately, the forward voltage drop V_F across the rectifier in the circuit of Figure 9.29 is a source of error. When correctly forward-biased, a silicon diode typically has a forward voltage drop V_F of 0.7 V. V_F increases slightly with increase in I_F, and if I_F falls to a very low level (below the knee of the characteristic curve) there can be taken into account in design calculations, its variations cannot.

To avoid the errors, the voltage follower feedback loop to the inverting terminal is taken from the cathode of the rectifier instead of from the amplifier output, as demonstrated in Figure 3. The result is that the half-wave rectified output precisely follows the positive half cycle of the input voltage. There is no V_F voltage drop from input to output because the inverting terminal follows the input to the noninverting terminal. With the inverting input referred to the rectifier cathode, the voltage to the meter always equals the voltage at the amplifier input. Such a circuit configuration is commonly referred to as a precision rectifier circuit.

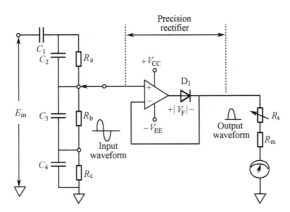

Figure 9.29 Precision rectifier amplifier

Note that capacitors C_2, C_3, and C_4 are connected across the attenuator resistors in Figure 9.29. These capacitors are normally employed in AC voltmeter attenuators to compensate for the input capacitance of the operational amplifier. Input capacitance problems also occur in oscilloscopes.

Low-level AC voltages too small for normal display must be accurately amplified before being rectified and applied to the meter. The circuit shown in Figure 9.30 combines amplification with half-wave rectification. Ignoring the rectifier diode for the moment, we see that the circuit is a noninverting amplifier with a gain factor of

$$A_V = \frac{R_2 + R_3}{R_3}$$

Inclusion of the diode causes the positive half cycles to be amplified and the negative half cycles ignored. Again, the configuration of the feedback loop guarantees that the amplification is precise and that no rectifier volt drop is involved in the measurement.

2. Full-wave Rectifier Meters

When full-wave rectification is used, both the positive and negative peaks of the waveform are converted into DC signals, as shown in Figure 9.31. Generally, the full-wave rectifier is of the bridge type. A bridge rectifier passes the positive, diodes D_1 and D_4 conduct, causing current to flow through the meter from the negative meter terminal to the positive. When the input goes negative, diodes D_1 and D_4 are reversebiased, and diodes D_2 and D_3 conduct current through the meter. Because of their arrangement, D_2 and D_3 force the current to again flow through the meter from the negative meter terminal to the positive. The resulting current flow is a series of positive half cycles without any intervening spaces.

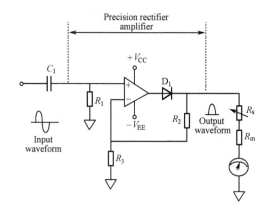

Figure 9.30 Precision rectifier amplifier

When a full-wave rectifier is used, the form factor is only 1.11, because

$$E_{avg} = \frac{2E_m}{\pi} = 0.636 E_m \qquad E_m = 1.414 E_{rms}$$

therefore

$$E_{avg} = 0.636 \times 1.41 E_{rms} = 0.9 E_{rms}$$

and

$$E_{rms} = \frac{E_{avg}}{0.9}$$

$$E_{avg} = 1.11 E_{rms}$$

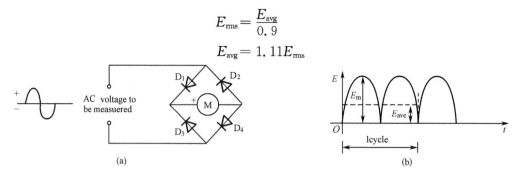

Figure 9.31 Full-wave bridge rectifier

As indicated earlier, the rectifier-type AC voltmeter is calibrated to read the effective, or rms, value of the measured average voltage.

The circuit in Figure 9.32 employs a full-wave bridge rectifier to convert the AC signal to DC. This circuit is basically a voltage-to-current converter with rectification. In this design, the voltage drop across R_3 is always equal to the input voltage.

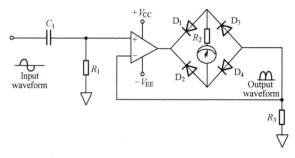

Figure 9.32 Full-wave voltage-to-current converter

Lesson 10   Digital Instruments

While it can be argued that analog instruments have earned their place in the laboratory, they are still limited by finite mechanical accuracy. Analog measurements are largely subjective, influenced by visual and psychological perception errors which may cause viewers to report slightly different readings for the same event. While much of the error can be attributed to parallax, which is caused by viewing the meter pointer and scale at an angle other than straight on, our psychological inner self tends to make us fudge on measurements even when we have interpreted them properly in the physical sense. If a measurement is slightly off the mark, we have the tendency to compensate for the error by allowing the recorded reading to creep closer to our perceived impression of the true value. [1] Aggravating the situation is the fact that the meter itself does not always register identical values in the same way twice, depending on which way the pointer happens to be moving. [2]

The digital readout, on the other hand, is a truly objective one, with all users perceiving the same indication. The direct numerical readout reduces the human error and the tedium involved in long periods of measurement. 'Cheating' is less prevalent because the numbers are before you, in black and white (or red on black, as the case may be). Digital electronics also brings with a wide range of peripheral benefits not available in analog instruments, including auto-ranging, waveform computation, and high common-mode rejection qualities.

10.1   Digital Displays

The first digital display devices were mechanical, made up of numbered wheels and driven by servo-controlled motors. Next on the scene were various formats consisting of miniature incandescent lamps. These devices evolved to a surprisingly satisfactory state of the art and gave rise to control techniques that now enable the popular light emitting diodes (LED) and liquid-crystal displays (LCD) to dominate instrumentation readouts.

The weak link in the incandescent lamp display is the fragile filament. Even when operated at low current, the incandescent bulb has a life expectancy that is measured in hours. Short life notwithstanding, the thermal response time of the filament also poses a problem in instrument readouts with a relatively long lag time. Although lamps are little used now, they established the popular seven-segment display format that is used by all digital readouts today.

Gas-discharge plasma readouts emit light via the ionization of neon-argon, and other noble gases. A small amount of mercury is often used to manipulate the ionization potential and to modify the spectrum. These displays have been made into many forms, some of which have external electrodes. Light output is very good and the definite on-off characteristics of plasma devices allow for convenient multiplexing. A long-enduring and very satisfactory gas-discharge readout was the Nixie tube made by Burroughs. These neon display tubes had nine internal cathodes, each shaped as a digit. The cathodes were stacked one behind the other in a line perpendicular to the line of sight. When a cathode was selected for excitement, it would glow in the shape of the number it represented. The fact that the nine digits were each in a different plane was somewhat disconcerting, but acceptable. A disadvantage of plasma displays is the high ionization potential required to ignite the lamp. It is on the order of 150 to 200V, which creates interfacing problems with today's low-voltage, solid-state circuitry. While no longer the dominant readout, plasma displays continue to find applications (Figure 10.1).

Truth table of segmented module

	1	2	3	4	5	6	7	8	9
a		×	×		×	×	×	×	×
b	×	×	×	×			×	×	×
c	×		×	×	×	×	×	×	×
d		×			×	×		×	×
e		×				×		×	
f				×	×	×		×	×
g		×	×	×	×	×		×	×

Display format of segmented module

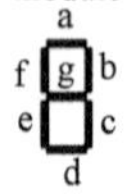

Figure 10.1 Seven-segment display

Vacuum fluorescent readouts were also a product of vacuum-tube technology. The fluorescent readout had a multiplicity of anodes that were coated with a light-emitting phosphor similar to the material found inside fluorescent lamps. While blue-green was the most popular color, others existed. The anode would fluoresce when bombarded by electrons from the cathode emitter. In most cases, the digit was formed using the seven-segment format described above. While these readout devices had the shortcomings inherent to vacuum tubes, including their fragile, short-lived filaments and high voltage requirements, they nevertheless had considerable aesthetic appeal and still find their way into modern applications, such was automotive and computer displays.

Electroluminescent display also make use of a light-emitting phosphor. Instead of the electron tube structure, though, these displays are essentially capacitors. A phosphor-rich film is sandwiched between two transparent, conductive plates. When AC voltage is applied to the plates, the capacitance displacement current excites the phosphor and causes it to glow.

Light emitting diode devices, developed as a result of continued advancement with semiconductor materials and fabrication, are the workhorses of digital display. They are bright, cost-effective, compatible with solid-state voltage and current levels, and long-lasting. Originally available only in red, LEDs now come in a wide range of colors, including red, orange, yellow, green, and blue.

Liquid crystal displays (LCDs) do not emit light, but instead alter the characteristics of the crystal material to reflect incident light. Optically, LCDs shine where LEDs fail; the brighter the ambient light, the brighter the LCD display. Conversely, LEDs wash out with increasing ambient light levels, and can scarcely be seen. On the other hand, the LCD display becomes shadowy at the low light levels at which the LED offers maximum readability. Because they are not light emitters, LCD displays offer the lowest power consumption of any electrically activated display. It is for this reason that: LCD displays are very popular with portable and battery-operated equipment.

10.2 Electronic Digital Counter

Common to all digital instruments is the electronic digital counter. The electronic counter, most often thought of as a device that totalizes input events, is the basic building block for all digital test and measurement applications.

A simplified block diagram of an electronic digital counter for frequency measurements is shown in Figure 10.2. A simple frequency counter consists of seven major functions: input conditioning, timerbase oscillator, timerbase dividers, main-gate flip-flop, main gate, counting register, and readout display. Central to the operation of the digital counter is the main gate.

The main gate is nothing more than a logic AND circuit. When both the input conditioning and main-gate flip-flop are logically true, the main gate opens for a period of time that is determined by the timerbase divider. While the main gate is open, the conditioned input signal pulses are passed through to the counting register, where they are tallied and then scaled for output by the display circuitry. [3] At the end of the counting period, the main gate is closed and the counter reset for the next sampling period.

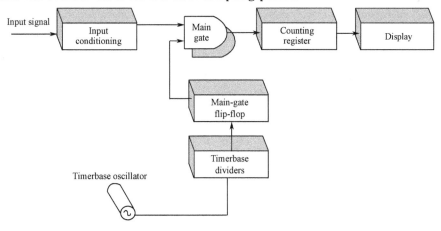

Figure 10.2 Electronic digital counter block diagram

10.3 Input Signal Conditioning

First, the input signal is conditioned to remove unwanted noise and make the signal level compatible with the internal digital processing circuitry. Because the frequency counter must accommodate a wide range of input signals, with varying degrees of voltage, noise component, and DC offset, the input must pass through many conditioning circuits before it is fully conditioned. [4] The individual components of the input conditioning function are shown in Figure 10.3.

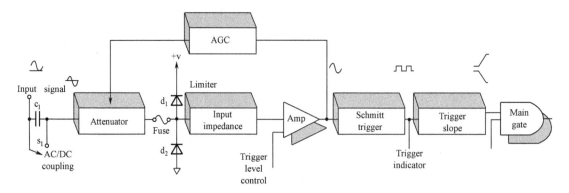

Figure 10.3 Signal conditioner block diagram

The input signal to be measured is first passed through an attenuator to reduce the amplitude of the signal, if necessary. The attenuator is a voltage-dividing RC network that typically provides switch-selectable input division by 1, 10, or 100. In some instruments, the RC network may be replaced with a potentiometer to provide continuously variable input attenuation.

An AC signal with a large DC component can shift the level of the signal outside the usable range of the instrument. Consequently, input coupling to the attenuator may be either DC or AC, depending on the signal input characteristics. With AC coupling, the input signal passes through a capacitor that blocks the DC voltage and establishes a ground reference for the contain a large offset voltage. AC coupling is of little value, however, if the duty cycle-the ratio of the input network may exceed the width of the input pulse. The result is an integration of the pulse that appears as an averaging voltage rather than a countable pulse. For the same reason, AC coupling should be avoided when measuring variable duty cycle signals, such as the kind found in switching power supplies and power controllers(Figure 10.4).

To prevent instrument damage from an accidental overload, two diodes (D_1 and D_2) are used to limit the input voltage in the event the input signal should exceed the range of the input attenuator setting. The input signal is then routed through an impedance-matching network into a wide-band amplifier.

The input to the amplifier is generally high impedance (one megohm), but some frequency counter permit a choice. A high-impedance gives the instrument better sensitivity, by reducing the amount of noise picked up by the input cable. A sensitivity of 20 mV is possible with a 50Ω input, as opposed to the typical 250mV sensitivity of the high-impedance input. The lower-impedance input also reduces the effect of pulse integration when using AC input coupling. Instrument loading, must be taken into consideration when choosing an input impedance.

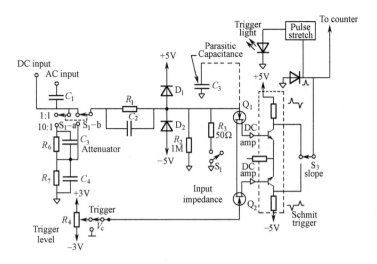

Figure 10.4 Input conditioning circuitry

The trigger level establishes a reference voltage, above which the input signal will trigger the frequency counter. If the trigger level (V_T) is set too high, as shown in Figure 10.5(b), the input signal will never fall below the lower hysteresis voltage, and no pulses will be counted. By lowering the trigger threshold, as shown in Figure 10.5(a), the input signal traverses the complete hysteresis range, and each input pulse will be counted. For negative input voltages, the trigger level (V_T) is set as shown in Figure 10.5(c).

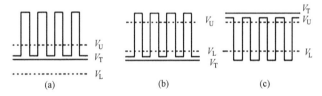

Figure 10.5 Triggler level

The AGC circuit adjusts the sensitivity of the amplifier to accommodate inputs of different magnitudes. The AGC smoothes out any rough spots that may be encountered in every input signals, so they should vary somewhat in amplitude, thus providing a constant output voltage to the Schmitt trigger. A trade-off exists between the response of the AGC and the minimum frequency signal that can be counted, which is usually about 50Hz with common AGC circuit tends to respond to high-frequency AM (amplitude modulated) signals. The AGC circuit tends to respond to the peak voltage levels and ignore the valleys. In an attempt to track a high-frequency AM input, the response time may not be fast enough to catch some low-level modulations, which leads to erroneous counting.

The output of the amplifier goes to the input of a Schmitt trigger. The Schmitt trigger provides the hysteresis necessary for proper triggering of the input signal and conditions any signal with a slow rise time, a slow fall time, or both (as in the case of low-frequency sinewave), to provide a rectangular output with sharp edges. The signal is then presented to one input of the main gate.

10.4 Timerbase Oscillator

The output pulse from the Schmitt trigger passes to the counting circuits only when output Q from the main-gate flip-flop is at a logic 1. The flip-flop changes state each time a negative-going output is received from the timerbase dividers. Therefore, if the timerbase divider is configured to supply an output signal once every second, the flip-flop output is alternately at level 1 for one second and at level 0 for one second. Consequently, the main gate is alternately open for one second and closed for one second. In other words, the main gate alternately passes the Schmitt trigger output pulses to the counting registers for one second, and then blocks them for one second. The timing of the main-gate flip-flop is controlled by the timerbase dividers, which are driven by the timerbase oscillator.

The timerbase oscillator is an accurate crystal oscillator that drivers the timerbase dividers. The timerbase divider scales the output of the timerbase oscillator (RTXO), temperature-compensated crystal oscillator (TCXO), and oven-controlled crystal oscillator.

Temperature is the natural enemy of any oscillator. The frequency stability of an oscillator is directly dependent on the temperatue stability of the tuned-circuit components. In a crystal oscillator, this is the crystal itself. Temperature effects can be minimized by thermostatic control, temperature compensation, or choice of an appropriate crystal type. The merits of these methods are summarized in Table 10.1.

Table 10.1 Crystal oscillator stability

TYPE	FREQUENCY	STABILITY	TEMP. RANGE
Double oven	2.5 MHz	$+1\times10^{-10}$	0 to +40℃
Single oven	5 MHz	$+1\times10^{-8}$	−20 to +65℃
TCXO	2-20 MHz	$\pm1\times10^{-7}$	−30 to +50℃
RXCO	4-21 MHz	$\pm2.5\times10^{-6}$	−10 to +60℃

Thermostatically controlled ovens with proportional control are the most stable of all oscillators. Single-oven types are able to control the oven temperature to ±0.0001℃ over a similar temperature range. When the oven temperature, a very low overall temperature, coincides with the crystal turnover temperature, a very low overall temperature coefficient is achieved. It generally takes 24 hours after the application of power to the instrument for an oven to attain its maximum specified accuracy, but an oven can usually attain an accuracy of 7 parts in 10^{-9} within 20 minutes.

When an oven is not practical, a crystal can be temperature-compensated (TCXO) with a series reactance whose temperature coefficient cancels the temperature coefficient of the crystal. Another alternative is to use a voltage-variable capacitor (varactor) with a temperature-dependent bias, as shown in Figure 10.6. The value of inductor is selected so that

$$\omega_s < \sqrt{\frac{1}{L_1 C_1}} < \omega_0$$

where ω_0 is the desired oscillation frequency and ω_s is the spurious frequency. Although this technique does not give as low a temperature coefficient as a thermostatically controlled oven, it is useful because of its lower power consumption and absence of any appreciable warm-up time.

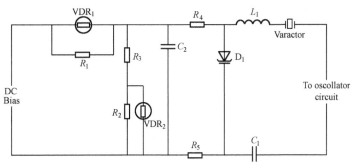

Figure 10.6 TCXO compensation

10.5 Timerbase Dividers

The main-gate flip-flop clock frequency is obtained by dividing down the timerbase oscillator frequency in the timerbase dividers. The timerbase dividers contains a bank of decade counters, each of which divides the input frequency by 10.

The output frequency from the final flip-flop of a decade counter is exactly one-tenth of the input triggering frequency. This means that the time period of the output waveform is exactly ten times the time period of the input waveform. The 1MHz output waveform of the first decade counter, which is driven by the 1MHz timerbase oscillator, has a time period of $10\mu s$. The time period of the output from the second decade counter (triggered by the output of the first decade country) is $100\mu s$, and that from the third decade counter is 1 ms, etc. With all six decade counters, the available time periods range from $10\mu s$ to 1s.

10.6 Counting Register

When the main-gate flip-flop opens the main gate, digital pulses from the Schmitt trigger stream into the counting register. The counting register is an array of decade counters that count and tally the digital pulses. Each display digit has its own decade counter, which consists of a latch circuit and a BCD to seven-segment decoder (Figure 10.7).

The decade counter is the pulse accumulator and totalizer. It is within this circuit that the pulses are counted and the count is converted to decimal format. The counters are cascaded to provide carryover as the addition proceeds. The output of the first decade (units) sends a pulse to the second decade counter (tens) each time the count rolls over. When the second counter has accumulated ten pulses from the first counter, it rolls over (zero) and outputs a pulse to the third (hundreds) decade counter, and so forth. The number of display digits determines the digital counter's resolution.

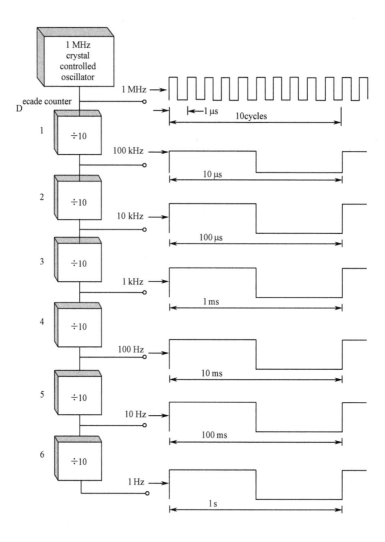

Figure 10.7 Timerbase oscillator and dividers

The latch circuit is used to provide isolation between the decade counter and readout display while counting is in progress. During counting, the display is disconnected from the decade counter output. At the end of the counting period, a signal to the latch loads the results of the tally into the final condition of the counting circuits. The BCD to seven-segment decoders convert the binary format of the decade counter output to seven-segment format for the display (Figure 10.8).

10.6.1 Digital Frequency Counters

The digital counter has come a long way since it first appeared on the commercial market. Even in its original form it was recognized as perhaps the most useful measuring instrument to emerge from the laboratory since the oscilloscope. The original application of the digital counter was that of a frequency counter.

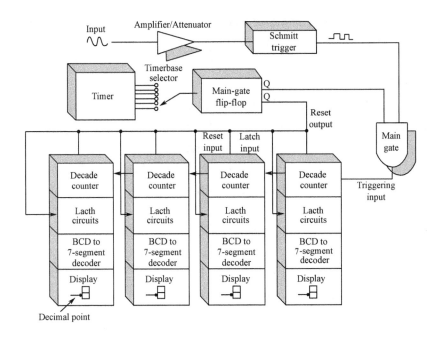

Figure 10.8 Counting register

When the main gate is controlled by an accurate time interval, the counter is in the frequency mode. The frequency mode is similar to the totalize mode described above, with the difference being in the way the gate control is operated.

For frequency measurements, the counting time interval is precisely controlled to be multiples of a second. If, for example, 45 500 pulses pass through the counting register while the main gate remains open for one second, then the input signal frequency is 45.5kHz. Depending on the number of digits in the readout, this number could be displayed directly or expressed as powers of ten.

To measure frequencies that would otherwise exceed the maximum count of the counting register, the time interval of the main gate can be reduced, with an appreciate change made to the scale factor. [5] A similar number of pulses (45 500) accumulating in the counting register within a 0.1 second period, for example, would represent a frequency of 455kHz; 25 000 pulses counted within 10μs is 2.5MHz, and so forth. Scaling can be done automatically by the positioning of the decimal point on the digital readout, as the timerbase period is switched from range to range.

Instead of manually switching to the appropriate timer range, some instruments have an automatic range selection system called auto ranger. [6] The auto ranger usually consists of a circuit that generates a voltage approximately proportional to the input frequency. Depending on the actual voltage level, one of several transistor switches is turned on to select the correct timerbase and decimal point position.

Using the frequency counter as a frequency meter, however, is exploiting but one of its several electronic counter functions. A counter can also indicate the period of an input signal, compare two signals in the ratio mode, indicate the time between two points on a waveform, or do reciprocal counting. The mode of operation is a simple reconfiguration of the basic counter building blocks shown in Figure 10.2.

10.6.2 Period Mode

The period mode of a signal is the reciprocal of its frequency. Period measurements are made by determining the amount of time a signal takes to complete one cycle of oscillation. In this mode, the counting register counts pulses from the timerbase oscillator, and the input signal is used to enable the main-gate flip-flop. The circuit configuration is shown in Figure 10.9. The number of timerbase pulses counted is directly proportional to the time period of the input signal.

For example, let's say the signal to be measured is 100kHz (10μs) and the timerbase reference oscillator is 1MHz (1μs). The main gate triggers on the input pulse and opens for a period of 10 microseconds, and during this time 10 pulses from the timerbase oscillator pass through the main gate. These 10 pulses are then processed and scaled by the counter to provide a readout indication of 10 microseconds.

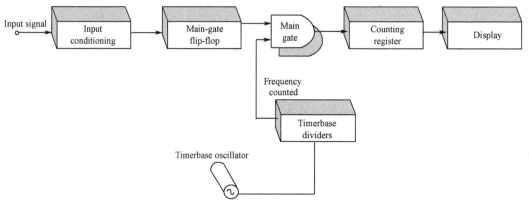

Figure 10.9　Period mode

10.6.3 Ratio Mode

A third mode of operation is the ratio mode. In the ratio mode, the ratio of two input signals is displayed on the counter readout. Frequency ratio may be measured, as shown in Figure 10.10, by disconnecting the timerbase oscillator from the gating circuitry and allowing the lower-frequency signal to control the main gate.

In effect, what happens is that the lower-frequency signal determines how long the gate is open and thus how long the gate passes the higher-frequency signal from the upper input. Therefore, the gate output is the ratio of the two signals. If the lower-frequency signal allows but one higher-frequency signal to pass, then the ratio is 1 : 1. Should the lower-frequency signal allow 10 higher-frequency cycles to pass, then the ratio is 10 : 1. This scheme will not work with reference frequencies that are lower than the signal to be referenced (i.e., when the frequency to Input A is less than the frequency at Input B), because the main gate would never be open long enough to pass even a signal pulse to the counting register.

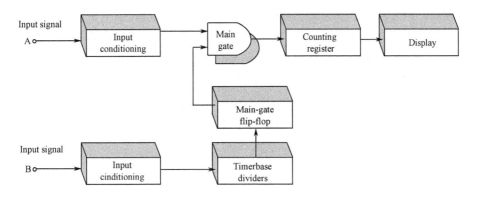

Figure 10.10 Ratio mode

10.6.4 Time Interval Mode and Phase Measurements

The time interval mode permits the counting of any number of events occurring between any two points in time. In effect, it is the equivalent of an electronic stopwatch, and uses start and stop inputs to control the opening and closing of the main gate. The equivalent circuit diagram is shown in Figure 10.11. During the counting period, the counting register counts pulses from the timerbase oscillator. Selection of the timerbase divider may alter the timerbase frequency in most instruments.

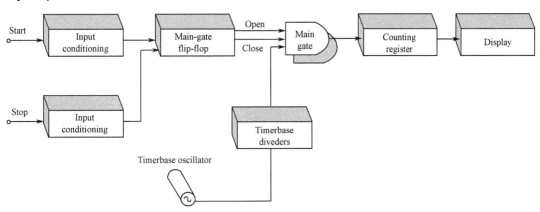

Figure 10.11 Time interval mode

A special application of the time interval mode is phase measurement. A frequency counter can be used to measure the phase difference between two equal-frequency signals by taking a time interval measurement between identical points on the waveform and calculating the phase from the measured time.

Consider the timing diagram in Figure 10.12. In this illustration, the values T_1 and T_2 are identical points on the reference-phase and unknown-phase waveforms. T_ϕ is the time difference between successive positive-going zero crossings of each signal. To measure T_ϕ, the frequency counter is set to operate in the START/STOP time interval mode, with the count starting at T_1 and ending at T_2.

The phase interval of time measurement is possible only with a more elaborate gate control than those associated with other modes previously discussed. As shown in Figure 10.13, the main gate has three inputs: one from the main-gate flip-flop and two from the signal conditioner stages, START and STOP inputs.

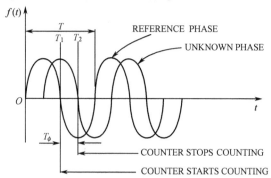

Figure 10.12   Phase measurement mode

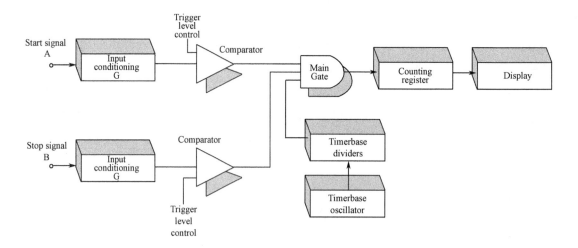

Figure 10.13   Phase mode timer control

The main-gate flip-flop feeds an accurate oscillator frequency from the timerbase oscillator to the counting register when the START signal triggers the main gate on. A similar trigger from the stop input halts the counting processes. The phase angle between the two signals is then calculated from the equation

$$F(\text{degrees}) = (T_\phi/T) \times 360$$

The START and STOP points are selected by triggering levels on the input waveforms.

10.7   Digital Voltmeter

The digital voltmeter (DVM) is perhaps the most prolific of all digital test instruments. Basically, the DVM circuitry is similar to the analog voltmeters discussed in Chapter1, with the expectation that the readout is displayed digitally rather than on a mechanical meter movement.

Although there are numerous techniques used to convert analog values to digital numbers, the basic principles of operation are the same: The voltage is unknown; the pulses are counted using a digital counter circuit, as described above, and the count is displayed on a 7-segment readout. Five methods of analog-to-digital conversion are commonly used. They are:

Voltage to frequency integrating
Single ramp
Dual-slope integrating
Successive approximation potentiometeric
Continuous balance potentiometeric

Words and Expressions

Activate ['æktiveit] *vt.* 激发
aesthetic [i:s'θetik] *adj.* 美学的
ambient ['æmbiənt] *adj.* 周围的
ambiguity [,æmbi'gju:iti] *n.* 模糊，含混不清
appeal [ə'pi:l] *n.* 要求
bombard ['bɔmbɑ:d] *vt.* 轰击
disconcert [,diskən'sə:t] *vt.* 使惊惶，使仓皇失措，破坏
duty cycle 占空比
elaborate [i'læbərət] *adj.* 精心制作的，详细阐述的，精细
electroluminescent [i'lektrəu'lu:mi'nesnt] *adj.* [物] 场致发光的，电致发光的
erroneous [i'rəunjəs] *adj.* 错误的
filament ['filəmənt] *n.* 灯丝
flip-flop [flipflɔp] *n.* 触发器
fluorescent [fluə'resənt] *adj.* 荧光的
fragile ['frædʒail] *adj.* 脆的
fudge [fʌdʒ] *vi.* 蒙混，捏造
incandescent [,inkæn'desnt] *adj.* 遇热发光的，白炽的
interpret [in'tə:prit] *v.* 解释
ionization [,aiənai'zeiʃən] *n.* 离子化，电离
oscillator ['ɔsileitə] *n.* 振荡器
parallax ['pærəlæks] *n.* 视差
perceive [pə'si:v] *vt.* 察觉
peripheral [pə'rifərəl] *adj.* 外围的
phosphor ['fɔsfə] *n.* 磷
plasma ['plæzmə] *n.* [物] 等离子体，等离子区
potentiometer [pə,tenʃi'ɔmitə] *n.* 电位计

prevalent ['prevələnt] adj. 盛行的
prolific [prə'lifik] adj. 多产的，丰富的，大量繁殖的
reciprocal [ri'siprəkəl] adj. 互惠的，相应的，倒数的，彼此相反的
spurious ['spjuəriəs] adj. 伪造的，假造的，欺骗的
tally ['tæli] vt. 计算
tedium ['ti:diəm, -djəm] n. 厌烦，沉闷
thermal ['θə:məl] adj. 热的
thermostatic [ˌθə:mə'stætik] adj. 恒温的

Notes

1. If a measurement is slightly off the mark, we have the tendency to compensate for the error by allowing the recorded reading to creep closer to our perceived impression of the true value.
大多数的测量误差来自视差，即由于观察仪器指针的角度不垂直而造成的测量误差，即使我们已经解释了这种生理特性，却还是想逃避自身责任。

2. Aggravating the situation is the fact that the meter itself does not always register identical values the same way twice, depending on which way the pointer happens to be moving.
如果测量值有轻微的偏差，我们总是倾向于补偿误差，将记录值向感觉的真值靠拢。更糟的是，事实上两次测量中仪器指针本身不会一直指向相同的位置，这是由于仪器指针是移动的。

3. While the main gate is open, the conditioned input signal pulses are passed through to the counting register, where they are tallied and then scaled for output by the display circuitry.
当主控门打开时，经过转换的输入脉冲通过主控门进入计数寄存器，并在此进行统计，然后通过读数屏输出。

4. Because the frequency counter must accmmodate a wide range of input signals, with varying degrees of voltage, noise component, and DC offset, the input must pass through many conditioning circuits before it is fully conditioned.
由于频率计数器必须提供一个宽范围的输入信号，使之能够适用于不同电压、噪声成分和直流偏移的情况，所以输入信号在充分变换前必须经过许多转换电路。

5. To measure frequencies that would otherwise exceed the maximum count of the counting register, the time interval of the main gate can be reduced, with an appreciate change made to the scale factor.
为了测量超过计数寄存器的最大计数值的频率，主控门被打开的时间间隔将减小，这个变化有个比例因子。

6. Instead of manually switching to the appropriate timer range, some instruments have an automatic range selection system called auto ranging.

某些仪器具有自动范围修正系统，该系统能够自动选择时间范围，以取代人工选择合适的时间范围。

Exercises

Translate the following passages into English or Chinese.

1. 大多数绝缘体的电阻率随温度的上升而下降。所以绝缘了的导体必须保持处于低温状态。
2. 一组相互连接的电阻器的等效电阻可以替换电路中的这组电阻器，而不会影响这组电阻器所在电路中电流值的大小。
3. As with the analog multimeter, the digital multimeter (DMM) is capable of measuring voltage, current, and resistance. Basically, the DMM is an electronic voltmeter that can be configured to express current and resistance measurements through a voltage. Current is derived by measuring the voltage drop across a known resistance and scaling the digital readout accordingly. Resistance is measured using the constant-current technique.
4. The time-interval probe has other advantages, including high input impedance (1 MΩ) and low input capacitance (10pF). Ordinary high-impedance probes have about 40pF of capacitance, and, at high frequencies, delays through the probe, cable, and circuitry can cause an inaccurate determination of time interval.

Reading Material

Rise-time Measurements

Rise-time measurements can also be made using the modified circuit configuration of Figure 10.11. When measuring rise time, the start and stop trigger levels are set at 10 percent and 90 percent, respectively, of the maximum voltage V_M (see Figure 10.14).

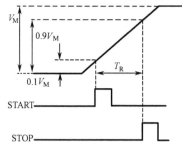

Figure 10.14 Rise-time measurements

Standard TTL gates, for example, have a nominal high output voltage of 2.4V. In this example, the rise time is measured by setting the START control to trigger at 0.24V and the STOP control to trigger at 2.16V. The intervening pulse count from the timerbase oscillator/timerbase divider is the rise time.

1. Reciprocal Frequency Counters

The reciprocal frequency counter differs from the ordinary frequency counter in that it uses separate registers to accumulate time- and event- counters, as noted in Figure 10.15. The contents of these registers are then processed to provide outputs that represent either period or frequency. It works like this: as long as the main gate is open, the event counter accumulates pulses from the external source, while the time counter accumulates pulses from an internal clock. The two counts may then be displayed or manipulated to achieve a wide range of measurements.

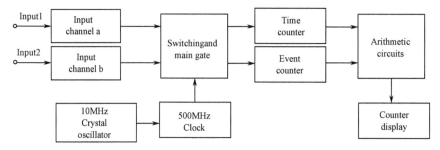

Figure 10.15 Reciprocal counter

Some reciprocal counters allow counting to be externally armed, as shown in Figure 10.16. External arming could be useful, for example, in measuring the frequency of a pulsed-RF signal, as shown in Figure 10.17. By setting the measurement-time control to the with of the RF burst, an accurate frequency measurement of the burst can be made.

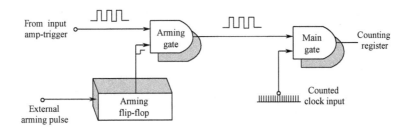

Figure 10.16　Armed input for reciprocal counter

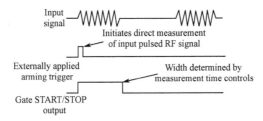

Figure 10.17　RF burst pulse counter

In a reciprocal counter, time-interval measurements are made by counting the number of pulses between independent START and STOP inputs, as shown in Figure 10.18. The resolution is determined by the frequency of the timerbase oscillator. A 10MHz clock, for example, could provide a 1×10^{-7} or 100-nanosecond, resolution.

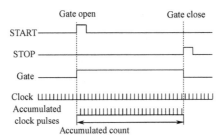

Figure 10.18　Time-Interval measurements with a reciprocal counter

2. Counting Errors

Although the notion of connecting a frequency counter to a frequency source to get an instantaneous digital readout down to the last hertz and beyond is intriguing, care must be exercised in the measurement. Everything is fine when the frequency counter deals with nothing but smooth, repetitive waveforms with high amplitude. However, erroneous readings can result if the wave has certain irregularities, involves transients, or is accompanied by RFI noise or other disturbances.

The input stage of a counter contains a Schmitt trigger circuit, which has an upper trigger point and a lower trigger point. The dead band separating the two is called the hysteresis. The effect of hysteresis is demonstrated in Figure 10.19. Assuming that the amplifier/attenuator stage is set for unity gain (1), the Schmitt output goes positive when the input sine wave passes the upper trigger point. When the sine wave goes below the lower trigger point, the Schmitt trigger output returns to its previous level.

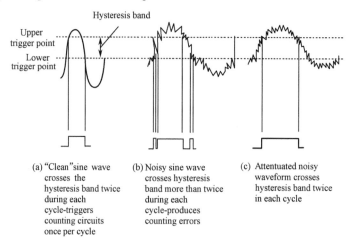

(a) "Clean" sine wave crosses the hysteresis band twice during each cycle-triggers counting circuits once per cycle

(b) Noisy sine wave crosses hysteresis band more than twice during each cycle-produces counting errors

(c) Attentuated noisy waveform crosses hysteresis band twice in each cycle

Figure 10.19 Hysteresis and counting errors

Now look at the effect of noise on the input signal, as illustrated in Figure 10.19. The noise spikes cause the signal to cross the hysteresis band more than twice in each cycle. Thus unwanted additional output pulses are generated that introduce errors in the frequency measurement. Such errors can be overcome by expanding the hysteresis band, but at the expense of not being able to detect low-level signals.

In typical frequency counters, the input circuitry is optimized for frequency counting by detecting zero crossings. This means that every time the waveform crosses over the zero reference line, the counter is triggered. That makes it difficult to measure rise times, propagation delays, and low-periodic-rate pulses because of the limited trigger-level range used (typically one volt or less). At best, the level control varies the center of the hysteresis band; at worst it is offset by several millivolts.

An alternative is to compensate for the ambiguity in the trigger point by using hysteresis compensation, in which a DC voltage equal to 1/2 the hysteresis band is added to or subtracted from the selected trigger level of reference voltage.

This method is most effective when used in conjunction with an oscilloscope. The main function of the oscilloscope is to reveal the true nature of the signal presented to digital counter. Armed with the knowledge of the waveform, certain techniques, such as rectification, can be used to eliminate sources of interference. Wherever feasible, connections to the digital counter should be made with coaxial cable. In such instances, it is also preferable to use the low-impedance (50Ω) input mode.

Rather than use hysteresis compensation, the time-interval prode solves the problem of trigger ambiguity by using an automatic calibration scheme. The user grounds the probe to be calibrated and then presses a front panel switch. That causes the reference voltage V_R to decrease in a stair-step fashion in 1mV steps until the device triggers. For negative slope calibration, the voltage increases in 1mV steps. The system then uses the value of V_R to adjust itself so that the actual trigger voltage corresponds to the trigger level selected by the user. Recalibration, when slopes or probes change, assures constant triggering accuracy. Some instruments allow the trigger voltage to be set from $-9.99V$ to $+9.99V$, in 10mV steps, by adjusting front panel thumbwheel switches.

Counting errors are also introduced by the main gate itself. In the system described in Figure 10.2, the timerbase could switch the main gate on or off while an input pulse from the main gate during the switching period may or may not succeed in triggering the counting circuits. Thus there is always a possible error of ± 1 cycle in the count of input cycles during the timing period. The accuracy of a digital frequency meter is usually stated as 1 count timerbase error.

So far in this discussion of digital and frequency counters, it has been assumed that the counting period is a one-time event. While this is generally true, most digital counters also provide for counting averaging. In the averaging mode, a counter can average the measurement reading over a number of periods or time intervals to give better resolution. Using the averaging mode, one can eliminate the uncertainty of the ± 1 count and arrive at a more precise measurement, particularly with waveforms that tend to generate erroneous displays, as described above.

Lesson 11 Computer-based Test Instruments

Personal-computer-based test instruments have fast become the quick and easy way for companies of all sizes to realize the advantages of computer-aided testing (CAT) without incurring the enormous expense of program development and main frame time. [1] Such instruments are able to meet the goals of reduced labor cost, increased productivity, and the elimination of human error in reading and processing measurements, by utilizing the power of software to perform many of the functions traditionally done by workbenches loaded with hardware. [2]

The general design of these instruments is around a desktop personal computer, such as the IBM PC or the IBM PC AT. PC instruments, as they have come to be known, are categorized as being either internal or external.

11.1 Internal Adapters

PC instruments designed for internal use are fabricated on one or more computer adapter boards. These adapter boards are physically identical to the video and I/O adapter cards required by the computer for normal operation. To install the instrument into the computer, the card is simply plugged into an available expansion slot on the motherboard, and the system is powered up as usual. Use of the test instrument is then controlled by software from the computer's keyboard. There are no knobs, switches, or indicators available to the user. The entire operation of the instrument is through the computer and its interfaces.

An example of an internal-adapter PC-test device is an analytical oscilloscope from R. C. Electronics called the COMPUTERSCOPE-IND IS-16. The IS-16 Data Acquisition package consists of a 16-channel analog-to-digital conversion board, an external instrument interface box, and appropriate software.

In operation, the IS-16 offers a 1-MHz aggregate sampling rate capability on 16 individual input channels with 12-bit resolution at input voltages within the range of $-10V$ to $+10V$. Fully automated keystroke commands (which programmers call hotkeys) provide the user with control over all features of the instrument, including channel selection, trigger control (internal to any channel, external, $+/-$ level or slop), sampling rate, and memory buffer size from 1KB to 64KB. In effect, the hotkeys are designated keys on the computer keyboard that act like the knobs and levers found on an oscilloscope's front panel.

Beyond the simulated mechanical aspects of an oscilloscope, the IS-16 employs a ring buffer that allows the capture of data in pretrigger intervals of virtually any length. Software commands permit timerbase expansion and contraction, left and right scrolling, independent vertical gain adjustment, and waveform storage and retrieval.

It is this last feature, waveform storage and retrieval, that sets the PC-based IS-16 oscilloscope apart from its mechanical counterpart. Input measurements can be stored in computer files or temporary computer memory (RAM) for archival purposes or further processing. The fact that the dynamic input becomes static in nature provides the user with the ability to modify the contents of an existing file in many useful ways, so that subsequent analysis of the signal may be effectively performed. The entire operation is done in software, following the acquisition of the input signal (Figure 11.1).

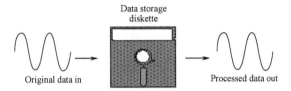

Figure 11.1 Computer diskettes save measured raw data for subsequent processing

The program begins by creating a verbatim copy of the input signal in a computer data file, using strings of binary words to represent instantaneous events and values in real time. Once saved, the entire measurement procedure can be duplicated down to the last detail by simply playing back the file record into the computer, in much the same way music is forever captured on a record disk or magnetic tape and played back upon demand. Each channel of data may be saved and processed separately by this method.

The saved data may now be processed by the computer's software to derive measurements not initially performed by the PC instrument. Table 11.1 gives a brief summary of the procedures that may be performed on the raw data. When one realizes that all these measurements can be made after the fact, the implications are staggering. One-time events that are elusive to conventional measurement can be done routinely in the leisure of an office setting by simply running the recorded data through the computer one more time using the appropriate software selection.

Table 11.1 Summary of Computerscope IS-16 computer-aided measurements

GLOSSARY
Signal averaging: Provides greater resolution by averaging the results over a user-specified period of time.
Full-wave rectification: Data points with voltage values below a userspecified threshold are converted to voltage values above the threshold, relative to the initial difference in absolute magnitude.
Data Inversion: The voltage value for each recorded data point is inverted in its relationship to a user-specified threshold.
Gain adjustment: Data points may be reassigned voltage values according to a new gain factor.
Integration: Channels may be integrated individually with a user-specified time constant, the results of which are adjusted in their temporal relationship so that no time lag is introduced.
Differentiation: The analog waveform is converted to its first derivative.
DC offset adjustment: The value for each recorded data point is increased or reduced by a user-specified constant, thus adjusting baseline voltage.
Algebraic functions: Each waveform may be added, subtracted, multiplied, or divided by a user-specified factor.
Digital filtering: Five differential digital frequency filtering functions—low pass, high pass, band pass, band stop, and notch—permit the user to determine the cutoff frequencies and roll-off efficiency independently for each input channel.
Three-point symmetric smoothing: A three-point symmetric smoothing function with double weighting on the middle point provides the ability to attenuate high-frequency noise in the data.

Furthermore, the results can build one on another. For example, in the first pass, noise and amplification factors may be adjusted, resulting in a data file that can now be averaged, integrated, or differentiated to see the effects of each, without ever having to change the physical setup of the test instrument or device under test.

11.2 External PC Instruments

Such performance is not limited to internal adapter devices or PC-based oscilloscopes. Similar results can be achieved using externally connected PC instruments.

In an externally connected PC instrument, the test instrument is housed in a conventional cabinet external to the computer's cabinet. Installed in the cabinet may be input-output connectors, selector switches, and maybe an LED or two. There is also a connector cable coming from the external instrument that plugs into the PC.

The connection between the external PC test instrument and the computer is in the form of an umbilical cord that feeds voltage, data, and control signals between the two devices. Without this connection, the test instrument is little more than a stand-alone device with features less than comparable to the instruments examined earlier in this book. When connected to the computer, however, it becomes a virtual powerhouse.

Inside the computer is an interface card that translates the signals coming from the external measuring instrument into digital pulses with voltage levels and timing requirements compatible with those on the computer bus (Figure 11.2).

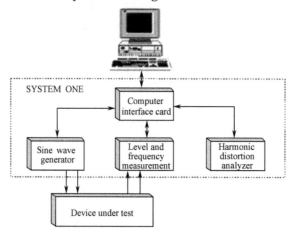

Figure 11.2 System one block diagram

An example of an external PC-test instrument is the System One audio analyzer from Audio Precision. The System One is designed as a computer-interfaced audio test station capable of performing more than 36 standard performance tests on audio amplifiers, magnetic tape systems, and related audio components.

The System One unit contains of an ultra-low-distortion (0.0008% at midband) sine-wave generator, a level-detector and frequency-measurement amplifier, a harmonic distortion analyzer, and a computer interface card.

The System One has no controls or displays other than power on/off. Located on the front panel of the instrument are the audio connectors required to interface the System One with the device under test. All user-related functions of control, such as selection of test type, level adjustment, frequency output, and display are accomplished via the keyboard and display of an IBM PC-compatible personal computer.

Like the COMPUTERSCOPE-IND IS-16, all equipment and test functions are controlled by predefined keystrokes on the PC keyboard. Because most audio measurements are sets of measurements rather than an individual spot measurement, the keyboard is programmed to execute an entire series of measurements at the touch of a key, when requested.

Generally, audio tests are performed using sweep frequencies that span beyond the range of the device under test. Some of the tests available are listed in Table 11.2.

Table 11.2 System One test functions

Amplitude
Phase
Noise
Common-mode rejection ratio
Crosstalk
Equalization curve
Frequency response
Frequency drift
Phase jitter
Tape noise
Tape frequency response
Head azimuth alignment
WOW & Flutter
THD amplitude
THD frequency
THD versus power output

The type of test performed, such as frequency response, is selected from a menu of standard tests. In these tests, the sweep range, rate, and amplitude are preset, according to SMPTE, DIN, CCIR, EIA, IHF, NAB, and other established audio standards. Each of these test patterns is recorded within the software of the System One application package for easy access through a screen menu.

The user may also change any of these parameters before initiating the sweep by calling up a control panel screen on the computer. The control panel is essentially a screen version of a mechanical quantities from the keyboard. There is no twisting of the knob or changing of the ranges. A frequency change is as simple as locating the line displaying the current frequency and entering the new value. While conventional hardware panel instruments are limited to the units of measurement for which the designer has the panel room and the meter scale space, the choices of the System One panel values can be changed to include volts, dBV, dBu, watts, dBm, or percentage for any specified impedance or function. If desired, the new parameters may be saved to a computer file for future use.

Once the testing procedure starts, the System One software begins its measurement of the unit under test. During the course of testing, the measurements are displayed on the computer's video monitor. The result is an X-versus-Y graph representation of the tests according to the response of the unit under test. Multiple line charts may be displayed by entering the proper values into the opening menus.

Test measurements are also saved to a computer file. As with the PC oscilloscope described above, these files may be played back for re-enactment of the tests performed or verification of test results. Unlike the COMPUTER-SCOPE, the System One does its testing in real time. Because of the nature of the tests involved, measurements are more accurate when programmed to be performed on the unit under test at the time it is connected to the test equipment than measurements derived from calculated values are. That is not to say that new measurements and conclusions cannot be derived from the data in the files created by the System One, because they can. [3] Therein lies the advantage of the PC-based test instrument over other types of testing devices.

11.3 Analog-to-Digital Conversion

Central to the performance of any PC-based test instrument is an analog-to-digital converter, or ADC. The ADC converts an analog input voltage into a resented by the binary word. For example, an input of 2.00V would be converted to 0011001001. The accuracy of the conversion is totally dependent on the number of bits in the binary word. In the above examples, the values are accurate to 10 places, or 10 bits.

Basically, there are two analog-to-digital conversion methods used for PC-based test instruments. They are the successive approximation register type and the flash converter.

11.3.1 Successive Approximation Register Conversion

Until the advent of monolithic electronics, the successive approximation register (SAR) method was considered the black sheep of the ADC family because of its slow conversion rate. But with today's technology—ultra-fast comparators, fast current switching, and ECL logic—very useful performance is obtainable from successive approximation converters.

An attractive feature of the successive approximation technique is that it requires very little hardware. Consequently, it is possible to integrate an ADC of this type into a single chip with 12-bit or higher resolution at very little cost.

The successive approximation converter, a block diagram of which is shown in Figure 11.3, uses a digital-to-analog converter (DAC) in a feedback loop. The digital-to-analog converter is generally a resistive network that generates an output voltage proportional to a digital code, using precision resistors in a ladder array. With a constant current passing through the resistor array, the output voltage is proportional to the total resistance, which is dependent on the number and value of the resistors inserted in the circuit by a binary code.

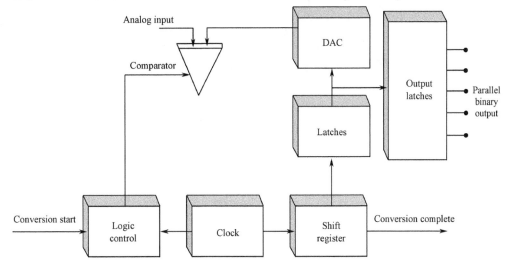

Figure 11.3 Successive approximation register ADC block diagram

In operation, the shift register sets a one (1) in the MSB latch of the approximation register and resets all other latches to zero (0). The digital converter's (DAC) analog output is then compared with the analog input voltage to determine which value is then compared with the analog input voltage to determine which value is higher. If the analog input voltage is less than the output voltage from the DAC, then the MSB latch is cleared to zero (0) and a one (1) is placed in the next-most-significant bit register. All this takes place within a single clock cycle.

During the succeeding clock cycle, the new DAC voltage is compared to the analog input voltage, and a determination is made as to which is greater than the digitally simulated output voltage from the DAC. When this is the case, the control logic retains the one value in the current latching register and places a one (1) into the next latch in line. On the next clock cycle, this new binary value, 0110000..., is compared to the input analog voltage and a decision made about whether to keep the newly inserted one or discard it.

During successive clock periods, this process is repeated over and over with bits of diminishing significance until the register is exhausted. The DAC output therefore becomes a progressively more accurate approximation of the analog input voltage, taking N clock periods to achieve N-bit resolution. At the end of the conversion period, the binary word is read into the computer as an input value and another conversion started. While conversion rates as high as 15MHz are possible using current design techniques, most successive approximation converters for use with PC instruments operate within the 1MHz to 5MHz range (Figure 11.4).

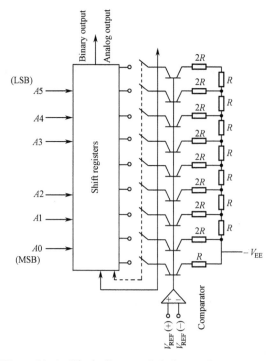

Figure 11.4 Block diagram digital-to-analog converter

11.3.2 Flash Converters

As economically attractive as the successive approximation converter may be, it cannot possibly address the issues of video frequency or digital sampling applications. For such applications, another type of ADC called the flash converter is used.

The flash converter performs a conversion and outputs data within a single clock cycle. Referring to Figure 11.5, note that the clock input is sensitive to the rising and falling clock edges. All significant operations are referenced to these edges.

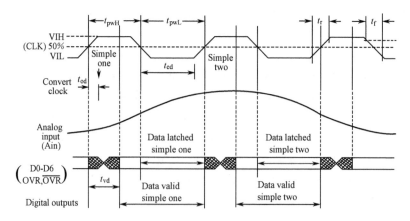

Figure 11.5 Flash converter timing

Analog-to-digital conversion is accomplished through a parallel array of voltage comparators. In the block diagram of Figure 11.6, 128 parallel comparators are used to achieve a resolution of 7bits.

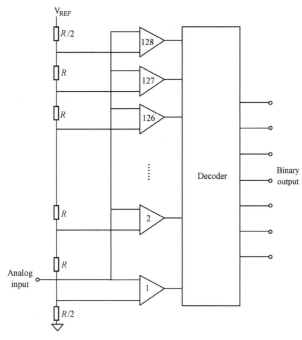

Figure 11.6 Flash-converter block diagram

The reference voltage to each of the 128 comparators is derived from a uniformly spaced voltage divider that spans both the positive and negative range of the input signal. All 128 sampling inputs, on the other hand, are commonly tied together and connected to the analog input voltage.

As the analog input voltage changes from one value to another, comparators are turned on or off according to the value of the analog input signal. These outputs are analyzed by a logic circuit, called a 127-of-7 encoder, and converted into a binary number representing the number of comparator gates turned on and the number turned off. On the falling edge of the clock pulse, this value is locked into the output register and is input to the computer for evaluation.

11.4 Computer Interface

Thus far we have indicated that the digital data bits from the PC instrument are input to the computer for processing, but little has been said about the method of transfer. In the case of the internal PC instrument, the data bits are transferred directly to the computer using the computer's internal data bus. The only interface to the outside world by the computer is through the test instrument's input/output ports.

In the design of external PC-test instruments, provisions must be made to interface the external box to the internal computer bus. In the case of the System One audio test station, the connection is made through a proprietary communications link to a digital interface card inserted into a computer expansion slot. The interface card then translates the digital input pulses into signals that the computer bus understands. Many external PC-based test equipment manufactures have taken this route, using an enormous variety of proprietary interface schemes.

In some cases, the external test instrument manufacturer attempts to standardize its interface, making it compatible with a wide range of computers and not limiting it to the IBM PC family. One such approach is the use of the RS-232 serial interface port available on every computer and many other types of devices.

An example of a typical RS-232 interfaced PC instrument is the SAV 10 Serial ASCII Voltmeter from Maron Production Inc. The SAV 10 is a multichannel ASCII voltmeter that monitors four analog voltage inputs, converts the information to digital values, and transmits them as ASCII encoded messages in a format compatible with standard data display terminals and computer RS-232 serial ports. The unit requires no control messages from a host computer and operates independently of the computer system. In effect, the SAV 10 is a satellite data-gathering device that inputs raw data into the computer for subsequent storage or processing.

11.5 General-purpose Interface BUS

As the need for electronic measurements has grown in number and complexity, more and more instrument users have come to realize that their traditional design and test procedures are inadequate.[4] The modern technician and engineer have new needs that can be satisfied only by assembling individual test instruments into interactive, automated systems. With more than one million personal computers in the hands of engineers and scientist, it was inevitable that instrument measurements and automated testing would find their way into the realm of desktop computing (Figure 11.7).

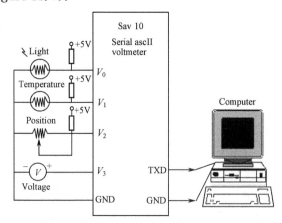

Figure 11.7 Typical SAV 10 test setup and computer interface

The first major step toward this goal was taken by Hewlett-Packard in the early 1970's with their introduction of the Hewlett-Packard Interface-Bus (HP-IB). The HP-IB is essentially a communications link that allows one instrument to talk to another over a standard electrical bus composed of many wires. Before HP-IB, connecting programmable test instruments to a computer desktop calculator or local area network (LAN) was a major undertaking. Different standards among instrument manufacturers created proprietary interfaces that tied the user to one manufacturer and denied use of test instruments not available from that supplier. The HP-IB standard once and for all defined an interface that made it possible to put together computer-controlled test systems.

In 1975, the HP-IB standard was adopted by the IEEE committee as the IEEE-488 standard. The IEEE-488 standard was updated in 1978 to IEEE-488-1978 and officially labeled the General Purpose Interface Bus, or GPIB.

The IEEE-488 interface is an asynchronous, handshaking-oriented parallel computer's LSI-11 bus (which is based on DEC's extremely popular PDP-11 minicomputer system). Central to the operation of any GPIB system is the controller. The controller is analogous to a telephone operator in that it receives, directs, and routes messages across the network. The GPIB uses talkers and listeners to communicate data over the IEEE-488 bus. Talkers send data to listeners via the controller. GPIB devices or instruments can assume one of three communications roles: talker, listener, or talker/listener (Figure 11.8).

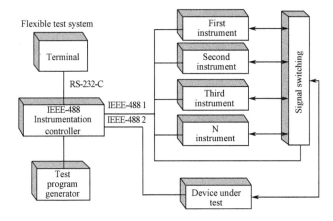

Figure 11.8 Mix and match instrument modules using the GPIB

A talker, which is normally referred to as a talk-only device, can only send data over the bus; it cannot receive data or commands. A common example of a talk-only device is a voltmeter that only sends measurements. Talkers are generally not desirable for IEEE-488 test systems because the controller cannot turn them off or on. Such devices, like the voltmeter, must be actuated by a separate trigger line or front-panel pushbutton. A system containing a talk-only device is generally a very simple setup with only two components, such as a voltmeter and a data printer. In this arrangement, the printer would periodically print out voltage readings as they were generated by the voltmeter. When a talk-only instrument is used in GPIB systems with more than one other device, however, the talk-only device could end up monopolizing the bus and interfering with commands issued by the controller and/or data on the bus coming from other instruments.

An instrument that is a listener, or a listen-only device, can only receive data over the bus and cannot transmit. Although listen-only devices are not common in IEEE-488 test configurations, they are practical for some non-measurement devices. A good example of a listen-only device is a power supply whose function is to receive voltage and current programming information over the bus rather than send information. Even in this case, however, it is beneficial for the power supply to have talker capabilities in the event of a failure. An out-of-spec warning signal transmitted over the GPIB bus by the power supply can give the controller an opportunity to disable the power source before the unit under test is damaged.

The preferred instruments for an IEEE-488 test system are talker-listeners, which can both send and receive information over the bus. Talker-listener capability can make the tasks of programming, troubleshooting, and maintaining IEEE-488 test systems much easier. Most new GPIB instruments are of the talker-listener variety. When the application requires stricter control, these instruments can usually be configured for talk-only or listen-only operation, using either a software command or a hardware switch.

Words and Expressions

actuate ['æktjueit] *vt.* 激发，开动
adapter board　适配板
albeit [ɔːl'biːit] *adv.* 虽然
analog-to-digital conversion　模/数转换
archival [ɑː'kaivəl] *adj.* 关于档案的
asynchronous [æ'siŋkrənəs] *adj.* 异步的
cluster ['klʌstə] *vt.* 聚集
dedicated ['dedikeitid] *adj.* 专门的
desktop ['desktɔp] *n.* 工作平台
distortion [dis'tɔːʃən] *n.* 失真
dynamic [dai'næmik] *adj.* 动态的
elusive [i'ljuːsiv] *adj.* 易忘的，难懂的
emulate ['emjuleit] *vt.* 仿效
expansion slot　扩展槽
execute ['eksikjuːt] *vt.* 执行
fabricate ['fæbrikeit] *vt.* 制作
generator ['dʒenəreitə] *n.* 发生器
hefty ['hefti] *adj.* 重的，肌肉发达的
interface ['intə(ː)ˌfeis] *n.* 接口

invoke [in'vəuk] vt. 激发
keystroke ['kiːstrəuk] n. 按键
latching register   锁存触发器
local area network (LAN)   局域网
monolithic [ˌmɔnə'liθik] n. 单片电路；单块集成电路
monopolize [mə'nɔpəlaiz] vt. 独占，垄断
motherboard ['mʌðəbɔːd] n. 母板
optoisolator [ˌɔptə'aisəˌleitə] n. 光绝缘体
parametric [ˌpærə'metrik] adj. 参量的
proprietary [prə'praiətəri] adj. 所有的，私人拥有的；n. 所有者，所有权
protocol ['prəutəkɔl] n. 协议
resolution [ˌrezə'luːʃən] n. 分辨率
retrieval [ri'triːvəl] n. 检索，巡检
scale [skeil] n. 刻度
simulate ['simjuleit] vt. 模仿
simultaneously [siməl'teiniəsli；(us) saim-] adj. 同时的
slot [slɔt] n. 长槽
succeeding [sək'siːdiŋ] adj. 以后的，随后的
supplier [sə'plaiə] n. 供应商
terminal ['təːminl] n. 终端
uniformly ['juːnifɔːmli] adv. 均匀地
verbatim [vəː'beitim] adj. 逐字的
workbench ['wəːkbentʃ] n. 工作台

Notes

1. Personal-computer-based test instruments have fast become the quick and easy way for companies of all sizes to realize the advantages of computer-aided testing (CAT) without incurring the enormous expense of program development and main frame time.
   各类公司迅速运用基于PC的测试仪器来实现计算机辅助测试（CAT），这种快捷的方法可以使工程师在程序开发和结构搭建上节约大量的时间。
2. Such instruments are able to meet the goals of reduced labor cost, increased productivity, and the elimination of human error in reading and processing measurements, by utilizing the power of software to perform many of the functions traditionally done by workbenches loaded with hardware.
   这样的仪器能够降低劳动成本，提高生产率，减少读取和处理中的人为失误，并通过使用软件来完成许多传统上由硬件搭建的结构所实现的功能。
3. Because of the mature of the tests involved, measurements are more accurate when

programmed to be performed on the unit under test at the time it is connected to the test equipment than are measurements derived from calculated values. That is not to say that new measurements and conclusions cannot be derived from the data in the files created by the System One.

在测试单元连接到测试设备上执行时，由于采用了直接测试，程序化的测量比由计算数值导出的测量更准确，这并不是说新的测量和结论不能从多功能音频分析器创建的数据中得出。

4. As the need for electronic measurements has grown in number and complexity, more and more instrument users have come to realize that their traditional design and test procedures are inadequate.

随着电子测量的应用在数量和复杂性上的发展，越来越多的仪器用户认识到传统的设计和测试过程的不足。

Exercises

Translate the following passages into English or Chinese.

1. 运动电子形成电流。当电流流过电路时，它会消耗一部分能量。电阻率越大，建立给定的电流密度所需的电场就越大。
2. 阻抗为无穷大表明：终端阻抗至少应为特定端口标称阻抗的 10 倍到 100 倍，而阻抗为零这一标记则表明它应该远小于该端口标称阻抗的 10%，若可能的话应小于其 1%。
3. By combining the various features of the GPIB with available GPIB test instruments, one can achieve a well-conceived test system capable of performing many measurements simultaneously, with data transfer between the GPIB devices for storage, print out, and evaluation.
4. The merits of automated testing are not to be debated here. That issue is best left to another treatment. What needs to be said is that, when considering PC-based measurement systems, consider the long-term investment and do not dismiss the advantages of ATE, should that possibility loom in the future. It is much easier to upgrade to ATE status than it is to start from scratch.

Reading Material

PC-based Workstation

There are, of course, applications for which the owner of a PC may want higher performance than that offered by the GPIB alone, with its odd assortment of devices. Such systems are generally referred to as PC-based workstations. In a PC-based workstation, instrument modules with matched performance and compatibility are clustered around a PC computer. The PC-based workstation allows data from multiple instruments to be rapidly gathered, stored, graphed, analyzed, and incorporated into report form on a single PC. Systems of this type are generally put together by one manufacturer to enhance performance and cooperation between the instrument modules involved.

An example of a PC-based workstation is the PC Instruments system introduced by Hewlett-Packard. PC Instruments are a series of nine dedicated instrument modules designed to operate over a dedicated interface bus by an HP Touchscreen II personal computer, an IBM PC/XT/AT, an AT&T PC 6300, or an HP Vectra PC. Among the instrument modules available for PC Instruments are a 50-MHz digitizing oscilloscope, a 12-bit dual-channel digital-to-analog converter, a 100-MHz universal counter, a 5-MHz function generator, and a digital multimeter.

Test data gathered from the instrument modules are fed directly to the PC for storage, display, and analysis. The instrument modules themselves have no meters or displays of their own. All outputs, including waveform displays, are through the PC screen, and up to eight different instrument outputs may be displayed simultaneously.

PC Instruments communicates with the PC via the PC Instruments Bus, or PCIB. PCIB is implemented with a single interface card that plugs into one expansion slot of the PC motherboard. Each PCIB interface card supports up to eight instrument modules. Installing a second PCIB into the computer permits support for an eight additional instrument modules.

The PCIB is unique in that it supports both parallel and serial communications (Figure 11.9). At first glance, the idea of having two communications channels in a single bus system appears to be redundant. On closer analysis, however, a cost saving is realized and the redundancy is reduced.

In a typical GPIB system there are instruments, such as voltage and current meters, that require the inputs and/or outputs to be isolated from earth ground and the computer so that floating measurements can be made. The isolation is typically provided by optoisolators communicating the data from the GPIB side to the instrument in serial fashion using a parallel-to-serial converter. Data coming from the instrument likewise must pass to the GPIB through an optoisolator using a serial-to-parallel converter. Consequently, every device that requires isolation also requires its own pair of optical isolators and respective converters.

In PC Instruments, the test instrument may choose between an isolated serial communications channel or the faster, nonisolated parallel communications bus. The parallel-to-serial and serial-to-parallel converters for the serial channel are placed on the interface card, thus reducing system cost while improving performance.

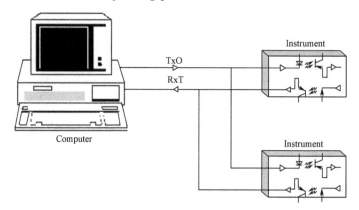

Figure 11.9　PCIB serial communications channel

The PCIB parallel communications channel (Figure 11.10) offers a high-speed data path for instruments that do not need isolation. Data can be transmitted at rates up to 100K bytes/second, subject to limitations of the host personal computer. The parallel communications channel consists of an 8-bit data path, two control lines, with appropriate ground returns, are part of the 26-conductor ribbon cable that connects the host personal computer to the modular instruments. The PCIB parallel communications channel is also upwardly compatible the GPIB, making PC Instruments a most versatile system.

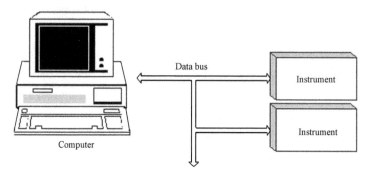

Figure 11.10　PCIB parallel communications channel

One of the most impressive aspects of PC Instruments is the soft front-panel system software, which provides an interactive graphics mechanism for the user to control instruments manually. The control panel can also be invoked directly from either DOS, BASIC, or any other language that produces the correct response, thus lending the system to computer automated testing using customized software routines.

The front-panel software emulates a benchtop stacked with traditional instruments, allowing the user to control one module while viewing them all. It is a simple matter to select a new instrument using simple touch-screen, mouse, or keyboard inputs.

The software for each installed instrument module is contained within the host computer at all times. The performance of each test module has been optimized for both individual performance and interactive cooperation. The software for the digitizing oscilloscope, for example, is the most complex in the current PC Instruments line, and is representative of what the system is capable of doing. The digitizing oscilloscope software package is broken into five major functions: hardware setup, data acquisition and display, user interface, measurements, and program control.

The hardware setup sets the vertical and horizontal sensitivities, delay time, coupling input, acquisition mode, and trigger configuration for the oscilloscope module. When invoked, the setup instructions are sent over the PCIB to the oscilloscope module for execution.

Data acquisition consists of reading a 251-byte data string from the oscilloscope module following a measurement procedure. The data is scaled according to the hardware settings described in the procedure above. The settings may be modified by the user using the front-panel graphics that are available on every screen. Consequently, change in the display of the data can be made on the fly.

The measurement package includes control of voltage and time markers and waveform analysis for automatic parametric measurements of rise time, fall time, period, frequency, pulse width, overshoot, preshoot, and peak-to-peak voltage. These measurements are based on a statistical analysis of the 251-byte data string from the oscilloscope module, using a histogram to determine the absolute and relative maximum, minimum, and percentage points, as well as the location of rising and falling edges. Because the data is displayed statistically, not physically, it is also subject to software processing in much the same way that signal processing was done with the R. C. Electronics IS-16 COMPUTERSCOPE described at the beginning of this chapter.

The PC Instruments software also allows for data exchange between instruments via the computer. For example, changes seen on the oscilloscope display can be adjusted using software feedback loops to the function generator, relay actuator, etc. One application may be expected to vary with frequency. By monitoring the output voltage of the device under test, using the oscilloscope's data string and software processing, adjustment to the output level of the function generator can be made to automatically compensate for signal fluctuations.

Automatic Test Equipment

The PC-based workstation, whether purchased as a system like the Hewlett-Packard PC Instruments of assembled from GPIB components, also lends itself well to automated testing environments. Automated testing systems, assembled from automatic test equipment (ATE), are a result of the computer being interfaced with digitally controlled stimulus and measurement instrumentation.

ATE is quickly replacing the benchtop setups of instrumentation that were manually controlled to make measurements. Once started, ATE can continue its operation with no outside intervention and can test thousands of parameters in seconds. ATE does not make mistakes when recording test data, does not compromise a test, does not forget tests, and does not get tired. Consequently, ATE dramatically improves throughput and measurement accuracy. It can even correct its own errors by comparing its measurement results to a known standard and compensating for the difference.

ATE test system may be divided into three classes: benchtop, dedicated, and general-purpose. For manufactures of semiconductors and large electronics systems, dedicated automatic test equipment is a necessary, albeit hefty, investment. But for thousands of small systems and subsystems manufacturers, the less costly benchtop PC-based workstation is the better solution.

A flexible ATE system usually consists of off-the-shelf instruments joined by an IEEE-488 or RS-232 bus to relieve the incompatibility problem that can occur in linking parts from different vendors. A major advantage of this approach is that it can easily be reconfigured for a variety of test applications by simply swapping one instrument for another.

Flexible ATE is generally fully programmable for applications in which testing is to be done with no operator intervention. Nonetheless, its components should also allow for operator intervention or manual operation when the situation so dictates. For example, manual operation is desirable when an instrument has to function off-line for maintenance or troubleshooting purposes, for minor testing tasks, or for writing and debugging the system test software.

A drawback to ATE is the expense incurred in writing the software. Because writing software can be the largest single task when configuring an ATE system, care must be exercised in the programming language selected and the structuring of the program. In some estimates, the cost of software development can run 90 percent of the cost of the system. Generally, an ATE system will cost four to seven times more than equivalent manual instrument.

Lesson 12   Industrial Bus

12.1   What is an Industrial Bus

What is an industrial bus? Traditionally, the industrial bus has been used to allow a central computer to communicate with a field device. The central computer was a mainframe or a MINI and the field device could be a discreet device such as a flow meter, or temperature transmitter or a complex device such as a CNC cell or robot.[1] As the cost of computing power came down, the industrial bus allowed computers to communicate with each other to coordinate industrial production.

As with human languages, many ways were devised to allow the computers and devices to communicate and, as with their human counterpart, most of the communications are incompatible with any of the other systems.[2] The incompatibility can be broken into two categories: the physical layer and the protocol layer.

Two popular industrial buses that use the RS-232 and RS-422/485 standards areModbus and Data Highway. Modbus was developed by Modicon for its line of PLCs, up to and including the 9 84 line of controllers. Modbus can be configured for either RS-232 or RS-485 in a 4-wire mode. Data Highway is the name of the industrial bus. An RS-485 port is available on some PLC-2, 3 and 5 controllers. Consult the manual provided with your controller to be certain of the type of bus supported. The industrial buses that adhere to the RS-232 and RS-422/485 standard are listed below along with products that are compatible with various industrial buses. B&B products support these buses at the physical layer only and are mainly used as repeaters, line extenders and isolators. B&B also offers a custom design service to solve particular problems that arise from industrial buses.

The physical layer and the protocol layer can be defined using the phone system as an example. Any spoken language can be carried over a phone line. As long as both the speaker and the listener(s) understand the language, communication is possible. The phone line is not concerned with the meaning of the signal that it carries, it is merely moving those signals from one point to another physically. This is the physical layer, the conduit in which communications pass from one point to another. On the other hand, the speaker and listener(s) are concerned with what is transported over the phone line. If the speaker is talking in Spanish and the listener(s) are only fluent in English, communication is not possible. Although the physical layer is working, the language or protocol is not correct, and communications cannot exist. The industrial world has developed a variety of different physical and protocol communications standards. A list of all of them would fill the rest of this article, so we will limit this discussion to industrial buses using the RS-232 and RS-422/485 standards for their physical layer.

The greatest difference between RS-232 and RS-422/485 is the way information is transmitted.

12.2 Data Line Isolation Theory

When it comes to protect data lines from electrical transients, surge suppression is often the first thing that leaps to mind. The concept of surge suppression is intuitive and there are a large variety of devices on the market to choose from(Figure 12.1). Models are available to protect everything from your computer to answering machine as well as those serial devices found in RS-232, RS-422 and RS-485 systems. [3]

Unfortunately, in most serial communications systems, surge suppression is not the best choice. The result of most storm and inductively induced surges is to cause a difference in ground potential between points in a communications system. The more physical area covered by the system, the more likely those differences in ground potential will exist.

The water analogy helps explain this. Instead of phenomenon water in a pipe, we will think a little bigger and use waves on the ocean. Ask anyone what the elevation of the ocean is, and you will get an answer of zero - so common that we call it sea level. While the average ocean elevation is zero, we know that tides and waves can cause large short-term changes in the actual height of the water. This is very similar to earth ground. The effect of a largeamount of current dumped into the earth can be visualized in the same way, as a wave propagating outwards from the origin. Until this energy dissipates, the voltage level of the earth will vary greatly between two locations.

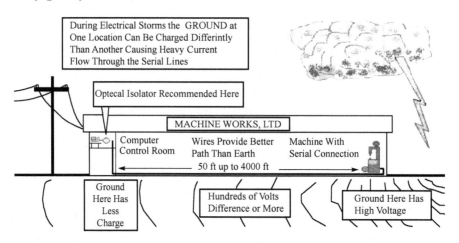

Figure 12.1 The method of surge suppression

Adding a twist to the ocean analogy, what is the best way to protect a boat from high waves? We could lash the boat to a fixed dock, forcing the boat to remain at one elevation. This will protect against small waves, but this solution obviously has limitations. While a little rough, this comparison isn't far off from what a typical surge suppressor is trying to accomplish. [4] Attempting to clamp a surge of energy to a level safe for the local equipment requires that the clamping device be able to completely absorb or redirect transient energy. [5]

Instead of lashing the boat to a fixed dock let the dock float. Now the boat can rise and fall with the ocean swells (until we hit the end of our floating dock's posts).

Instead of fighting nature, we're simply moving along with it. This is our data line isolation solution.

Isolation is not a new idea. It has always been implemented in telephone and Ethernet equipment. For asynchronous data applications such as many RS-232, RS-422 and RS-485 systems, optical isolators are the most common isolation elements. With isolation, two different grounds (better thought of as reference voltages) can exist on opposite sides of the isolation element without any current flowing through the element(Figure 12. 2).[6] With an optical isolator, this is performed with an LED and a photosensitive transistor. Only light passes between the two elements.

Another benefit of optical isolation is that it is not dependent on installation quality. Typical surge suppressors used in data line protection use special diodes to shunt excess energy to ground. The installer must provide an extremely low impedance ground connection to handle this energy, which can be thousands of amps at frequencies into the tens of megahertz. A small impedance in the ground connection, such as in 1. 8m of 18 gauge wire, can cause a voltage drop of hundreds of volts - enough voltage to damage most equipments. Isolation, on the other hand, does not require an additional ground connection, making it insensitive to installation quality.

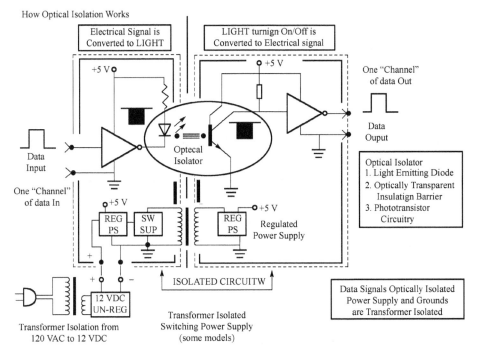

Figure 12. 2 An optical circuit

Isolation is not a perfect solution. An additional isolated power supply is required to support the circuitry. This supply may be built in as an isolated DC-DC converter or external. Simple surge suppressors require no power source. Isolation voltages are limited as well, usually ranging from 500V to 4000V. In some cases, applying both surge suppression and isolation is an effective solution.

When choosing data line protection for a system it is important to consider all available options. There are pros and cons to both surge suppression and optical isolation, however isolation is a more effective solution for most systems. If in doubt, choose isolation.

Words and Expressions

mainframe ['meɪnfreɪm] n. 主机，大型机
transmitter [trænz'mɪtə] n. 传感器，传送者，变送器，发送器，传递器
counterpart ['kauntəpɑːt] n. 副本，极相似的人或物，配对物
incompatible [ˌɪnkəm'pætəbl] adj. 不相容的，矛盾的
consult [kən'sʌlt] vt. & vi. 商量，商议，请教，参考，考虑
conduit ['kəndɪt] n. 导管，水管，水道
leap [liːp] vt. & vi. 跳跃，飞跃 n. 跳跃，飞跃
surge [səːdʒ] n. 巨涌，汹涌，澎湃 vi. 汹涌，澎湃，振荡，滑脱，放松
　　　　vt. 使汹涌奔腾，急放
suppression [sə'preʃən] n. 抑制
intuitive [ɪn'tjuːɪtɪv] adj. 直觉的
elevation [ˌelə'veɪʃən] n. 上升，高地，正面图，海拔
dump [dʌmp] vt. 倾卸 n. 堆存处
visualize ['vɪʒuəlaɪz] vt. 形象，形象化，想象 vi. 显现
twist [twɪst] n. 扭，扭曲，螺旋状 vt. 捻拧，扭曲，绞，搓，编织
　　　　vi. 扭弯，扭曲，缠绕，扭动
clamp [klæmp] n. 夹子，夹具，夹钳 vt. 夹住，夹紧
redirect [ˌriːdɪ'rekt, -daɪ-] v.（信件）重寄，使改道，使改方向
transient ['trænzɪənt] adj. 短暂的，瞬时的 n. 瞬时现象
photosensitive [ˌfəutəu'sensɪtɪv] adj. 感光性的
megahertz ['megəhəːts] n. 兆赫（MHz）
gauge [geɪdʒ] n. 标准尺，规格，量规，量表 vt. & vi. 测量
pros [prəu] adv. 正面地
cons [kən] adv. 反对地，反面 n. 反对论

Notes

1. The central computer was a mainframe or a MINI and the field device could be a discreet device such as a flow meter, or temperature transmitter or a complex device such as a CNC cell or robot.
   本句说明了两点，一点是 central computer 是什么，一点是 field device 是什么样的设备。在描述 field device 时，举例进行了说明。discreet device 译为"智能设备"。
2. As with human languages, many ways were devised to allow the computers and devices to communicate and, as with their human counterpart, most of the communications are

incompatible with any of the other systems.

本句分成两部分去理解。第一部分将计算机和设备之间的通信和人类语言类比；第二部分将系统兼容情况与人类相比，这两部分内容用 and 连接。其中，with 表示"相对照，相比"，counterpart 表示"相似物"。

3. Models are available to protect everything from your computer to answering machine as well as those serial devices found in RS-232, RS-422 and RS-485 systems.

   本句的基本句型是，models are available to protect everything from…to…as well as…。这样，from…to…就很清楚了，它描写的是 protect 的范围。available 表示"可以得到的，可以利用的"。

4. Wile a little rough, this comparison isn't far off from what a typical surge suppressor is trying to accomplish.

   本句是包含 while 的转折句，while 表示"尽管"。rough 有很多意思，这里修饰的是 comparison，表示"粗略（的比较）"，far off from 表示"远离，远非"，本句出现的是 is not far off from，表示的意思正好相反。

5. Attempting to clamp a surge of energy to a level safe for the local equipment requires that the clamping device be able to completely absorb or redirect transient energy.

   本句的主语是动名词短语 attempting to clamp a surge of energy to a level safe for the local equipment，谓语是 require，that 引导宾语从句。动词不定式短语 to clamp a surge of energy to a level safe for the local equipment 是动名词 attempting 的宾语。因为本句的谓语是 require，that 从句里的谓语动词就要用虚拟语气。因为谓语后省略了 should，所以用 be able to，而不是 is able to。

6. With isolation, two different grounds (better thought of as reference voltages) can exist on opposite sides of the isolation element without any current flowing through the element.

   本句中的主干是 two different grounds call exist on opposite sides of the isolation element，with 表示"以……的方式"，without 用在动名词前有"不……，未……"的意思，表示前面的动作发生时没有后面的情况出现。

Exercises

Translate the following passages into English or Chinese.

1. All IBM PC and compatible computers are typically equipped with two serial ports and one parallel port. Although these types of ports are used for communicating with external devices, they work in different ways.

2. A serial port sends and receives dada one bit at a time over one wire. While it takes eight times as long to transfer each byte of data this way, only a few wires are required.

3. 如同人类语言一样，人们用许多方式来设计计算机同设备之间的通信，也同人类相似，大部分通信与任何别的系统之间的通信是不兼容的。

4. 对于异步数据传送的应用，如许多 RS-232、RS-422 和 RS-485 系统，光学隔离器是最常用的隔离元件。通过隔离，两个不同的接地处于隔离元件的两边，而不让任何电流穿过隔离元件。

Reading Material

Serial Communications Systems

What do you do when you need to make a dedicated low speed data connection between two places? Need to monitor a PLC on the factory floor from an office area, connect a host PC to a time clock or alarm system, or make a connection between buildings?

The asynchronous serial connection has been the workhorse of low bandwidth communications for decades. For control, monitoring and low volume data transfer asynchronous serial provide a low cost, low development solution. The RS-232 serial port is prevalent on PC's as well as scores of industrial, scientific and consumer devices making it a convenient starting point for communications. Since RS-232 itself is only suited to short connections, many applications require that it be adapted to fit requirements. This article summarizes the choices a system designer has when selecting a serial communications system (Table 12.1).

Table 12.1 COM Type

COM Type	Pros	Cons
RS-232	Low cost Widely available	Limited distance Poor noise immunity
RS-422	Good noise immunity	
Long distance	May require additional isolation to prevent ground loops	
RS-485	Good noise immunity Long distance Multidrop capability	May require additional isolation to prevent ground loops
Current Loop	High noise immunity Built-in isolation	Low speedCompatibility problems Rarely used in new designs
Fiber Optics	Ideal noise immunity Long distance	More care required in installation Higher initial cost
RF Wireless	High mobility	High cost Sensitive to environmental variables

RS-232 or, more currently, EIA-232 uses a single ended, bipolar voltage signal. Voltages typically swing from -12V to +12V with respect to signal ground. Suitable for low noise environments and distances below 30.5m, RS-232 is commonly used for the desktop modem and mouse. An increase in modem speeds has spurred an effort to increase RS-232 data rate by chip vendors. Transceivers capable of 460kbit and higher are now available, although the actual throughput gains of running higher data rates than 115.2kbit on Interrupt based systems is questionable at best.

RS-422 is suited to longer distance communications, up to 1200m without repeaters. Using a balanced differential pair results in higher noise immunity than EIA-232. The differential voltage provides a valid signal down to 200mV. Two wires are required for each signal in addition to a signal ground conductor. RS-422 is most commonly used for point-to-point communications, although up to 10 receivers may be connected to a single transmitter.

RS-485 is also suited to longer distance communications, up to 1200m without repeaters. Again, a balanced differential pair is used for higher noise immunity than ElA-232. Voltage levels are identical to RS-422. In addition, RS-485 offers amultidrop capability, up to 32 nodes can be connected. The multidrop feature also allows "two-wire" (in addition to signal ground) half-duplex data connection to be made.

Current Loop is the oldest method of connecting serial devices, dating back to teletype machines. Typically a loop current of 20mA indicates a marking condition and 0mA represents a space. Unfortunately, there is no true standard for current loop, so switching thresholds, voltage requirements and connections vary widely. A well designed current loop system has high noise immunity, and is inherently optically isolated. However, speeds are generally low and the lack of a standard makes connectivity between manufacturers spotty.

Fiber optic communications is growing in popularity as another low bandwidth serial solution. While costs are still higher than copper solutions, fiber optic links benefit from optimum isolation, noise immunity, and distances up to several miles. Installation of fiber optic cabling requires more care than copper, and repairing damaged cabling is difficult.

RF wireless has become more affordable in recent years and the adoption of spread spectrum technology has further improved performance. Modules to convert RS-232 signals to RF can be used for low to medium data rates. Range is limited, typically several hundred feet, although units are available that reach several miles with appropriate antennas. Higher power units are also available but require an FCC site license to operate. The range and performance of RF wireless is highly dependent on the physical and electrical environment and costs are high. If mobility is required or wire isn't possible, wireless has become a viable solution.

RS-232 uses a single - ended, bipolar voltage to move data between two points. RS-422/485 uses a balance differential pair to accomplish this same task. The advantage of using RS-422/485 in an industrial environment is greater noise immunity. This allows a greater distance between the transmitter and receiver. There is a downside to the greater distances provided by RS-422/485 - the difference of potential between end points.

Industrial buses cover a large area. Often different areas of the network are supplied by different power sources. Even though all of the sources are grounded, a voltage difference can exist between the grounds of the se voltage sources. This voltage difference can upset the data line in an RS-422/485 bus by pushing the signal voltage out of range and, in some cases, an excess voltage can damage equipment. Another source of excess voltage potential can be caused by intermittent sources. Power line surges and lightning are causes of this type of disturbance, but other causes, such as large electric motors starting and stopping, can temporarily affect the ground reference voltage. The solution to this problem is to employ RS-422/485 devices that provide isolation between different parts of the network. Additional protection can be achieved by using a fiber optic link between the network and areas known for voltage problems such as a power house or a water treatment plant.

Two popular industrial buses that use the RS-232 and RS-422/485 standards areModbus and Data Highway. Modbus was developed by Modicon for its line of PLC, up to and including the 984 line of controllers. Modbus call be configured for either RS-232 or RS-485 in a 4-wire mode. Data Highway is the name of the industrial bus produced by Allen. Bradley and is used on some SLC 500 controllers. An RS-485 port is also available on some PLC-2, PLC-3 and PLC-5 controllers. Consult the manual provided with your controller to be certain of the type of bus supported. The industrial buses that adhere to the RS-232 and RS-422/485 standard are listed below along with products that are compatible with various industrial buses. B&B products support these buses at the physical layer only and are mainly used as repeaters, line extenders and isolators. B&B also offers a custom design service to solve particular problems that arise from industrial buses.

Lesson 13 Programmable Logic Controller

Programmable Logic Controllers (PLCs), also referred to as programmable controllers, are in the computer family. They are, used in commercial and industrial applications. A PLC monitors inputs, makes decisions based on its program, and controls outputs to automate a process or machine.[1] This course is meant to supply you with basic information on the functions and configurations of PLCs.

13.1 Basic PLC Operation

PLCs consist of input modules or points, a Central Processing Unit (CPU), and output modules or points(Figure 13.1). An input accepts a variety of digital or analog signals from various field devices (sensors) and converts them into a logic signal that can be used by the CPU.[2] The CPU makes decisions and executes control instructions based on program instructions in memory. Output modules convert control instructions from the CPU into a digital or analog signal that can be used to control various field devices (actuators). A programming device is used to input the desired instructions. These instructions determine what the PLC will do for a specific input. An operator interface device allows process information to be displayed and new control parameters to be entered. Pushbuttons (sensors), in this simple example(Figure 13.2), connected to PLC inputs, can be used to start and stop a motor connected to a PLC through a motor starter (actuator).

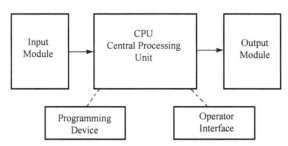

Figure 13.1 The configuration of PLC

13.2 Hard-Wired Control

Prior to PLCs, many of these control tasks were solved by contactors or relay controls. This is often referred to as hardwired control. Circuit diagrams had to be designed (Figure 13.3). Electrical components specified and installed, and wiring lists created.[3] Electricians would then wire the components necessary to perform a specific task. If an error was made the wires had to be reconnected correctly. A change in function or system expansion required extensive component changes and rewiring.

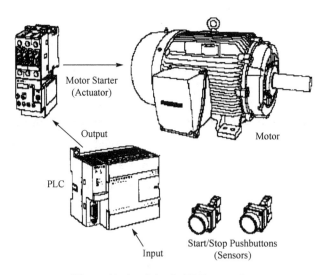

Figure 13.2 A basic PLC operation

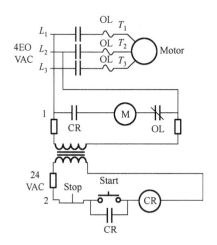

Figure 13.3 Circuit diagram

13.3 Advantages of PLCs

The same, as well as more complex tasks, can be done with a PLC. Wiring between devices and relay contacts is done in the PLC program. Hard-wiring, though still required to connect field devices, is less intensive. Modifying the application and correcting errors are easier to handle. It is easier to create and change a program in a PLC than it is to wire and rewire a circuit.

Following are just a few of the advantages of PLCs.

(1) Smaller physical size than hard-wire solutions.

(2) Easier and faster to make changes.

(3) PLC s have integrated diagnostics and override functions.

(4) Applications can be immediately documented.

(5) Applications can be duplicated faster and less expensively.

13.4 Siemens PLCs

Siemens makes several PLC product lines in the SIMATIC S7 family. They are: S7-200, S7-300, and S7-400.

1. S7-200

The S7-200 is referred to as a micro PLC because of its small size. The S7-200 has a brick design which means that the power supply and I/O terminal are on-board. The S7-200 can be used on smaller, stand-alone applications such as elevators, car washes, or mixing machines. It can also be used on more complex industrial applications such as bottling and packaging machines.

2. S7-300 and S7-400

The S7-300 and S7-400PLCs are used in more complex applications that support a greater number of I/O points. Both PLCs are modular and expandable. The power supply and I/O consist of separate modules connected to the CPU. Choosing either the S7-300 or S7-400 depends on the complexity of the task and possible future expansion. Your Siemens sales representative can provide you with additional information on any of the Siemens PLCs.

13.5 CPU

CPU is a microprocessor system that contains the system memory and is the PLC decision-making unit. The CPU monitors the inputs and makes decisions based on instructions held in the program memory. The CPU performs relay, counting, timing, data comparison, and sequential operations.

13.6 Programming Devices

The program is created in a programming device (PG) and then transferred to the PLC. The program for the S7-200 can be created using a dedicated Siemens SIMATIC S7 programming device, such as a PG 720 or PG 740, if STEP 7 Micro/WIN software is installed.

A personal computer (PC), with STEP 7 Micro/WIN installed, call also be used as aprogramming device with the S7-200.

13.7 Software

A software program is required in order to tell the PLC what instructions it must follow. Programming software is typically PLC specific. A software package for one PLC, or one family ofPLCs, such as the S7 family, would not be useful on other PLCs. The S7-200 uses a Windows. based software program called STEP 7-MicroWN32. The PG 720 and PG 740 have STEP 7. Micro WIN32 software pre-installed. Micro WIN32 is installed on a personal computer in a similar manner to any other computer software.

13.8 Connector Cables PPI (Point-to-Point Interface)

Connector cables are required to transfer data from the programming device to the PLC. Communication can only take place when the two devices speak the same language or protocol. Communication between a Siemens programming device and the S7-200 is referred to as PPI protocol. An appropriate cable is required for a programming device such as a PG 720 or PG 740. The S7-200 uses a 9-pin D-connector. This is a straight-through serial device that is compatible with Siemens programming devices (MPI port) and is a standard connector for other serial interfaces.

A special cable, referred to as a PC/PPI cable, is needed when a personal computer is used as a programming device. This cable allows the serial interface of the PLC to communicate with the RS-232 serial interface of a personal computer. DIP switches on the PC/PPI cable are used to select an appropriate speed (bit rate) at which information is passed between the PLC and the computer.[4]

Words and Expressions

starter ['stɑːtə] n. 启动器，启动钮
actuator ['æktjueitə] n. 操作（执行）机构，执行器
reconnect [ˌriːkə'nekt] vt. & vi. 再接合
override [ˌəuvə'raid] vt. 不顾，无视，蔑视
bottling n. 装瓶
packaging n. 包装
PPI (Point-to-Point Interface) 点对点接口
compatible [kəm'pætəbl] adj. 协调的，一致的，兼容的
depend on 依靠，依赖
counting 计数器
timing 计时器，定时器
PG 编程器（编程设备）

Notes

1. A PLC monitors inputs, makes decisions based on its program, and controls outputs to automate a process or machine.

   本句是个简单句。主语是 PLC，谓语由 monitors、makes、controls 组成，分别描写 3 个行为，不定式短语 to automate a process or machine 作为目的状语。based on its program 修饰 makes decisions，告诉我们结果是"根据程序"得到的。

2. An input accepts a variety of digital or analog signals from various field devices (sensors) and converts them into a logic signal that can be used by the CPU.

本句结构很简单，由 and 连接两个并列句，后一个并列句里 that 引导的定语从句修饰 signal。an input 里的 an 是不定冠词，泛指 input，而不是指"一个"input。field devices 解释为现场设备。

3. Circuit diagrams had to be designed, electrical components specified and installed, and wiring lists created.

   本句是英语常见的动词省略现象。本句的前半句谓语 had to be designed 是完整的。后面两个部分，分别以 electrical components 及 wiring lists 作为主语，谓语部分的 had to be 被省略。这句话补充完整应该是 Circuit diagrams had be to designed, electrical components had to be specified and installed, and wiring lists had to be created.

4. DIP switches on the PC/PPI cable are used to select an appropriate speed (bit rate) at which information is passed between the PLC and the computer.

   本句中有一个以 which 引导的定语从句。在这个定语从句中，介词 at 前置，和引导词在一起，实际指的是 at the speed，这个 speed 指的是 PLC 和计算机之间信息传输的速度。

Exercises

Translate the following passages into English or Chinese.

1. A super capacitor, so named because of its ability to maintain a charge for a long period of time, protects data stored in RAM in the event of a power loss.
2. One of the most confusing aspects of PLC programming for first-time users in the relationship between the device that controls a status bit and the programming function that uses a status bit.
3. PLC 通过输入模块接收来自不同现场设备（传感器）的各种数字或模拟信号，然后将它们转换为 CPU 可以使用的逻辑信号。
4. PLC 通过输出模块将来自 CPU 的控制指令转换为现场控制的各个设备（执行机构）所能使用的数字或逻辑信号。

Reading Material

Terminology

The language of PLCs consists of a commonly used set of terms; many of which are unique to PLCs. In order to understand the ideas and concepts of PLCs, an understanding of these terms is necessary.

1. Sensor

A sensor is a device that converts a physical condition into an electrical signal for use by the PLC. Sensors are connected to the input of a PLC. A pushbutton is one example of a sensor that is connected to the PLC input. An electrical signal is sent from the pushbutton to the PLC indicating the condition (open/closed) of the pushbutton contacted.

2. Actuator

Actuator converts an electrical signal from the PLC into a physical condition. Actuator is connected to the PLC output. A motor starter is one example of an actuator that is connected to the PLC output. Depending on the output PLC signal the motor starter will either start or stop the motor.

3. Discrete Input

A discrete input(Figure 13.4), also referred to as a digital input, is an input that is either in an ON or OFF condition. Pushbuttons, toggle switches, limit switches, proximity switches, and contact closures are examples of discrete sensors which are connected to the PLCs discrete or digital inputs. In the ON condition a discrete input may be referred to as a logic 1 or a logic high. In the OFF condition a discrete input may be referred to as a logic 0 or a logic low(Figure 13.5).

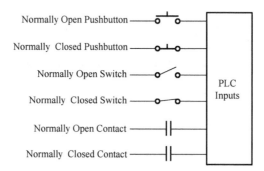

Figure 13.4 A discrete input

A Normally Open (NO) pushbutton is used in the following example. One side of the pushbutton is connected to the first PLC input. The other side of the pushbutton is connected to an internal 24 VDC power supply. Many PLCs require a separate power supply to power the inputs. In the open state, no voltage is present at the PLC input. This is the Off condition. When the pushbutton is depressed, 24 VDC is applied to the PLC input. This is the On condition.

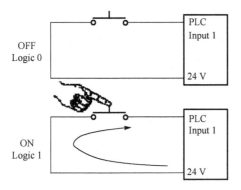

Figure 13.5 A logic 0 or a logic 1

4. Analog Input

An analog input is an input signal that has a continuous signal. Typical analog inputs may vary from 0 to 20mA, 4 to 20mA, or 0 to 10V. In the following example, a level transmitter monitors the level of liquid in a tank. Depending on the level transmitter, the signal to the PLC can either increase or decreases or the level increases or decreases.

5. Discrete Output

A discrete output is an output that is either in an ON or OFF condition. Solenoids, contactor coils, and lamps are examples of actuator devices connected to discrete outputs. Discrete outputs may also be referred to as digital outputs. In the following example, a lamp can be turned on or off by the PLC output it is connected to.

6. Analog Output

An analog output is an output signal that has a continuous signal. The output may be as simple as a 0~10 VDC level that drives an analog meter. Examples of analog meter outputs are speed, weight, and temperature. The output signal may also be used on more complex applications such as a current-to-pneumatic transducer that controls an air-operated flow-control valve(Figure 13.6).

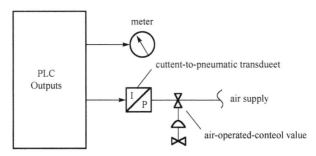

Figure 13.6 An analog output of PLC

7. Programming

A program consists of one or more instructions that accomplish a task. Programming a PLC is simply constructing a set of instructions. There are several ways to look at a program such as ladder logic, statement lists, or function block diagrams.

8. Ladder Logic

Ladder logic (LAD) is one programming language used with PLCs. Ladder logic uses components that resemble elements used in a line diagram format to describe hard-wired control.

9. Ladder Logic Diagram

The left vertical line of a ladder logic diagram represents the power or energized conductor. The output element or instruction represents the neutral or return path of the circuit. The fight vertical line, which represents the return path on a hard-wired control line diagram, is omitted. Ladder logic diagrams are read from left-to-fight, top-to-bottom. Rungs are sometimes referred to as networks. A network may have several control elements, but only times referred to as networks. A network may have several control elements, but only one output coil.

In the example program shown example I0.0, I0.1 and Q0.0 represent the first instruction combination(Figure 13.7). If inputs I0.0 and I0.1 are energized, output relay Q0.0 energizes. The input devices could be switches, pushbuttons, or contact closures. I0.4, I0.5, and Q1.1 represent the second instruction combination. If either input I0.4 or I0.5 are energized, output relay Q0.1 energize.

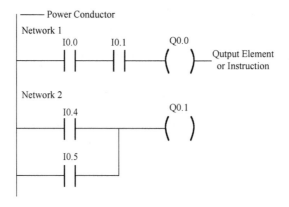

Figure 13.7 A ladder logic diagram

Lesson 14   Remote Sensing

Remote Sensing is a process of obtaining information about land, water, or an object, without any physical contact between the sensor and subject of analysis. The term remote sensing most often refers to the collection of data by instruments carried aboard aircraft or satellites. Remote sensing systems are commonly used to survey, map, and monitor the resources and environment of the earth. They also have been used to explore other planets.

There are several different types of remote sensing devices. Many systems take photographs with cameras, recording reflected energy in the visible spectrum. Other systems record electromagnetic energy beyond the range of human sight, such as infrared radiation and microwaves. Multispectral scanners produce images across both the visible and the infrared spectrum.

14.1   Sensors

The most familiar form of electromagnetic energy is visible light, which is the portion of electromagnetic spectrum to which human eyes are sensitive.[1] When film in a camera is exposed to light, it records electromagnetic energy. For more than 50 years, photographic images obtained from airborne cameras have been used in urban planning, forest management, topographic mapping, soil conservation, military surveillance, and many other applications.[2]

Infrared sensors and microwave sensors record invisible electromagnetic energy. The heat of an object, for example, can be measured by the infrared energy it radiates. Infrared sensors create images that show temperature variations in an area —a difficult or impossible task using conventional photography. Thermal infrared sensors can be used to survey the temperatures of water, locate damaged underground pipelines, and map geothermal and geologic structures.[3]

Microwave sensors, such as radar, transmit electromagnetic energy toward objects and record how these objects reflect the energy. Microwave sensors operate at very long electromagnetic wavelengths capable of penetrating clouds, are useful when cloud cover prohibits imaging with other sensors.[4] By scanning an area with radar and processing the data in a computer, scientists can create radar maps. The surface of Venus, which is entirely shrouded by dense clouds, has been mapped in this way. Radar imagery is also used in geologic mapping, estimating soil moisture content, and determining sea-ice conditions to aid in ship navigation.

Multispectral scanners provide date electronically for multiple portions of the electromagnetic spectrum. Scientists often use computers to enhance the quality of these images or to assist in automated information-gathering and mapping. With computers, scientists can combine several images obtained by multispectral scanners operating at different frequencies. [5]

14.2 Satellites

Satellites have proved extremely useful in the development of remote sensing systems. In 1972 the United States launched Landsat-1, the first in a series of satellites designed specifically for remote sensing. Today, Landsat-5 produces images of most of the earth's surface every 16 days. Each Landsat image covers more than 31,000 sq km (11,970 sq mi). Objects as small as 900 sq m (9688 sq ft) can be seen in the images produced by Landsat's Thematic Mapper, a type of multispectral scanner. Landsat date are used for applications such as mapping land use, managing forested land, estimating crop production, monitoring grazing conditions, assessing water quality, and protecting wildlife.

Between 1990 and 1996, almost 50 remote sensing satellites were placed into orbit. Since 1986, France's SPOT satellites have provided images showing objects as small as 100 sq m (1076 sq ft) and have produced stereoscopic images useful for topographic mapping. Earth-observing satellites have also been launched by the European Space Agency and Japan, Russia, India, and other nations.

Meteorological satellites, such as other operated by the U. S. National Oceanic and Atmospheric administration, provide images for use in weather forecasting as well as in oceanic and terrestrial applications. Remote sensors on weather satellites can track the movement of clouds and record temperature changes in atmosphere.

14.3 Outlook

Remote sensing is changing rapidly. Some satellites carry instruments that can images of objects as small as an automobile and constantly improving technology promises even better resolution in the near future. Computer-assisted image-analysis techniques are leading to many new applications for remote sensing. In the late 1990s, the U. S. National Aeronautics and Space Administration is scheduled to launch the Earth Observing System, a key program in its Mission to Planet Earth, which involves launching a series of satellites to study environmental changes on the planet.

Words and Expressions

remote sensing   遥感
scanner ['skænə] n. 扫描仪
shroud [ʃraud] vt. 遮蔽，隐蔽，覆盖
spectrum ['spektrəm] n. 频谱
stereoscopic [ˌsteriə'skɔpik] adj. 实体镜的，立体的
survey [sə(:)'vei] n. 调查，视察；vt. 观测，测量，俯视
Venus ['vi:nəs] n. 金星（罗神）维纳斯

Notes

1. The most familiar form of electromagnetic energy is visible light, which is the portion of electromagnetic spectrum to which human eyes are sensitive.
   我们最熟悉的电磁能就是可见光,它是电磁波谱中人眼能够感觉的那部分。

2. For more than 50 years, photographic images obtained form airborne cameras have used in urban planning, forest management, topographic mapping, soil conservation, military surveillance, and many other applications.
   50多年来,用空中摄影机获取的照片图像一直用于城市规划、森林管理、地形测量与绘图、土壤保护、军事监察和其他很多应用中。

3. Thermal infrared sensors can be used to survey the temperatures of water, locate damaged underground pipelines, and map geothermal and geologic structures.
   热红外传感器可用于水温的测量、已损坏的地下管道定位、地热和地质结构图的绘制。

4. Microwave sensors operate at very long electromagnetic wavelengths capable of penetrating clouds, a useful when cloud cover prohibits imaging with other sensors.
   微波传感器工作在波长很长的电磁波段,它能穿透云层,当云层覆盖使别的传感器不能成像时,它却能成像它的这个特点是非常有用了。

5. Scientists often use computers to enhance the quality of these images or to assist in automated information-gathering and mapping. With computers, scientists can combine several images obtained by multispectral scanners operating at different frequencies.
   科学家常用计算机来提高图像的质量或借以进行自动信息收集和绘图。利用计算机,科学家还能把工作在不同频率上的扫描仪得到的不同频谱的几个图像进行组合。

Exercises

Translate the following passages into English or Chinese.

1. 雷达可以确定远距离物体的许多特性,如物体的位置、运动速度和方向形状等。雷达能够探测视力范围以外的物体,并能在各种气象条件下工作,这使得它成为很多工业部门的重要且用途广泛的一种工具。

2. 太空探测器上的雷达能穿透金星大气层浓密的云层,它提供的图像质量比雷达从地球上获得的图像质量好得多,这将帮助我们的科学家更好地探索外太空。

3. A new use for GPS location involves the integration of a GPS receiver with an intelligent compact disc player capable of displaying road maps and other graphical information. Upon receiving the GPS location data, the CD player can pinpoint the location visually on one of the road maps contained on disk.

4. In the near term, before the actualization of a fully digital telecommunications world, devices like modems will still be necessary to provide an essential like between the old analog world and the upcoming digital one.

Reading Material

GPS

1. Introduction

Global Positioning System (GPS) is a space-based radionavigation system, consisting of 24 satellites and ground support. GPS provides users with accurate information about their position and velocity, as well as the time, anywhere in the world and in all weather conditions.

2. History and Development

GPS, formally known as the Navstar Global Positioning System, was initiated in 1973 to reduce the proliferation of navigation aids. GPS is operated and maintained by the United States department of Defense. By creating a system that overcame the limitations of many existing navigation systems, GPS became attractive to a broad spectrum of users. GPS has been successful in classical navigation application, and because its capabilities are accessible using small, inexpensive equipment, GPS has also been used in many new applications.

3. How GPS Works

GPS determines location by computing the difference between the time a signal is sent and the time it is received. GPS satellites carry atomic clocks that provide extremely accurate time. The time information is placed in the codes broadcast by the satellites so that a receiver can continuously determine the time the signal was broadcast. The signal contains data that a receiver uses to computer the locations of the satellites and to make other adjustments needed for accurate positioning. The receiver uses the time difference between the time of signal reception and the broadcast time to compute the distance, or range, from the receiver to the satellite. The receiver must account for propagation delays, or decreases in the signal's speed caused by the ionosphere and the troposphere. With information about the ranges to three satellites and the location of the satellite when the signal was sent, the receiver can compute its own three-dimensional position.

An atomic clock synchronized to GPS is required in order to compute ranges from these three signals. However, by taking a measurement from a fourth satellite, the receiver avoids the need for an atomic clock. Thus, the receiver uses four satellites to compute latitude, longitude, altitude, and time.

4. The Parts of GPS

GPS comprises three segments: the space, control, and user segments. The space segment includes the satellites and the delta rockets that launch the satellites from Cape Canaveral, in Florida. GPS satellites fly in circular orbits at an altitude of 20,100km (12,500mi) and with a period of 12 hours. The orbits are tilted to the earth's equator by 55 degrees to ensure coverage of polar regions. Powered by solar cells, the satellites continuously orient themselves to point their solar panels toward the sun and their antennae toward the earth. Each satellite contains four atomic clocks.

The control segment includes the master control station at Falcon Air Force Base in Colorado Springs, Colorado, and monitor stations at Falcon Air Force Base and on Hawaii, ascension Island in the Atlantic Ocean, Diego Garcia Atoll in the Indian Ocean, and Kwajalein Island in the South Pacific Ocean. These stations monitor the GPS satellites. The control segment uses measurements collected by the monitor stations to predict the behavior of each satellite's orbit and clock. The prediction data is uplinked, or transmitted, to the satellites for transmission to the users. The control segment also ensures that the GPS satellite orbits and clocks remain within acceptable limits.

The user segment includes the equipment of the military personnel and civilians who receive GPS signals. Military GPS user equipment has been integrated into fighters, bombers, tankers, helicopters, ships, submarines, tanks, jeeps, and soldier's equipment. In addition to basic navigation activities, military applications of GPS include target designation, close air support, 'smart' weapons, and rendezvous.

With more than 500 000 GPS receivers, the civilian community has its own large and diverse user segment. Surveyors use GPS to save time over standard survey methods. GPS is used by aircraft and ships for route navigation and for airport or harbor approaches. GPS tracking systems are used to route and monitor delivery vans and emergency vehicles. In a method called precision farming, GPS is used to monitor and control the application of agricultural fertilizer and pesticides. GPS is available as an in-car navigation aid and is used by hikers and hunters. GPS is also used on the space shuttle because the GPS user does not need to communicate with the satellite, GPS can serve an unlimited number of users.

5. GPS Capabilities

GPS is available in two basic forms: the standard positioning service (SPS) and the precise positioning service (PPS). SPS provides a horizontal position that is accurate to about 100m (about 330ft); PPS is accurate to about 20m (about 70ft). For authorized users—normally the United States military and its allies—PPS also provides greater resistance to jamming and immunity to deceptive signals.

Enhanced techniques such as differential GPS (DGPS) and the use of a carrier frequency processing have been developed for GPS. DGPS employs fixed stations on the earth as well as satellites and provides a horizontal position accurate to about 3m (about 10ft). Surveyors pioneered the use of a carrier frequency processing to compute positions to within about 1cm (about 0.4in). SPS, DGPS and carrier techniques are accessible to all users.

The availability of GPS is currently limited by the number and integrity of the satellites in orbit. Outages due to failed satellites still occur and affect many users simultaneously. Failures can be detected immediately and users can be notified within seconds or minutes depending on the user's specific situation. Most repairs are accomplished within one hour. As GPS becomes integrated into critical operations such as traffic control in the national airspace system, techniques for monitoring the integrity of GPS on-board and for rapid notification of failure are being developed and implemented.

6. The Future of GPS

As of March 1994, 24 GPS satellites were in operation. Replenishment satellites are ready for launch, and contracts have been awarded to provide satellites into the 21st century. GPS applications continue to grow in land, sea, air, and space navigation. The ability to enhance safety and to decrease fuel consumption will make GPS an important component of travel in the international airspace system. Airplanes will use GPS for landing at fogbound airports. Automobiles will use GPS as part of intelligent transportation systems. Emerging technologies will enable GPS to determine not only the position of a vehicle but also its altitude.

Lesson 15 Multi-sensor Data Fusion

15.1 Introduction

Multi-sensor data fusion (MDF) is an emerging technology to fuse data from multiple sensors in order to make a more accurate estimation of the environment through measurement and detection. Applications of MDF cross a wide spectrum, including the areas in military services such as automatic target detection and tracking, battlefield surveillance, etc. and the areas in civilian applications such as environment surveillance and monitoring, monitoring of complex machinery, medical diagnosis, smart building, food quality characterization and even precision agriculture.[1] Techniques for data fusion are integrated from a wide variety of disciplines, including signal processing, pattern recognition, and statistical estimation, artificial intelligence, and control theory.[2] The rapid evolution of computers, proliferation of micro-mechanical/electrical systems sensors, and the maturation of data fusion technology provide a basis for utilization of data fusion in everyday applications.

15.2 Technical Background

Multi-sensor data fusion provides an approach improving of performance of single sensor. In general, the data of measurement from a single sensor is limited to achieve a high quality. If a number of sensors can be used to perform the same measurement and the data of measurements from these sensors can be combined some way, the resultant has a great potential to outperform over each single measurement with increased accuracy.

In general, as its name implies, MDF is a technique by which data from a number of sensors are combined through a centralized data processor to provide comprehensive and accurate information. This technology has a powerful potential to track changing conditions and anticipate impacts more consistently than a single data source could do traditionally even this single data source might be a highly reliable one. Thus, MDF makes it possible to create a synergistic process, in which, the consolidation of individual data creates a combined resource with a productive value greater than the sum of its parts.[3]

Although the concept is not new, MDF technology is still in its infancy. This technology has undergone rapid growth that started in the late 1980s and has continued to the present. The U.S. Department of Defense (DoD) conducted much of the early research on this technology and explored its usefulness in military surveillance and land-based battle management systems. The application of data fusion technology to commercial endeavours (e.g. robotics and general image processing) and non-military government projects (e.g. weather surveillance and NASA missions) is also growing rapidly. In its current state, the technology can combine sensor data of many types, including radar, infrared, sonar, and visual information.

In data analysis and processing of measurement and instrumentation, pattern recognition techniques are necessary. Pattern recognition is used to develop data fusion algorithms. Artificial neural networks, which have been developed based on studies about the mechanism of human brain, are the top option over other conventional statistical pattern recognition methods. [4] Linn and Hall surveyed more than fifty data fusion systems. Only three of these systems used the neural network method. This low number may indicate an underestimation of the importance of neural networks in the field of data fusion.

Artificial neural networks have been widely used in solving complex problems, such as pattern recognition, fast information processing and adaptation. Artificial neural networks are structured based on studies about the mechanism and the structure of human brain. The architecture and implementation of a neural network models a simplified version of the structure and activities of the human brain. The vast processing power inherent in biological neural structures has inspired the study of the structure itself as a model for designing and organizing man-made computing structure. A MDF scheme integrated with neural network pattern recognition is a promising structure to achieve high quality data analysis and processing in measurement and instrumentation.

15.3 Method

Three basic alternatives can be used for MDF.

(1) Direct fusion of sensor data.

(2) Representation of sensor data via feature vectors, with subsequent fusion of the feature vectors.

(3) Processing of each sensor to achieve high-level inferences or decision, which is subsequently combined.

Each of these approaches is motivated and influenced by some type of applications and utilizes different fusion methods. They can not be generalized to all data fusion methods. This paper will focus on the feature-based data fusion method.

With the measurement data from each sensor, the features can be quantitatively extracted from each sensor's output. The extracted feature vectors can be combined as the following matrix:

$$\bm{F} = [f_1, f_2, \cdots, f_n]^\mathrm{T} \quad (n \geqslant 2) \tag{15-1}$$

where f_i is the feature vector from the ith sensor and n is the number of sensors.

With each of the feature vectors as the input, a pattern recognition technique, typically a neural network is needed to differentiate the measurement data into appropriate classes. The output vectors of the n neural networks are necessary to be merged for data fusion.

Suppose $O_1^i, O_2^i, \cdots, O_m^i (i=1,2,\cdots,n)$ are the m outputs of the neural network for the ith sensor. The merged output should be in the form:

$$O_j^M = c_1 O_j^1 + c_2 O_j^2 + \cdots + c_n O_j^n \quad (j = 1, 2, \cdots, n) \tag{15-2}$$

where $c_1, c_2, \cdots, c_n$ are the merging coefficients which can be determined by assigning values or by optimization.

Figure 15.1 shows the diagram of the MDF scheme with neural networks. This scheme should be implemented as follows.

(1) Extract features from the measurement data of each of the n sensors.
(2) Classify the feature vectors through a neural network classifier.
(3) Merge the m outputs of the neural network classifiers for all the n sensors.
(4) Post process to produce a fused classification result.

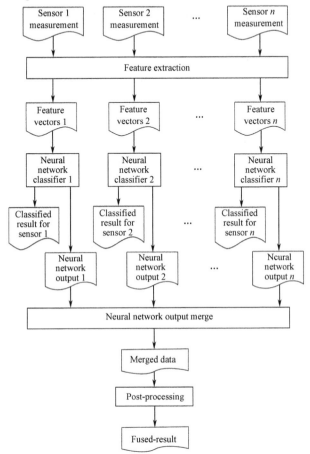

Figure 15.1 Scheme diagram of multisensor data fusion with neural network classifiers

15.4 Application

We have developed and are going to develop a number of applications of MDF for high quality data analysis and processing in measurement and instrumentation. The areas of these applications have few articles in engineering literature.

In 1977, Gros wrote a book that is the first book to devote exclusively to Multi-sensor integration and data fusion applied to NDT (Non-Destructive Testing) or NDI (Non-Destructive Inspection). This book provides detailed case studies and practical guidelines for readers wishing to explore NDT/NDI data fusion. Since then research and application of NDT/NDI data fusion have been conducted in various areas. However, a general data fusion system model capable of handling various applications is very difficult, if not impossible, to design. Various data fusion models under the general concept are necessary for each specific area of research and application.

The NDI ultrasound and eddy current imaging methods were studied for corrosion detection for aging aircraft panels. In order to increase the accuracy of the detection from the individual sensor, a NDI data fusion method was developed for the same application.

We developed another NDI data fusion method integrated with artificial neural network classifiers. Ultrasonic and eddy current image data run through neural network classifiers to identify the corrosive spots on the same aircraft panel specimen. In order to evaluate the overall performance of the NDI imaging in corrosion detection, the fusion of the classified corrosion data from the two different imaging sensors is of the interest. The MDF is expected to have a complete picture of the method and more accurate corrosion detection.

Figure 15.2 shows the diagram of the NDI imaging data fusion scheme with neural network classifiers. This scheme can be implemented as follows:

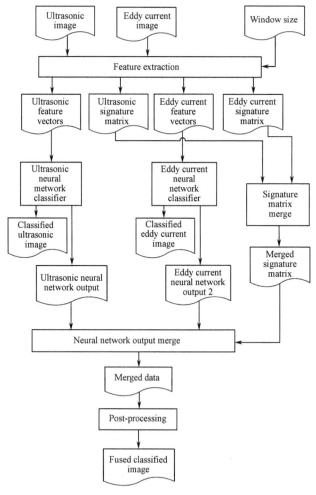

Figure 15.2 Scheme diagram of corrosion NDI imaging data fusion with neural network classifiers

(1) Extract statistical features from the ultrasonic and eddy current images with a window that goes over the images with a specified size: 2×2, 3×3,···, 12×12.[5]

After feature extraction for each image, there are two sets of new data: feature vectors and signature matrix. The signature matrix is used to record what each pixel indicates in the image:

0: indicates that this pixel in the image shows no corrosion;

1: indicates that this pixel in the image may shows certain degree of corrosion but this pixel will not be included in neural network classification since the windowed feature extraction skipped it due to around it there are too many non-corrosive pixels;[6]

2: indicates that this pixel in the image shows certain degree of corrosion and it will be included in neural network classification.

(2) Generate outputs of the neural network classifiers and produce classified ultrasonic and eddy current images.

(3) Merge the output vectors of the neural network classifiers for different sensors with the merged signature matrix.

The purpose of the signature matrix is to help find the relationship between the output sequences of neural network classifiers for different sensors. The merged signature matrix is formed as:

0: when both are 0 at the same pixel in ultrasonic and eddy current signature matrices;

1: when either 0 or 1 at the same pixel in ultrasonic or eddy current signature matrices;

2: when at least a 2 at the same pixel in ultrasonic or eddy current signature matrices.

(4) Produce a fused classified image.

This scheme implements two merging functions: merge of feature signature matrices and merge of outputs and neural network classifiers for different sensors. The fused image can outperform over each individual image from the two different imaging sensors because the fused product integrates what the two different sensors captured.

The products of this scheme are: a classified ultrasonic image, a classified eddy current image, and a fused classified image. To evaluate the performance of the data fusion model, the three images were compared with a benchmark X-ray corrosion image of the same specimen (Figure 15.3). Figure 15.4 is an ultrasonic image of the corrosive aircraft panel specimen. Figure 15.5 is the eddy current image of the specimen. Figure 15.6 is the neural network classified ultrasonic image with a match rate of 50.61% with the X-ray data at 4×4 window size feature extraction. Figure 15.7 is the neural network classified eddy current image with a match rate of 57.70% with the X-ray data at 4×4 window size feature extraction. Figure 15.8 is the fused classified image of Figure 15.6 and Figure 15.7. The match rate goes up to 65.49%.

Table 15.1 summarizes the match rates of the data fusion of the neural network classifiers with the X-ray data at different window sizes. The results indicate that data fusion consistently enhances neural network corrosion detection with ultrasonic and eddy current image data individually.

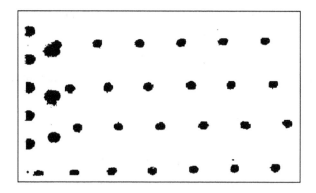

Figure 15.3　X-ray image of a corrosive aircraft panel specimen

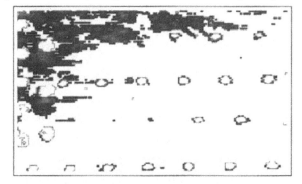

Figure 15.4　Ultrasonic image of a corrosive aircraft panel specimen

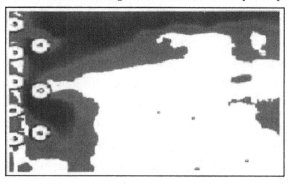

Figure 15.5　Eddy current image of a corrosive aircraft panel specimen

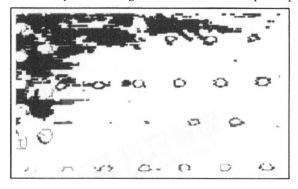

Figure 15.6　Neural network classified ultrasonic image

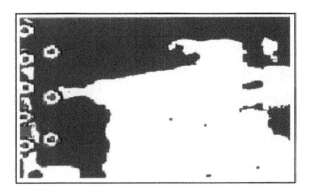

Figure 15.7 Neural network classified eddy current image

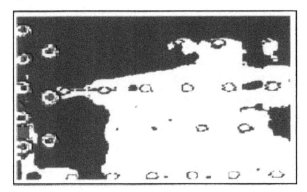

Figure 15.8 Fused classified image from ultrasonic and eddy
current neural network classified results

Table 15.1 Match rates corrosion neural network classification with X-ray
with image feature extractionat different window sizes

Window size	Ultrasonic	Eddy current	Data fusion
2×2	50.61%	57.70%	65.49%
3×3	50.61%	57.70%	65.49%
4×4	50.61%	57.70%	65.49%
5×5	50.61%	57.70%	65.49%
6×6	50.61%	57.70%	65.49%
7×7	50.61%	57.70%	65.49%
8×8	50.61%	57.70%	65.49%
9×9	50.61%	57.70%	65.49%
10×10	50.61%	57.70%	65.49%
11×11	50.61%	57.70%	65.49%
12×12	47.45%	57.25%	62.52%
Average	48.68%	57.33%	64.32%

Words and Expressions

fusion [ˈfjuːʒn] *n.* 融合，熔解，合并
surveillance [səːˈveiləns] *n.* 监视，监督
monitor [ˈmɔnit] *vt.* 监控
synergistic *adj.* 协作的，互相作用的
consolidation [knˈsɔliˈdeiʃn] *n.* 巩固，合并，统一
endeavour [inˈdevə (r)] *n.* 尽力，竭力
inference [ˈinfərəns] *n.* 推理，推断，推论

Notes

1. Applications of MDF cross a wide spectrum, including the areas in military services such as automatic target detection and tracking, battlefield surveillance, etc. and the areas in civilian applications such as environment surveillance and monitoring, monitoring of complex machinery, medical diagnosis, smart building, food quality characterization and even precision agriculture.
   MDF 的应用领域广泛，包括如自动目标检测与跟踪、战场监视等军用方面，以及如环境监视与监控、复杂机械监控、医疗诊断、智能建筑、食品质量检验和精细农业民用方面。

2. Techniques for data fusion are integrated from a wide variety of disciplines, including signal processing, pattern recognition, and statistical estimation, artificial intelligence, and control theory.
   数据融合技术是多门学科的综合，包括信号处理、模式识别、统计估计、人工智能和控制理论。

3. Thus, MDF makes it possible to create a synergistic process, in which, the consolidation of individual data creates a combined resource with a productive value greater than the sum of its parts.
   因而，MDF 使得创造一个协同处理的过程成为可能，在这个过程中，单个数据的合并创造出一个其总值大于部分之和的联合数据源。

4. Artificial neural networks, which have been developed based on studies about the mechanism of human brain, are the top option over other conventional statistical pattern recognition methods.
   人工神经网络是根据对人脑机制的研究而开发的理念，应用该理念进行统计模式识别要远胜过其他常规方法。

5. Extract statistical features from the ultrasonic and eddy current images with a window that goes over the images with a specified size: 2×2, 3×3, ..., 12×12.
   以一个指定大小的窗口：2×2, 3×3, ..., 12×12，扫描超声波和涡流图像，提取统计特征。

6. 1: indicates that this pixel in the image may shows certain degree of corrosion but this pixel will not be included in neural network classification since the windowed feature extraction skipped it due to around it there are too many non-corrosive pixels.

1：表示图像中该像素可能有一定程度的腐蚀，但该像素不会包含在神经网络分类中，这是由于其周围有太多的未腐蚀像素，因而窗口特征提取时跳过了它。

Exercises

Translate the following passages into English or Chinese.

1. 一个人工智能可以完全融合所有5种人类感知信息：视觉、听觉、嗅觉、味觉和触觉。
2. 人工神经网络已广泛用于解决各种复杂问题，如模式识别、快速信息处理和自适应等。
3. Image fusion can be broadly defined as the process of combining multiple input images into a smaller collection of images, usually a single one, which contains the 'relevant' information from the inputs.
4. The aim of image fusion is to integrate complementary and redundant information from multiple images to create a composite that contains a 'better' description of the scene than any of the individual source images.

Reading Material

Multi-sensor Image Fusion

Extraordinary advances in sensor technology, microelectronics and communications have brought a need for processing techniques that can effectively combine information from different sources into a single composite for interpretation. As many sources produce images, image processing has become one of the most important domains for fusion and has been used in many fields such as machine vision, medical diagnosis, military applications and remote sensing.

Image fusion can be broadly defined as the process of combining multiple input images into a smaller collection of images, usually a single one, which contains the 'relevant' information from the inputs. The aim of image fusion is to integrate complementary and redundant information from multiple images to create a composite that contains a 'better' description of the scene than any of the individual source images. This fused image should increase the performance of the subsequent processing tasks such as segmentation, feature extraction and object recognition. The different images to be fused can come from different sensors of the same basic type or they may come from different types of sensors. The sensors used for image fusion need to be accurately coaligned so that their images will be in spatial registration. A simple image fusion method is to take the average of the source images pixel by pixel. However, along with simplicity come several undesired side effects including reduced contrast. In recent years, many researchers recognized that multi-scale transforms are very useful for analyzing the information content of images for the purpose of fusion. Various methods based on multi-scale decompositions (MSD) have been proposed. Examples of this approach include the Laplacian pyramid, the gradient pyramid, the ratio-of-low-pass pyramid and the morphological pyramid. Since wavelet theory has emerged as a well developed yet rapidly expanding mathematical foundation for a class of multi-scale representations, it has also been used in multi-sensor image fusion. At the same time, some sophisticated image fusion approaches based on multi-scale representations began to emerge and receive increased attention. Most of these approaches were based on combining the multi-scale decompositions (MSD's) of the source images. The basic idea is to perform a multi-scale transform on each source image, then construct a composite multi-scale representation from these. The fused image is obtained by taking an inverse multi-scale transform. In general, discrete wavelet transform (DWT) based fusion method is superior to the previous pyramid-based methods. Although these methods often perform satisfactorily, their multi-resolution decompositions and the consequent fusion results are shift-variant because of an underlying down-sampling process. When there is slight movement or mis-registration of the source images, their performance will deteriorate quickly.

The traditional algorithms based on MSD techniques are mainly pixel-based approaches where each individual coefficient of the MSD decomposition (or possibly the coefficients in a small xed window) is treated more or less independently. However, in most image fusion applications, we are not interested in individual pixels but rather in the objects and regions they represent. It therefore seems reasonable to incorporate object and region information into the fusion process. Because the pulse-coupled neural network (PCNN) has great advantage in image segmentation, we present a new region -based image fusion scheme using PCNN, which combines aspects of feature and pixel-level fusion. The basic idea is to segment all different input images by PCNN and to use this segmentation to guide the fusion process.

As we observed before, an important pre-processing step in image fusion is image registration, which ensures that the information from each source is referring to the same physical structures in the real-world. Throughout this paper, it will be assumed that all sources images have been registered. Comprehensive reviews on image registration can be found in Barbara, Z. and Flusser's paper.

1. Multi-focus image fusion applications

We consider situations where two or more objects in the scene are at different distances from the camera. As is typical with most inexpensive cameras, the image thus obtained will not be in focus everywhere, i. e. , if one object is in focus, another one will be out of focus. However, by fusing images with different focus points, an image that is in focus everywhere can be obtained. As showed in Figure 15. 9, each image contains multiple objects at different distances from the camera. The focus in Figure 15. 9(a) is on the clock, while that in Figure 15. 9(a) is on the student. We artificially produce an all in focus reference image showed as Figure 15. 9(c).

It can be seen from Figure 15. 9 that Figure 15. 9(c) is a combination of the good-focus clock Figure 15. 9(a) and the good focus student Figure 15. 9(b). The left part of Figure 15. 9(a) is clearer than Figure 15. 9(b) in visual ect. Here, we take the clarity of character "8" in the clock as the symbol of the whole region it belongs to. Take its edge into account, the intensity difference between the object and its background, like character "8" and clock dial, is more obvious in Figure 15. 9(a) than in Figure 15. 9(b).

(a)   (b)   (c)

Figure 15. 9   The muti-focus source images and the reference one

2. Different sensor image fusion applications

Concealed weapon detection is an increasingly important topic in the general area of law enforcement, and image fusion has been identi? eds a key technology to enable progress on this topic. Figure 15. 10 shows a pair of visual and 94GHz millimeter-wave (MMW) images. The visual image Figure 15. 10(a) provides the outline and the appearance of the people while the MMW image Figure 15. 10(b) shows the existence of a gun. From the fused image Figure

15.10(c), there is considerable evidence to suspect that the person on the right has a concealed gun beneath his clothes. This fused image may be very helpful to police officer, for example, who must respond promptly.

(a)　　　　　　　　　　(b)　　　　　　　　　　(c)

Figure 15.10　The different sensor images and fused one

Lesson 16 Wireless Power Transfer via Coupled Magnetic Resonances

In the early 20th century, before the electrical-wire grid, Nikola Tesla devoted much effort toward schemes to transport power wirelessly. [1] However, typical embodiments (e. g., Tesla coils) involved undesirably large electric fields. The past decade has witnessed a surge in the use of autonomous electronic devices (laptops, cell phones, robots, PDAs, etc.). As a consequence, [2] interest in wireless power has reemerged. Radiative transfer, although perfectly suitable for transferring information, poses a number of difficulties for power transfer applications: The efficiency of power transfer is very low if the radiation is omnidirectional [3], and unidirectional [4] radiation requires an uninterrupted line of sight and sophisticated tracking mechanisms. A recent theoretical paper presented a detailed analysis of the feasibility of using resonant objects coupled through the tails of their nonradiative fields for midrange energy transfer. Intuitively, [5] two resonant objects of the same resonant frequency tend to exchange energy efficiently, while dissipating relatively little energy in extraneous offresonant objects. In systems of coupled resonances (e. g., acoustic, electromagnetic, magnetic, nuclear), there is often a general 'strongly coupled' regime of operation. If one can operate in that regime in a given system, the energy transfer is expected to be very efficient. Midrange power transfer implemented in this way can be nearly omnidirectional and efficient, irrespective of the geometry of the surrounding space, with low interference and losses into environmental objects.

The above considerations apply irrespective of the physical nature of the resonances. [6] Here, we focus on one particular physical embodiment: magnetic resonances. Magnetic resonances are particularly suitable for everyday applications because most of the common materials do not interact with magnetic fields, so interactions with environmental objects are suppressed even further. We were able to identify the strongly coupled regime in the system of two coupled magnetic resonances by exploring nonradiative (near-field) magnetic resonant induction at megahertz frequencies. At first glance, such power transfer is reminiscent of the usual magnetic induction [7]; however, note that the usual nonresonant induction is very inefficient for midrange applications.

16.1 Overview of theFormalism

Efficient midrange power transfer occurs in particular regions of the parameter space describing resonant objects strongly coupled to one another. Using coupled-mode theory to describe this physical system, we obtain the following set of linear equations:

$$\dot{a}_m(t) = (i\omega_m - \Gamma_m)a_m(t) + \sum_{n \neq m} ik_{nm}a_n(t) + F_m(t) \qquad (16\text{-}1)$$

Where the indices denote the different resonant objects. The variables $a_m(t)$ are defined so that the energy contained in object m is $|a_m(t)|^2$, ω_m is the resonant angular frequency of that isolated object, and Γ_m is its intrinsic decay rate (e. g., due to absorption and radiated losses). In this framework, an uncoupled and undriven oscillator with parameters ω_0 and Γ_0 would evolve in time as $\exp(i\omega_0 t - \Gamma_0 t)$. The $k_{mn} = k_{nm}$ are coupling coefficients between the resonant objects indicated by the subscripts, and $F_m(t)$ are driving terms. We limit the treatment to the case of two objects, denoted by source and device, such that the source (identified by the subscript S) is driven externally at a constant frequency, and the two objects have a coupling coefficient k. Work is extracted from the device (subscript D) by means of a load (subscript W) that acts as a circuit resistance connected to the device, and has the effect of contributing an additional term Γ_W to the unloaded device object's decay rate Γ_D. The overall decay rate at the device is therefore $\Gamma'_d = \Gamma_D + \Gamma_W$ The work extracted is determined by the power dissipated in the load, that is, $2\Gamma_W |a_D(t)|^2$. Maximizing the efficiency h of the transfer with respect to the loading Γ_W, given equation (16-1), is equivalent to solving an impedancematching problem. One finds that the scheme works best when the source and the device are resonant, in which case the efficiency is

$$\eta = \frac{\Gamma_W |a_D|^2}{\Gamma_S |a_S|^2 + (\Gamma_S + \Gamma_W) |a_D|^2} = \frac{\frac{\Gamma_W}{\Gamma_D} \frac{k^2}{\Gamma_S \Gamma_D}}{\left[\left(1 + \frac{\Gamma_W}{\Gamma_D}\right) \frac{k^2}{\Gamma_S \Gamma_D}\right] + \left[\left(1 + \frac{\Gamma_W}{\Gamma_D}\right)^2\right]} \quad (16\text{-}2)$$

The efficiency is maximized when $\Gamma_W / \Gamma_D = [1 + (k^2/\Gamma_S \Gamma_D)]^{1/2}$. It is easy to show that the key to efficient energy transfer is to have $k^2/\Gamma_s \Gamma_D > 1$. This is commonly referred to as the strong coupling regime. Resonance plays an essential role in this power transfer mechanism, as the efficiency is improved by approximately ω^2/Γ_{D2} ($\sim 10^6$ for typical parameters) relative to the case of inductively coupled nonresonant objects.

16.2 Theoretical Model for Self-resonant Coils

Our experimental realization of the scheme consists of two self-resonant coils. One coil (the source coil) is coupled inductively to an oscillating circuit; the other (the device coil) is coupled inductively to a resistive load (Figure 16.1). Selfresonant coils rely on the interplay between distributed inductance and distributed capacitance to achieve resonance. The coils are made of an electrically conducting wire of total length l and cross-sectional radius a wound into a helix of n turns, radius r, and height h. To the best of our knowledge, there is no exact solution for a finite helix in the literature, and even in the case of infinitely long coils, the solutions rely on assumptions that are inadequate for our system.[8] We have found, however, that the simple quasi-static model described below is in good agreement (within $\sim 5\%$) with experiment.

A is a single copper loop of radius 25 cm that is part of the driving circuit, which outputs a sine wave with frequency 9.9 MHz. S and D are respectively the source and device coils referred to in the text. B is a loop of wire attached to the load (light bulb). The various k_s

represent direct couplings between the objects indicated by the arrows. The angle between coil D and the loop A is adjusted to ensure that their direct coupling is zero. Coils S and D are aligned coaxially. The direct couplings between B and A and between B and S are negligible.

Figure 16.1 Schematic of the experimental setup

We start by observing that the current must be zero at the ends of the coil, and we make the educated guess that the resonant modes of the coil are well approximated by sinusoidal current profiles along the length of the conducting wire. We are interested in the lowest mode, so if we denote by s the parameterization coordinate along the length of the conductor, such that it runs from $-l/2$ to $+l/2$, then the time-dependent current profile has the form $\lambda_0 \cos(\pi s/l) \exp(i\omega t)$. It follows from the continuity equation for charge that the linear charge density profile is of the form $\lambda_0 \sin(\pi s/l) \exp(i\omega t)$, so that one-half of the coil (when sliced perpendicularly to its axis) contains an oscillating total charge (of amplitude $q_0 = \lambda_0 l/\pi$) that is equal in magnitude but opposite in sign to the charge in the other half.

As the coil is resonant, the current and charge density profiles are $\pi/2$ out of phase from each other, meaning that the real part of one is maximum when the real part of the other is zero. Equivalently, the energy contained in the coil is at certain points in time completely due to the current, and at other points it is completely due to the charge. Using electromagnetic theory, we can define an effective inductance L and an effective capacitance C for each coil as follows:

$$L = \frac{\mu_0}{4\pi |I_0|^2} \iint dr dr' \frac{J(r)J(r')}{|r-r'|} \tag{16-3}$$

$$\frac{1}{C} = \frac{1}{4\pi\varepsilon_0 |q_0|^2} \iint dr dr' \frac{\rho(r)\rho(r')}{|r-r'|} \tag{16-4}$$

Where the spatial current $J(r)$ and charge density $\rho(r)$ are obtained respectively from the current and charge densities along the isolated coil, in conjunction with the geometry of the object. As defined, L and C have the property that the energy U contained in the coil is given by

$$U = \frac{1}{2}L |I_0|^2 = \frac{1}{2C}|q_0|^2 \tag{16-5}$$

Given this relation and the equation of continuity, the resulting resonant frequency is $f_0 = 1/[2\pi(LC)^{1/2}]$. We can now treat this coil as a standard oscillator in coupled-mode theory by defining $a(t) = [(L/2)^{1/2}]I_0(t)$.

We can estimate the power dissipated by noting that the sinusoidal profile of the current distribution implies that the spatial average of the peak current squared is $|I_0|^2/2$. For a coil with n turns and made of a material with conductivity s, we modify the standard formulas for ohmic (R_o) and radiation (R_r) resistance accordingly:

$$R_o = \sqrt{\frac{\mu_0 \omega}{2\sigma}} \frac{l}{4\pi a} \tag{16-6}$$

$$R_r = \sqrt{\frac{\mu_0}{\varepsilon_0}} \left[\frac{\pi}{12} n^2 \left(\frac{\omega r}{c}\right)^2 + \frac{2}{3\pi^3} \left(\frac{\omega h}{c}\right)^2 \right] \tag{16-7}$$

The first term inequation (16-7) is a magnetic dipole radiation term (assuming $r \ll 2\pi c/\omega$, where c is the speed of light); the second term is due to the electric dipole of the coil and is smaller than the first term for our experimental parameters. The coupled-mode theory decay constant for the coil is therefore $\Gamma = (R_o + R_r)/2L$, and its quality factor is $Q = \omega/2\Gamma$.

We find the coupling coefficient k_{DS} by looking at the power transferred from the source to the device coil, assuming a steady-state solution in which currents and charge densities vary in time as $\exp(i\omega t)$:

$$P_{DS} = \int dr E_S(r) J_D(r) = -\int dr [\dot{A}_S(r) + \nabla \phi_S(r)] J_D(r)$$

$$= -\frac{1}{4\pi} \int dr dr' \left[\mu_0 \frac{J_S(r')}{|r'-r|} + \frac{\rho_S(r')}{\varepsilon_0} \frac{r'-r}{|r'-r|^3} \right] J_D(r') \equiv -i\omega M I_S I_D \tag{16-8}$$

WhereM is the effective mutual inductance, ϕ is the scalar potential, $\mathbf{A}$ is the vector potential, and the subscript S indicates that the electric field is due to the source. We then conclude from standard coupled-mode theory arguments that $k_{DS} = k_{SD} = k = \omega M/[2(L_S L_D)^{1/2}]$. When the distance D between the centers of the coils is much larger than their characteristic size, k scales with the D^{-3} dependence characteristic of dipole-dipole coupling. Both k and Γ are functions of the frequency, and k/Γ and the efficiency are maximized for a particular value of f, which is in the range 1 to 50 MHz for typical parameters of interest. Thus, picking an appropriate frequency for a given coil size, as we do in this experimental demonstration, plays a major role in optimizing the power transfer.

16.3 Comparison withExperimentally Determined Parameters

The parameters for the two identical helical coils built for the experimental validation of the power transfer scheme are $h = 20$ cm, $a = 3$ mm, $r = 30$ cm, and $n = 5.25$. Both coils are made of copper. The spacing between loops of the helix is not uniform, and we encapsulate the uncertainty about their uniformity by attributing a 10% (2 cm) uncertainty to h. The expected resonant frequency given these dimensions is $f_0 = (10.56 \pm 0.3)$ MHz, which is about 5% off from the measured resonance at 9.90 MHz.

The theoretical$Q$ for the loops is estimated to be ~ 2500 (assuming $\sigma = 5.9 \times 10^7$ m/ohm), but the measured value is $Q = 950 \pm 50$. We believe the discrepancy is mostly due to the effect of the layer of poorly conducting copper oxide on the surface of the copper wire, to which the current is confined by the short skin depth (~ 20 mm) at this frequency. We therefore use the experimentally observed Q and $\Gamma_S = \Gamma_D = \Gamma = \omega/2Q$ derived from it in all subsequent computations.

We find the coupling coefficient k experimentally by placing the two self-resonant coils (fine-tuned, by slightly adjusting h, to the same resonant frequency when isolated) a distance D apart and measuring the splitting in the frequencies of the two resonant modes. According to coupled-mode theory, this splitting should be $\Delta\omega = 2[(k^2-\Gamma^2)^{1/2}]$. In the present work, we focus on the case where the two coils are aligned coaxially, although similar results are obtained for other orientations.

16.4 Measurement of the Efficiency

The maximum theoretical efficiency depends only on the parameter $k/[(L_S L_D)^{1/2}] = k/\Gamma$, which is greater than 1 even for $D = 2.4$ m (8 times the radius of the coils). Thus, we operate in the strongly coupled regime throughout the entire range of distances probed.

As our driving circuit, we use a standard Colpitts oscillator whose inductive element consists of a single loop of copper wire 25 cm in radius (Figure 16.1); this loop of wire couples inductively to the source coil and drives the entire wireless power transfer apparatus. The load consists of a calibrated light bulb and is attached to its own loop of insulated wire, which is placed in proximity of the device coil and inductively coupled to it. By varying the distance between the light bulb and the device coil, we are able to adjust the parameter Γ_W/Γ so that it matches its optimal value, given theoretically by $[1+(k^2/\Gamma^2)]^{1/2}$. (The loop connected to the light bulb adds a small reactive component to Γ_W, which is compensated for by slightly retuning the coil.) We measure the work extracted by adjusting the power going into the Colpitts oscillator until the light bulb at the load glows at its full nominal brightness.

We determine the efficiency of the transfer taking place between the source coil and the load by measuring the current at the midpoint of each of the self-resonant coils with a current probe (which does not lower the Q of the coils noticeably). This gives a measurement of the current parameters I_S and I_D used in our theoretical model. We then compute the power dissipated in each coil from $P_{S,D} = \Gamma L |I_{S,D}|^2$, and obtain the efficiency from $\eta = P_W/(P_S+P_D+P_W)$. To ensure that the experimental setup is well described by a two-object coupled mode theory model, we position the device coil such that its direct coupling to the copper loop attached to the Colpitts oscillator is zero. We were able to transfer several tens of watts with the use of this setup, fully lighting up a 60W light bulb from distances more than 2m away.

As a cross-check, we also measured the total power going from the wall power outlet into the driving circuit. The efficiency of the wireless transfer itself is hard to estimate in this way, however, as the efficiency of the Colpitts oscillator itself is not precisely known, although it is expected to be far from 100%. Still, the ratio of power extracted to power entering the driving circuit gives a lower bound on the efficiency. When transferring 60W to the load over a distance of 2m, for example, the power flowing into the driving circuit is 400W. This yields an overall wall-to-load efficiency of 15%, which is reasonable given the expected efficiency of 40 to 50% for the wireless power transfer at that distance and the low efficiency of the Colpitts oscillator.

Words and Expressions

wireless power transfer（WPT） 无线电能传输
laptop [ˈlæptɒp] n. 便携式计算机；笔记本计算机
resonant [ˈrezənənt] adj. 共振的；谐振的
perpendicular [ˌpɜːpənˈdɪkjələ(r)] adj. 垂直的；成直角的
validation [ˌvælɪˈdeɪʃən] n. 生效；批准；验证
helical [ˈhelɪkl] adj. 螺旋的；螺旋形的
encapsulate [ɪnˈkæpsjuleɪt] v. 简述；概括；压缩
Colpitts oscillator 考毕兹振荡器
dissipate [ˈdɪsɪpeɪt] v. （使）消散，消失
yield [jiːld] v. 出产（作物）；产生（收益、效益等）
quality factor 品质因数

Notes

1. Before the electrical-wire grid, Nikola Tesla devoted much effort toward schemes to transport power wirelessly.

   在电网出现之前，尼古拉·特斯拉便致力于无线电力传输的构想。

   devote much effort to 为……付出很多努力

   注：Nikola Tesla 是世界上第一个从事无线电能传输技术研究的科学家。他在 1893 年的芝加哥世界博览会上，利用无线电能传输原理，在不用导线的情况下点亮了一盏电灯。

2. I regret to inform you he died as a consequence of his injuries.

   我很遗憾地通知你，他因伤势太重不治身亡。

   As a consequence 因此，……

3. omni－【词头】（形容词）全的

   如 omnidirectional，（接收或发射信号）全向的。

4. uni－【词头】（形容词）单的

   如 unidirectional，单向的。

5. Intuitively, two resonant objects of the same resonant frequency tend to exchange energy efficiently, while dissipating relatively little energy in extraneous offresonant objects.

   从直观上看，同一共振频率的两个共振物体往往能有效地交换能量，而在共振物体的外部耗散的能量相对较少。

   intuitively 直觉地，直观地

6. The above considerations apply irrespective of the physical nature of the resonances.

   无论共振的物理性质如何，上述考虑都适用。

irrespective of 不考虑；不管。同义词有 regardless of

7. The space is reminiscent of bars in Berlin and New York. 这个地方会让人想起柏林和纽约的酒吧。

   be reminiscent of　让人想起……

8. To the best of our knowledge, there is no exact solution for a finite helix in the literature, and even in the case of infinitely long coils, the solutions rely on assumptions that are inadequate for our system.

   to the best of our knowledge　据我们所知

   据我们所知，文献中没有关于有限螺旋线的精确解，即使在无限长线圈的情况下，解也依赖于对于系统不充分的假设。

Exercises

Translate the following passages into English or Chineses.

1. Work is extracted from the device (subscript D) by means of a load (subscript W) that acts as a circuit resistance connected to the device, and has the effect of contributing an additional term Γ_W to the unloaded device object's decay rate Γ_D.
2. When the distance D between the centers of the coils is much larger than their characteristic size, k scales with the D^{-3} dependence characteristic of dipole-dipole coupling.
3. 所有这些让人想起了凯恩斯经济学。
4. 据我们所知，这是世界银行行长第一次在清华大学发表讲话。

Reading Material

Wireless Charging System for Electric Vehicle

It's known that electric vehicle (EV) development is highly affected by the demands of users who pay much attention on safety and convenience of charging. Compared with prevailed plug-in charging systems, wireless power transfer (WPT) system can be safer and more convenient, which could greatly promote popularization of EV. Inductively coupling power transfer technology (ICPT), magnetic resonance coupling (MRC) wireless power transfer technology, microwave wireless power transfer technology and laser energy transfer technology are major means of WPT. In EV charging applications, ICPT and MRC are the prime technologies, and MRC technology is more suitable for EV mid-range wireless charging system.

In 2007, team of Professor Marin Soljacic in MIT put forward the MRC wireless power transfer technology. After that, institutions, universities, and enterprises around the world begun to study the technology. Research results show that MRC canrealize high power (kW level) wireless power transmission to distances several times of its own coil diameter. MRC possesses many favorable qualities, such as small impact on obstacles in transmission and tolerance of coil misalignment, which are applicable for EV charging.

1. Principle of Mid-range Wireless Charging

As illustrated inFigure 16.2, the MRC WPT system consists of four coils, in which S and D are self-resonant coils. To explain the MRC phenomenon, the Marin Soljacic team adopted the coupled-mode theory. It suggests that the resonant coil S and D are strong-coupled and their power transfer relationship could beexpressed as equation (1) and equation (2).

$$\frac{da_D(t)}{dt} = -(j\omega_S + \Gamma_S)a_S(t) + jK_{SD}a_S(t) + F_S(t) \tag{1}$$

$$\frac{da_D(t)}{dt} = -(j\omega_D + \Gamma_S + \Gamma_L)a_D(t) + jK_{SD}a_S(t) \tag{2}$$

Where, ω_S and ω_D are the self-resonant frequency of coil S and coil D; Γ_S and Γ_D is loss rates caused by radiation and internal resistance; $F_S(t)$ is the signal loaded on the coil S; K_{SD} is the coupling coefficient between coil S and coil D. When the system is tuned resonant, $\omega_S = \omega_D = \omega$.

On the other hand, some other researchers utilize the equivalent circuit model including coupled inductor to describe MRC and analyze it with simplified three-coefficient model. However, in actual system, the four coils are usually placed in parallel and the coupling coefficients between coilA and D, coil S and B shall not be ignored. Three coefficients cannot cover all necessary information to adequately describe the system, thus the six-coefficient model shown in Figure 1 should be adopted.

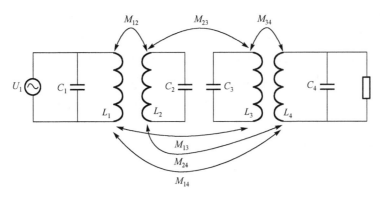

Figure 16.2 WPT subsystem model with four coupling coils

L_1 to L_4 are the equivalent inductances from coil 1 to coil 4. C_1 to C_4 are the compensation capacitors. The existence of C_1, C_4 depend on working condition, they are connected with L_1 and L_4. U_1 is the driving voltage of L_1. R is the load. M_{12}, M_{13}, M_{14}, M_{23}, M_{24}, M_{34} are mutual inductances between coils. Tomanifest the six-coefficient model in matrix, Equation (3) could be used.

$$\begin{bmatrix} U_1 \\ 0 \\ 0 \\ 0 \end{bmatrix} = \begin{bmatrix} Z_1 & j\omega M_{12} & j\omega M_{13} & j\omega M_{14} \\ j\omega M_{12} & Z_2 & j\omega M_{23} & j\omega M_{24} \\ j\omega M_{13} & j\omega M_{23} & Z_3 & j\omega M_{34} \\ j\omega M_{14} & j\omega M_{24} & j\omega M_{34} & Z_4 \end{bmatrix} \begin{bmatrix} I_1 \\ I_2 \\ I_3 \\ I_4 \end{bmatrix} \tag{3}$$

Z_1 to Z_4 are the equivalent self-impedances of coil1 to coil4. I_1 to I_4 are the corresponding currents. ω is the operating angular frequency.

2. Wireless Charging Circuit

The transmitter terminal, WPT subsystem and the receiver terminal constitute the EV charging system, as shown inFigure 16.3. In detail, the transmitter terminal comprises power-frequency rectifier (FB1, C_{p1}), Buck converter (Q_1, L_{p1}, VD_1, C_{p2}), full-bridge inverter and the inverter series LC filter circuit (L_{p3}, C_{p3}); the WPT subsystem includes driving coil (L_1), resonant coils (L_2, L_3), receiving coil (L_4) and compensation capacitors (C_1, C_2, C_3, C_4); the receiver terminal consists of the receiver compensation circuit (C_4, L_{s5}), high-frequency rectifier filter circuit (FB2, L_{s7}, C_{s7}) and a diode (VD_2).

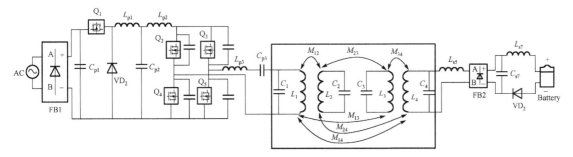

Figure 16.3 Model of charging system

The system uses the BUCK converter and the full-bridge inverter to maintain the stability of output current by adjusting the output voltage of BUCK circuit and the output current of the full-bridge inverter when the output power change. The LC (L_{p3}, C_{p3}) filter circuit lowers the current harmonics, eliminating the influences on the WPT subsystem when transmitted power changes. In addition, C_4 and L_{s5} form the receiving compensation circuit and could further reduce current harmonics.

In the simulation model, operating frequency of the charging system is set to 150 kHz and parameters of the WPT subsystem are the measured from wire wound coils, shown in Table 16.1.

Table 16.1 Parameters of the WPT subsystem

(Unit: H)

L_1	L_2	L_3	L_4	M_{12}	M_{13}	M_{14}	M_{23}	M_{24}	M_{34}
16.35	547.14	55.92	7.78	25.16	2.88	0.93	6.01	1.96	16.15

Set the simulation model operating frequency scanning from 100 kHz to 170 kHz and the curve showing the power-efficiency-frequency relationship could be obtained. Figure 16.4 provides that the efficiency-frequency curve is in the critical coupled single-peak condition while the power-frequency presents double-peak characteristic. In actual systems, the higher operating frequency point is selected, since the output power would plunge when frequency shift occurs with the lower frequency point.

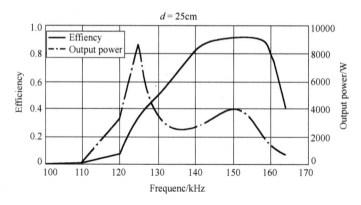

Figure 16.4 Frequency characteristic curve of EV charging system

The discrete incremental PID control method is applied in the 8A constant current output simulation. As illustrated in Figure 16.5, the output current is sampled to compare with reference current, and the error determines the operating frequency shift with PID control method.

Both the resistant load and battery load simulation are conducted and the results are provided in Figure 16.6. When the load is resistor, the overshoots of system output is 3%, the response time is less than 1ms and the accuracy of system output is higher than 2%. However, when the load is battery, the overshoots of system output rise up to 6% and the accuracy of system output is 3%. It could be summarized that wireless charging system has short response time, high precision and could meet the demand of the EV battery charging.

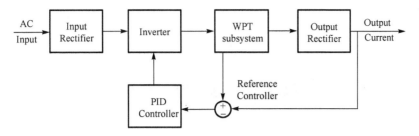

Figure 16.5 Block diagram of the constant current output simulation

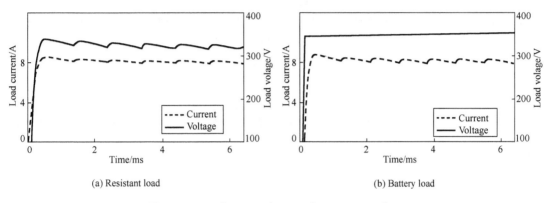

(a) Resistant load

(b) Battery load

Figure 16.6 Output voltage and current waveform

参考译文

第 1 课 周 期 信 号

1.1 时域描述

许多常被看成信号的函数都是时间的函数,这可以用时频信号处理理论来处理。一个周期信号可以看成每隔 T 秒重复其本身的信号,其中 T 称为信号波形的周期。周期波形的理论假定这种精确的重复延伸到整个时间轴上(不管是过去还是将来)。实际上,信号不能无限地重复它本身。不过,像电源整流器输出电压这样的波形,在平滑之前,还是重复本身很多次的,将其作为严格的周期信号进行分析,会产生颇有价值的结果。另一方面,像心电图这样的波形是准周期的,而且可以因为某种需要,把它当成真正的周期信号来处理。在通信信道中,一个真正的重复信号是没有什么用的,因为接收到第一个周期的波形后,就不会有更多的信息传送过来了。讨论周期信号的主要原因之一是:它对于深入理解处理周期信号和随机信号这种分析方法会有很大帮助。

周期信号的完整的时域描述将详细指明其在每个瞬间的精确值。在一些例子中,使用数字表达很容易确定其精确值。幸运的是,在许多例子中,描述信号波形时,有用的只是某些特定方面,或者只用近似的数学公式来表达它。在特定情况下,所涉及的物理量如下。

(1)信号平均值。
(2)信号达到的峰值。
(3)信号在 a、b 值时的时间比例。
(4)信号周期。

如果希望通过数学表达式得到近似的波形,可以使用多项式、泰勒级数及傅里叶级数。一个 n 次多项式为

$$f(t) = a_0 + a_1 t + a_2 t^2 + a_3 t^3 + \cdots + a_n t^n \tag{1-1}$$

可以用来拟合 $n+1$ 阶实际曲线。随着多项式幂的加大,拟合精度也逐渐提高。应注意到,在拟合点范围之外,正确的信号波形与多项式之间的误差一般很大,而且多项式本身也不是周期的。尽管一个多项式逼近可以以一定数量的点拟合实际波形,但泰勒级数逼近可以针对一个固定点提供一条光滑的连续曲线。选择泰勒级数的系数,可以使得级数和它的派生项在某点更吻合实际波形。泰勒级数中,幂的值决定了扩展的派生项的指数,以及在选定点区域内,级数和实际波形吻合的精确度。在某点的区域内,泰勒级数近似函数的一般形式为

$$f(t) = f(a) + (t-a) \cdot \frac{\mathrm{d}f(a)}{\mathrm{d}t} + \frac{(t-a)^2}{2!} \cdot \frac{\mathrm{d}^2 f(a)}{\mathrm{d}t^2} + \cdots + \frac{(t-a)^n}{n!} \cdot \frac{\mathrm{d}^n f(a)}{\mathrm{d}t^n} \tag{1-2}$$

一般来说,在选定点的区间内,级数与实际波形很吻合,但是在区间外,这种情况就会迅速恶化。因此,在波形的一个限定区间,可以用多项式和泰勒级数描述信号波形以期达到很高精度。在所选择区间外,精度通常会迅速降低,尽管可以通过补充一些项,使之有所改善(只要 t 位于序列的收敛域内)。这种方法提供的近似式在形式上从来不是周期变化的,因此也不能认为是描述周期信号的理想形式。

相比较而言,在一段较长时间内,通过傅里叶级数逼近可更好地表达信号波形。当信号是周期性的时候,傅里叶级数描述的精度可以始终得以保证,因为这个信号可以表示成一系列的正弦函数

之和，而正弦函数本身是周期性的。在详细分析表示信号的傅里叶级数方法之前，先介绍一下它的背景知识——频域描述方法。

1.2 频域描述

频域分析的基本概念是：任何复杂波形都可以看成许多具有适当振幅、周期和相对相位的正弦波之和。一个连续的正弦函数($\sin\omega t$)在频率(rad/s)上被认为是单频波。信号的频域描述涉及这样一些基本函数间的断点问题，这就是傅里叶级数分析方法。

有许多理由可以说明为什么在信号分析时信号成分中正弦波起到如此重要的作用。在一段较长的时间内，用一簇周期函数近似描述一个信号波形已经得到证明。后面将证明使用这种方法可以使得真实信号与它的近似波形间的误差降到最低。正弦函数在信号分析中如此重要的另一个原因是它们在物理上被广泛采用，并且易于进行数字处理。一个庞大的而且极其重要的机电系统，又称线性系统，加上任何频率的正弦干扰后都会得到正弦响应。通过物理上正弦函数的应用表明，用正弦函数进行分析，将简化信号与潜在的物理因素间的关系或信号与其系统或设备的物理属性间的关系等问题。最后，正弦函数可以形成一簇函数，这些函数称为"正交函数"。它的十分特殊的属性和优点将在下面进行介绍。

1.3 正交函数

1.3.1 矢量和信号

通过分析信号和矢量之间的相似之处，引入用来描述信号的正交函数概念。矢量用大小和方向来描述，如力和速度。假如有两个矢量 $\boldsymbol{V}_1$ 和 $\boldsymbol{V}_2$，几何学上通过在 $\boldsymbol{V}_1$ 末端到 $\boldsymbol{V}_2$ 上构造直角，我们沿着矢量 $\boldsymbol{V}_2$ 来定义矢量 $\boldsymbol{V}_1$，有

$$\boldsymbol{V}_1 = C_{12}\boldsymbol{V}_2 + \boldsymbol{V}_e \tag{1-3}$$

其中，矢量 $\boldsymbol{V}_e$ 是近似误差。很明显，当这个误差矢量沿着 $\boldsymbol{V}_2$ 方向被拉为直角时，它是最小的长度。因此，我们说沿着矢量 $\boldsymbol{V}_2$，$\boldsymbol{V}_1$ 是通过 $C_{12}\boldsymbol{V}_2$ 给定的，这里 C_{12} 的选择原则是使误差矢量尽可能小。正交矢量系中熟知的例子就是在坐标几何学里三个相互垂直的坐标轴的使用。

在这里，矢量分析的基本思想可以推广到信号分析里去。假定，我们希望在一定的区间($t_1 < t < t_2$)内，通过另一个信号或函数 $f_2(t)$ 去接近信号 $f_1(t)$，即

$$f_1(t) \approx C_{12} f_2(t) \qquad \text{其中} \quad t_1 < t < t_2$$

通过选择 C_{12} 可获得最佳逼近函数。若定义误差函数为

$$f_e(t) = f_1(t) - C_{12} f_2(t) \tag{1-4}$$

很明显，选择 C_{12} 的目的是在选定区间内使 $f_e(t)$ 的平均值达到最小。这种误差标准的缺点是在不同时刻出现的正负误差趋近于互相抵消。如果我们选择使均方差最小，而不是误差本身(相当于使平均误差的平方根最小化，或均方根误差)最小，这个缺点可以避免。用 ε 表示 $f_e^2(t)$ 的平均值，可以得到

$$\varepsilon = \frac{1}{(t_2 - t_1)} \int_{t_1}^{t_2} f_e^2(t) \mathrm{d}t = \frac{1}{(t_2 - t_1)} \int_{t_1}^{t_2} [f_1(t) - C_{12} f_2(t)]^2 \mathrm{d}t \tag{1-5}$$

对 C_{12} 求微分，然后令所得表达式为 0，就可以得到使 ε 最小的 C_{12} 的值，即

$$\frac{\mathrm{d}}{\mathrm{d}C_{12}} \left\{ \frac{1}{(t_2 - t_1)} \int_{t_1}^{t_2} [f_1(t) - C_{12} f_2(t)]^2 \mathrm{d}t \right\} = 0$$

去掉括号，交换积分和微分的次序，得出

$$C_{12} = \int_{t_1}^{t_2} f_1(t) f_2(t) \mathrm{d}t \bigg/ \int_{t_1}^{t_2} f_2^2(t) \mathrm{d}t \tag{1-6}$$

1.3.2 利用正交函数集描述信号

假定我们在一定区间上通过函数 $f_2(t)$ 已经得到一个近似的信号 $f_1(t)$,并且均方差达到最小,但是现在我们希望改善波形的近似性。这需要证明,根据一组互相正交函数 $f_2(t),f_3(t),f_4(t)$ 等来描述信号将会达到一个非常满意的效果。假设初始近似式是

$$f_1(t) \approx C_{12}f_2(t) \tag{1-7}$$

通过插值,使误差进一步减小,即

$$f_1(t) \approx C_{12}f_2(t) + C_{13}f_3(t) \tag{1-8}$$

其中,$f_2(t)$ 和 $f_3(t)$ 在有效区间内是正交的。由于并入了附加项 $C_{13}f_3(t)$,均方差进一步降低了。误差函数为

$$f_e(t) = f_1(t) - C_{12}f_2(t) - C_{13}f_3(t) \tag{1-9}$$

在区间 $t_1 < t < t_2$ 内,均方差为

$$\varepsilon = \frac{1}{t_2 - t_1} \cdot \int_{t_1}^{t_2} [f_1(t) - C_{12}(t)f_2(t) - C_{13}(t)f_3(t)]^2 \mathrm{d}t \tag{1-10}$$

为了求出使均方误差仍保持最小的 C_{12} 值,先对 C_{12} 求偏微分,再交换微分与积分次序,我们再次得到

$$C_{12} = \int_{t_1}^{t_2} f_1(t)f_2(t)\mathrm{d}t \Big/ \int_{t_1}^{t_2} f_2^2(t)\mathrm{d}t \tag{1-11}$$

换言之,如果 $f_2(t)$ 与 $f_3(t)$ 在所选择的时间区间内正交,在并入用 $f_3(t)$ 表示的附加项以改善逼近程度时,系数 C_{12} 无须修正。同理,如果信号只是通过 $f_3(t)$ 逼近,那么 C_{13} 值也不会改变。

这个重要结论可以推广到包括用整个正交函数集表示信号的情况。在进行逼近时,任何系数值与集合中用了多少函数没有关系,因此,即使包含更多的项,这些系数也不会改变。利用一个正交函数集描述信号,类似于在三维空间中利用三个互相垂直的轴描述矢量,这就引出了"信号空间"的概念。精确的信号表达往往需要多于三个正交函数,因此我们必须把一些区间 ($t_1 < t < t_2$) 上的信号在多维空间上用一个点来表示。

总之,有许多正交函数集可用来近似描述信号波形,如所谓的勒让德多项式和沃尔什函数等,正弦函数集是其中最常用的。在 t 时刻上,多项式函数集不是周期的,但是它可用来描述周期波形的一个循环;在选定区间外,真实信号和近似式之间的误差会理所当然地迅速增加。然而,用正弦函数描述一个周期信号的一个循环在任何时候都是等效的,因为这些函数都是正交的。

1.4 傅里叶级数

傅里叶级数的基本理论是把复杂的周期波分解成许多简谐的正弦波,这些正弦波构成一个正交函数集。如果有一个周期为 T 的周期信号 $f(t)$,它可用级数表示为

$$f(t) = A_0 + \sum_{n=1}^{\infty} A_n \cos n\omega_1 t + \sum_{n=1}^{\infty} B_n \sin n\omega_1 t \tag{1-12}$$

式中,$\omega_1 = 2\pi/T$。这样,$f(t)$ 可以认为是由稳定值 A_0 及许多不同频率的正弦波或余弦波叠加而成的。最小频率 ω_1(rad/s)称为基频,这种频率的波与信号的周期相等。频率为 $2\omega_1$ 的波称为二次谐波,频率为 $3\omega_1$ 的波称为三次谐波,依次类推。必须给 $f(t)$ 加上一定的限制,如狄利克雷条件,才能使用傅里叶级数。在整个周期上的积分 $\int |f(t)| \mathrm{d}t$ 必须有上下限,并在限定的区间上不可有太多的断点。幸运的是,实际中的信号波形都能满足这些条件。

1.4.1 系数的计算

问题已转化为计算系数 A_0、A_n 和 B_n，使用前面所述的最小方差准则，为描述方便起见，有

$$\left.\begin{aligned} A_0 &= \frac{1}{2\pi}\int_{-\pi}^{\pi} f(x)\mathrm{d}x \\ A_n &= \frac{1}{\pi}\int_{-\pi}^{\pi} f(x)\cos nx\,\mathrm{d}x \\ B_n &= \frac{1}{\pi}\int_{-\pi}^{\pi} f(x)\sin nx\,\mathrm{d}x \end{aligned}\right\} \quad (1\text{-}13)$$

尽管在大多数例子中，关于原点对称的区间积分起来很方便，但还是应该选择长度等于信号波形周期的区间。

许多实际波形要么是时间的偶函数，要么是奇函数。如果 $f(t)$ 是偶函数，根据定义有 $f(t)=f(-t)$；如果 $f(t)$ 是奇函数，则有 $f(-t)=-f(t)$。如果偶函数乘以奇函数 $\sin n\omega_1 t$，结果也是奇函数。这样每个 B_n 的被积函数都是奇函数。当一个奇函数在关于 $t=0$ 对称的区间上积分时，结果为零。因此，所有的系数 B 都是零，级数中只剩下余弦项了。同理，如果 $f(t)$ 是奇函数，那么系数 A 必为零，级数中只有正弦项。直观上很明显，偶函数只能由其他的偶函数构成，反之，奇函数也一样。

我们已经看到，在偶函数或奇函数的例子中，通过消去它的正弦项或余弦项，傅里叶级数得到简化。在具有"半波对称"的波形例子中，出现的简化形式也不同。在数学上，"半波对称"存在的条件是

$$f(t) = -f(t+T/2) \quad (1\text{-}14)$$

换句话说，在波形上任何相差 $T/2$ 的两个值数量相等，符号相反。一般来说，只有奇的简谐波具有"半波对称"，因此具有这种对称的任何复杂波形都不包含偶的简谐波成分；相反，包含有二次、四次或其他次谐波的波形都不能表现出"半波对称"的特点。

通常，我们总是在整个周期上积分得出系数值，然而在奇函数或偶函数的例子中，比较简单的是，只要在半个周期上积分，然后乘以 2 就可以得出系数值了。另外，如果波形既是偶函数或奇函数，又表现出"半波对称"，那么只要在 1/4 个周期上积分，然后乘以 4 就可以得出系数值了。在具有上述函数的例子中，这些比较近似算法是合适的，因为在这种情况下，积分是重复的，而在一个周期内还有"半波对称"时，积分要重复两次。

1.4.2 时间原点的选择及波形功率

对于一个具体的波形而言，如果它是偶函数或奇函数，那么计算该波形傅里叶级数系数时，可以通过适当的选择时间原点来减小其计算工作量。这种转变只是把只含有正弦项的傅里叶级数转化为只包含余弦项的傅里叶级数，而在任一频段的振幅正如我们所料，是不会改变的。对于既不是偶函数又不是奇函数的复杂波形来说，在其傅里叶级数里一定包含正弦项和余弦项。

随着波形时间原点的改变，傅里叶级数的正弦和余弦系数也将改变，但是任何两个系数 A_n 和 B_n 的平方和仍然是常量，也就是电子工程师非常熟悉的波形平均功率是不变的。

由上述思想很自然地得出两种三角形式的傅里叶级数。若有两个基波为

$$A_1 \cos \omega_1 t \quad \text{和} \quad B_1 \sin \omega_1 t$$

使用三角恒等式，傅里叶级数有两种形式来表示，即

$$A_1 \cos \omega_1 t + B_1 \sin \omega_1 t = \sqrt{(A_1^2 + B_1^2)} \cos\left(\omega_1 t - \arctan \frac{B_1}{A_1}\right)$$

$$= \sqrt{(A_1^2 + B_1^2)} \sin\left(\omega_1 t + \arctan\frac{B_1}{A_1}\right) \tag{1-15}$$

这样在一个特殊的频段上，正弦和余弦部分可表示成单一的余弦或正弦波形式，只是相位有所改变。如果这种过程应用到所有傅里叶级数的谐波中去，就得到以下两种形式

$$f(t) = A_0 + \sum_{n=1}^{\infty} C_n \cos(n\omega_1 t - \phi_n) \quad \text{或} \quad f(t) = A_0 + \sum_{n=1}^{\infty} C_n \sin(n\omega_1 t + \theta_n) \tag{1-16}$$

其中

$$C_n = \sqrt{A_n^2 + B_n^2}, \quad \phi_n = \arctan(B_n/A_n), \quad \theta_n = \arctan(A_n/B_n) \tag{1-17}$$

最后，要注明的是用任何波形表达的正弦平均功率是

$$(A_n^2 + B_n^2)/2 = C_n^2/2 \tag{1-18}$$

A_0 项的功率简单地表示成 A_0^2，总的波形平均功率为

$$P = A_0^2 + \frac{1}{2}\sum_{n=1}^{\infty} C_n^2 \tag{1-19}$$

但是，功率 P 可表示为 $[f(t)]^2$ 在一个周期上的平均值。再次使用 $[f(t)]^2$，是因为它常被认为一个电压波通过 1Ω 电阻所产生的功率。因此有

$$P = A_0^2 + \frac{1}{2}\sum_{n=1}^{\infty} C_n^2 = \frac{1}{T}\int_{-T/2}^{T/2} [f(t)]^2 \, dt \tag{1-20}$$

这个结果是一个常用形式，称为帕塞瓦尔定理。它表明了整个波形功率是傅里叶级数中部分功率之和。然而重要的是，由于不同部分的波都是从一个正交函数集中推导出来的，所以它是唯一正确的。

第 2 课　非周期信号

2.1　概述

在第 1 课中，我们已经看到周期信号是如何被表示成一组正弦谐波的叠加。这样一个由许多离散的频率所组成的信号的频谱称为一条"线"频谱。虽然周期信号分析可能有很大的实用性，但是大多数信号不是这个类型。首先，自身多次重复的信号总是处于"开"和"关"的状态。换句话说，它们不可能经常被假定为全部时间（过去、现在和将来）总存在，重要的是要了解它们在有限时间上的频谱。其次，除了时间限制的问题外，有一类重要的信号波形（其中包括随机信号）是在自然中简单而不重复的，因此其不能够被包含许多的谐波频率的傅里叶级数所表示。然而，幸运的是，我们可以将傅里叶级数作为研究这类信号频谱的出发点。

2.2　傅里叶级数的指数形式

我们已经知道（1.4.2节），一个周期信号的傅里叶级数有两种表达方式：或者表达为适当振幅和频率的一组正弦波和余弦波，或者表达为由振幅和相关相位角所定义的一组正弦曲线形式的波。现在将要讨论的是傅里叶级数的第三种形式，即指数形式，因为它对推导非周期信号的频谱是很有帮助的。

简单三角函数表达式为

$$f(t) = A_0 + \sum_{n=1}^{\infty} A_n \cos n\omega_1 t + \sum_{n=1}^{\infty} B_n \sin n\omega_1 t \tag{2-1}$$

而指数形式的表达式可以写成

$$f = \cdots + a_{-2}\exp(-j2\omega_1 t) + a_{-1}\exp(-j\omega_1 t) + a_0 + a_1\exp(j\omega_1 t) + a_2\exp(j2\omega_1 t) + \cdots$$

$$= \sum_{m=-\infty}^{\infty} a_m \exp(jm\omega_1 t) \tag{2-2}$$

其中，m 可以取任何的整数。虽然这两种表达式看起来不同，但是事实上如果使用下面的恒等式

$$\cos x = (e^{jx} + e^{-jx})/2 \tag{2-3}$$

和

$$\sin x = -j(e^{jx} + e^{-jx})/2 \tag{2-4}$$

它们其实是一样的。将式(2-3)和式(2-4)代入式(2-1)并移项整理可以很容易地得到指数形式的表达式。两种形式的系数关系如下：

$$a_0 = A_0; \ a_m = (A_m - jB_m)/2,\text{当}\ m\ \text{为正整数时} \tag{2-5}$$

$$a_m = (A_m + jB_m)/2,\text{当}\ m\ \text{为负整数时} \tag{2-6}$$

这些结果表明指数形式的系数一般是比较复杂的，而且它们是共轭的(也就是说，系数 a_n 的虚部和系数 a_{-n} 的虚部数值相等、符号相反)。虽然，复杂系数的概念乍一看是难于理解的，但是，应该记住系数的实部表示余弦波相关频率的大小，而虚部表示正弦波幅值的大小。特别地，如果系数 a_n 和 a_{-n} 为实数，那么频率 $n\omega_1$ 仅由一个余弦项组成；如果系数 a_n 和 a_{-n} 为只有虚部，则频率 $n\omega_1$ 仅由一个正弦项组成；如果系数 a_n 和 a_{-n} 是一般的情形，即系数 a_n 和 a_{-n} 是复数，则频率 $n\omega_1$ 同时存在余弦项和正弦项。

2.3 傅里叶变换

2.3.1 简介

傅里叶变换(又称傅里叶积分)为非重复的信号波形分析提供了数学工具，而傅里叶级数为重复的信号分析提供了数学工具。假定信号波形周期(即它的基频)不变，当脉冲宽度减小时，我们可以看到周期脉冲波形线谱的变化。假如脉冲的持续时间保持固定，但脉冲之间的间隔逐渐增大，这样就会引起周期逐渐增大。在这样的条件下，我们将会得到单一矩形脉冲，它旁边的脉冲已经移动到前面或后面的无限远处。在这种情况下，基频趋于零，它的谐波将近占满整个空间，而振幅极小。进一步，我们将得到一个连续的频谱。

在数学上，这种情形可以用修正的傅里叶级数的指数形式表达为

$$f(t) = \sum_{m=-\infty}^{\infty} a_m \exp(-jm\omega_1 t) \tag{2-7}$$

其中，系数 a_m 等于 f(t) 与 $exp(jm\omega_1 t)$ 的乘积的定积分的平均值，即

$$a_m = \frac{1}{2\pi}\int_{-\pi}^{\pi} f(x)\exp(-jmx)dx$$

$$= \frac{1}{T}\int_{-T/2}^{T/2} f(t)\exp(-jm\omega_1 t)dt \tag{2-8}$$

在趋于无限大的情形下，每个单独的系数都变成无穷小，则上述公式似乎就不再有用。然而 $a_m \cdot T$ 的乘积不是当 $T \to \infty$ 时就消失了，因此我们现在来引出一个新的变量 G。当 $T \to \infty$ 且 $\omega_1 \to 0$ 时，G 趋

于表示为一个 ω 函数的连续变量。由于变量 G 是连续的频率变量函数，我们现在写出上面方程的第二种形式为

$$G(\omega) = \int_{-\infty}^{\infty} f(t) e^{-j\omega t} dt \tag{2-9}$$

代入一簇无限谐波之和的表达式中，我们得到

$$f(t) = \sum_{-\infty}^{\infty} \frac{G(\omega)}{T} \exp(jm\omega_1 t) = \sum_{-\infty}^{\infty} G(\omega) \frac{\omega_1}{2\pi} \exp(jm\omega_1 t) \tag{2-10}$$

再用连续变量 ω 替换掉 G，并且基频 ω_1（现在是无穷小）写成 $d\omega$。无穷项和变成了上下限中的一个积分，则这个方程可改写为

$$f(t) = \frac{1}{2\pi} \int_{-\infty}^{\infty} G(\omega) e^{j\omega t} d\omega \tag{2-11}$$

这两个等式称为傅里叶积分，它表明了一个非重复的时域波形与它的连续光谱之间的关系。

掌握这两个方程的含义是很重要的。第一个方程告诉我们 f(t) 波形的能量是在频率范围 $(-\infty, \infty)$ 上连续分布的。第二个方程说明，实际上该波形可以由 $G(\omega)^2$ 的相关值与形如 $e^{-j\omega t}$ 的无限指数函数集加权合成。

连续频率函数 $G(\omega)$ 也许值得稍微进一步探讨其实际意义。的确很难想象，这些单个脉冲是由无限个无穷小振幅的波组成，这样一个波形的能量在频域中是连续分布的。下列众所周知的情形可能使 $G(\omega)$ 更为容易理解：有两根梁，其中一根梁仅在几个点上有载荷，而另一根梁是在全长上均匀地加载，如石头或混凝土。简言之，第一种情况是在离散点上加载，正如仅包含离散频率的周期信号波形。然而，如果人们问为什么载荷连续分布的梁上承载在一个点上，回答肯定是那个点（或任何其他点）上所承受载荷为无穷小。明智的方法是先弄清在一小段距离上平均载荷是多少，再回答出每米多少千克质量。同样地，一条连续频率的光谱表明在任何点频率上的部分是无穷小的，而且计算以那一点为中心的一小段频率所包含的能量。因此，函数 $G(\omega)$ 可认为是频率密度函数。

2.3.2 连续频谱的例子

为了举例说明傅里叶积分的应用，我们现在正式地计算一下图 2.1 所示单脉冲波形的频谱。积分的上下限是有限的，有

$$G(\omega) = \int_{-\tau}^{\tau} 1 \times e^{-j\omega t} dt = -\frac{1}{j\omega} [e^{-j\omega t}]_{-\tau}^{\tau} = 2\tau \left(\frac{\sin \omega \tau}{\omega \tau} \right) \tag{2-12}$$

根据前面已知知识，$(\sin x)/x$ 的这个函数的图形如图 2.1(b) 所示。当 ω 等于 π/τrad/s 的整数倍时，$\sin \omega\tau = 0$，图形通过零点。似乎很奇怪，脉冲在这样的频率下没有能量，然而这不难说明。例如，如果我们想知道在脉冲中包含多少如 ω=π/τ 或 f=1/2τ 这样的频率，一般惯例是脉冲函数乘以适当的正弦曲线波形的函数并且在超过 2τ 宽度上做积分，则得到的结果一定是零，因为在任何超过 2τ 宽度的正弦曲线波形的积分总是等于零的。

这个脉冲波形的频谱说明许多有关时限信号的重要性质。如果脉冲宽度 2τ 非常大，它的频谱能量分布在以 ω=0 为中心的区间上，在极限情况下 $(\sin x)/x$ 函数变成了零频率处的一条直线。换言之，脉冲波形已经变成了无限宽的固定电平。反之，如果脉冲极其短，则比较高的频率逐渐地在它的频谱中表现出来，在一个无限狭窄的脉冲的限制情形下，频谱变得平坦并且扩展到各个频率段。这些结果表现在图 2.2 中，而且说明了一个重要的理论，一个窄时限波形将会占据一个很宽的频率段，反之亦然。总之，当一个连续或重复的信号在"开"和"关"离散化时，结果与预期的一样。上述的脉冲波形会被看成不断"开"和"关"的平稳信号。"开""关"突然变化会产生新的频率，它将

引起单线频谱变宽。无论时限信号表示的是一个通信通道中的电子波形，还是一个限定区间上的采集或观测数据，它的属性都是很重要的。

既然已经定义了傅里叶变换方程，那么就有可能计算各种不同的非重复信号的频谱。工作的困难在于被计算积分的图形随着选择信号波形的改变而变化。我们必须用已熟悉的连续频谱的例子来分析问题，如图2.3所示的指数衰减的正弦曲线波形。由数学分析，这个波形可由方程 $f(t)=e^{-\alpha t}\sin\omega_0 t, t\geqslant 0$ 来描述。因此它的频谱为

$$g(\omega)=\int_{-\infty}^{\infty}f(t)e^{-j\omega t}dt=\int_{0}^{\infty}e^{-\alpha t}\sin\omega_0 t e^{j\omega t}dt \tag{2-13}$$

如果 $\sin\omega_0 t$ 用下面等式替换的话，这个积分是相当简单的，即

$$\sin\omega_0 t=\frac{1}{2j}[\exp(j\omega_0 t)-\exp(-j\omega_0 t)] \tag{2-14}$$

从而得到

$$G(\omega)=\frac{\omega_0}{\alpha^2+\omega_0^2-\omega^2+j2\alpha\omega} \tag{2-15}$$

首先要注意的是，这里的 $G(\omega)$ 是复数，然而在图2.1所示的单脉冲的情况下，它纯粹是个实数。这种区别的原因是脉冲波形是关于 $t=0$ 对称的，因此它的频谱只包含余弦部分。然而，衰减的正弦曲线波是一点也不关于 $t=0$ 对称的，因此只能由正弦项和余弦项共同组合，并由它复杂的频谱来反映这种不对称。

2.3.3 傅里叶积分方程的对称性

两个傅里叶积分方程具有明显的对称性，即

$$G(\omega)=\int_{-\infty}^{\infty}f(t)e^{-j\omega t}dt \tag{2-16}$$

$$f(t)=\frac{1}{2\pi}\int_{-\infty}^{\infty}G(\omega)e^{j\omega t}d\omega \tag{2-17}$$

的确，除了第二个方程有乘数 $(1/2\pi)$ 和在指数上符号的不同外，两个方程在形式上是相同的。如果我们考虑一个时间偶函数，例如，前面已经讨论过的单脉冲波形，则方程在时域和频域具有极好的对称性，有一个仅包含余弦的偶函数的频谱。在这种情形下，如果在第一个方程中用 t' 代替 $(-t)$，则这个方程变为

$$G(\omega)=\int_{-\infty}^{\infty}f(-t')e^{-j\omega t'}(-dt')=-\int_{-\infty}^{\infty}f(-t')e^{j\omega t'}dt' \tag{2-18}$$

但是如果 $f(t)$ 是一个偶函数，即 $f(-t')=f(t')$，则有

$$G(\omega)=\int_{-\infty}^{\infty}f(t')e^{j\omega t'}dt' \tag{2-19}$$

现在，除了相差一个乘数 $1/2\pi$ 之外，式(2-19)和第二个方程在形式上是相同的。可以推出，由于方波的频谱是 $(\sin x)/x$ 的图形，因此频谱能量均匀地分布在 $(-\omega',\omega')$ 的范围内，其他地方则没有能量。图2.4中说明了这种在时域和频域间的对称性，而且解释了用于傅里叶变换和反变换的傅里叶积分方程的一般描述。

2.3.4 傅里叶变换的局限性

到现在为止，似乎傅里叶变换适用于任何非重复的信号，但是事实上在它的应用中有一些限制和困难。从下面方程可以得到一个重要的限制，即

$$G(\omega) = \int_{-\infty}^{\infty} f(t) e^{-j\omega t} dt \tag{2-20}$$

明显地，只有当这个方程的右边有限的时候，才存在傅里叶变换。此外，由于 $e^{-j\omega t}$ 的幅值是有限的，要使傅里叶变换存在，必须满足

$$\int_{-\infty}^{\infty} |f(t)| dt < \infty \tag{2-21}$$

实际的许多波形，如图 2.5 所示的连续正弦波和余弦波，以及所谓的单位阶跃函数，都不满足后一个条件，严格地说都不能进行傅里叶变换。然而，幸运的是，可以把它们作为进行傅里叶变换的有限的几个波形。例如，我们可以计算在 $-T < t < T$ 区间内的正弦波或余弦波的傅里叶变换：T 趋于非常大，且在极限情况下对连续函数自身做变换。在研究单位阶跃函数时，首先考察如图 2.5 所示的指数衰减阶跃信号，然后令 T 变得非常小，在极限情况下，我们得到了如图 2.5(a) 所示的单位阶跃函数频谱。对于指数衰减阶跃信号，我们有

$$G(\omega) = \int_0^\infty e^{-\alpha t} e^{-j\omega t} dt = \frac{1}{-(\alpha + j\omega)} \left[e^{-\alpha t} e^{-j\omega t} \right]_0^\infty \tag{2-22}$$

当 $t = \infty$ 时，如果 $\alpha > 0$，则 $e^{-\alpha t}$ 等于零，因而

$$G(\omega) = \frac{1}{-(\alpha + j\omega)}(0 - 1) = \frac{1}{\alpha + j\omega}, \text{对于 } \alpha > 0 \tag{2-23}$$

在使单位阶跃函数趋向于零来得到其傅里叶变换之前，先谈一个涉及奇异点的问题。它常发生在函数值没有定义的某点：这样的点称为奇异点。如果我们让上述的例子中的 $\alpha \to 0$，则在 $\omega = 0$ 时，$G(\omega) \to \infty$，因此在 $\omega = 0$ 处有一个奇异点。阶跃函数波形在 $t = 0$ 处没有合适的定义，因为在这个点处并不连续。

奇异点概念经常出现在信号理论中。从数学的观点来看，奇异点的出现也许引起或不会引起特别的麻烦，但是在任何情形下都应该十分注意，在奇异区域中要特别考虑函数的性质。

现在我们回到单位阶跃函数的傅里叶变换问题，令 $\alpha \to 0$，乍看上去，似乎 $G(\omega)$ 趋近于 $1/j\omega$，但必须考虑在 $\omega = 0$ 时出现奇异点的特殊情况。假定我们使 α 足够小，但不为零，在频率范围上由于 $\omega \gg \alpha$，所以 $G(\omega) \approx (1/j\omega)$；但是当 ω 实际为零时，由于 $\omega \ll \alpha$，所以有 $G(\omega) \approx (1/\alpha)$。随着 α 变得越来越小，只剩下一个等于 $1/j\omega$ 的连续频谱和以 $\omega = 0$ 为中心、振幅为 $1/\alpha$ 的尖峰脉冲。这实际上代表"直流"或阶跃波平均值的频谱线。这样，单位阶跃函数具有由无数的正弦函数集组成的频谱，这个正弦函数的振幅与频率成反比，其频率和零点的频率一起表示为波形的平均值。

实际上，我们通过乘以一个使傅里叶积分收敛的因子 $e^{-\alpha t}$，来得出阶跃函数的频谱特性。尽管由于正数 α 随着时间 t 无限增加，$e^{-\alpha t}$ 的值变得越来越小，从而使得这类"收敛因子"不适用于整个时间上的信号波形，但是上述实用的方法还是广泛应用于其他波形当中。所以，这种方法只能成功地用在某确定时刻有非零值而在此之前可视为零的波形；通常，将信号最先出现非零值的时刻规定为 $t = 0$。如果这个收敛因子应用于 $t = 0$ 之前 $f(t)$ 为零的例子中，我们可以定义傅里叶变换的另一种形式为

$$G_1(\omega) = \int_0^\infty f(t) e^{-\alpha t} e^{-j\omega t} dt = \int_0^\infty f(t) e^{-(\alpha + j\omega)t} dt \tag{2-24}$$

当然，严格来说，G_1 不仅是 ω 的函数，也是关于因子 α 的函数，而且 α 的选择应使得在任何特殊情况下，积分都是收敛的。为了方便起见，我们引入新的变量 $s = (\alpha + j\omega)$，因此关于 s 的函数 G_1 为

$$G_1(s) = \int_0^\infty f(t) e^{-st} dt \tag{2-25}$$

为此，下面我们要详细说明信号 $f(t)$ 的拉普拉斯变换。

2.4 拉普拉斯变换

2.4.1 与傅里叶变换的关系

拉普拉斯变换与傅里叶变换的关系很密切。正如我们所看到的，傅里叶级数是把一个信号描述成正弦项和余弦项之和，而正弦项和余弦项又可以用形式为 $e^{j\omega t}$ 的一对虚指数项来表示。通过引入收敛因子，就有可能得出一定信号的波谱，这种信号用傅里叶积分是无法计算的。只要我们把引入的这个积分因子 $e^{-\alpha t}$ 乘以一个复杂的信号函数，再积分或许可以计算出这种信号，然后我们令 $e^{-\alpha t}$ 趋近于零即可。但是，正如傅里叶积分所表现的形式那样，是一个带有形如 $e^{j\omega t}$ 指数项的无限集，因此拉普拉斯变换等式的形式表明我们应该用具有无限集的指数项 $e^{s t}$ 的拉普拉斯变换表示 $f(t)$，这里的 s 是一个称为复频的复数。这些项不仅引入了傅里叶方法中的正弦波和余弦波，而且还引入了正弦波和余弦波的增长和衰退，以及指数的增长和衰退。

由于实际中拉普拉斯变换可以分析在时间上是振荡和非振荡函数的信号，因此它对信号 $f(t)$ 的限制要比傅里叶积分少。另一方面，它仍然是有限制的，因为变量的实部 α 必须是收敛的。例如，如果 $f(t)$ 含有一个渐增的指数部分 e^{7t}，只有当 $e^{-\alpha t}$ 项抵消它的增长，也就是说 α 必须至少为 7 时，积分才将收敛。但是，在 $f(t)$ 中，不论 α 的值如何变化，在 t 足够大时，$\exp(t^2)$ 项主导着 $e^{-\alpha t}$ 的变化，因此拉普拉斯变换不能使用。已经涉及的另一个局限是拉普拉斯变换不能处理无限延伸到过去的信号，只能在区间 $0<t<\infty$ 上进行积分运算。原因是 α 在区间 $0<t<\infty$ 上能收敛，而在区间 $-\infty<t<0$ 上不能收敛。

2.4.2 拉普拉斯变换的应用

作为第一个例子，考虑如图 2.6 所示的指数衰减，它的拉普拉斯变换式为

$$G(s)=\int_0^\infty f(t)e^{-st}dt=\int e^{-(s+\alpha)t}dt=\frac{1}{s+\alpha}[e^{-(s+\alpha)t}]_0^\infty=\frac{1}{(s+\alpha)} \tag{2-26}$$

第二个例子我们考虑图 2.6 中的正弦波，它的拉普拉斯变换式为

$$G(s)=\int_0^\infty \sin\omega_0 t\exp(-st)dt=\frac{\omega_0}{(s+j\omega_0)(s-j\omega_0)} \tag{2-27}$$

得出这些结果并不容易，因为尽管能够相当容易地想象一个由许多连续正弦波合成的信号，但从图 2.6 中很难考虑可能波形的范围。当然，变量 $s=(\alpha+j\omega)$ 可以是实数、纯虚数或复数，如果我们取出它的虚部 ($s=j\omega$)，那么就像在傅里叶变换中一样，实际考虑的信号是由正弦波组成的。在图 2.6 的指数衰减的例子中，令 $s=j\omega$ 时，有频谱

$$G(\omega)=1/(\alpha+j\omega) \tag{2-28}$$

它的模和相位如图 2.7 所示。结果表明，正弦波的相对振幅和相位需要合成为指数曲线。在区间 $-\infty<t<0$ 上，它们都被抵消为零，但是在 $0<t<\infty$ 上，相加可得到我们想要的波形。为了研究图 2.6 中的频谱，如果我们令 $s=j\omega$，就得到

$$G(j\omega)=\frac{\omega_0}{(j\omega+j\omega_0)(j\omega-j\omega_0)} \tag{2-29}$$

在这个例子中，$G(j\omega)$ 表明了两个奇异点，即在 $\omega=\pm\omega_0$ 时，频谱上的幅值为无穷大，尽管实际上我们选择了单位幅值的正弦波作为时间函数。令 $s=j\omega$，我们试图计算出一个永远连续的波的傅里叶变换，而它的积分是不收敛的，这样困难就应运而生。在这样的例子中，在拉普拉斯变换式中替换 s 是不容易做到的。

我们已经知道怎样进行时间函数的拉普拉斯变换了，它是一个复频变量 s 的函数，但是没有提

到关于 s 的时间函数反变换过程的构造。拉普拉斯变换难点在于反变换的逆，它没有给出直接计算公式。一般地，拉普拉斯反变换定义为

$$f(t) = \frac{1}{2\pi j}\int_{\sigma-j\omega}^{\sigma+j\omega} G(s)e^{st}ds \tag{2-30}$$

而这个等式的一般形式类似于相应的傅里叶积分等式，实际中，上下限 $\sigma\pm j\omega$ 的确定较为困难。对此，等式的积分计算要熟悉复变函数及余项的微积分学。这里采用的方法是在许多教材中处理拉普拉斯变换时普遍采用的方法，即提供一个比较常用的变换表供以后使用。

拉普拉斯变换表不可能涵盖在实际中可能遇到的所有函数，而去研究反变换的不同技巧超过了本书的范围，下面只介绍反变换中十分常用的一个简单方法。这种方法要求各函数之和的拉普拉斯反变换等于它们反变换的和，这称为部分分式法。在大多数例子中，一个函数由许多简单的线性因子组成，部分分式法就是要我们把一个函数表示为许多比较简单的函数之和，而每个简单的函数都出现在拉普拉斯变换表中。

当许多信号不能通过傅里叶变换进行分析时，可以尝试用复频表示该信号，再通过拉普拉斯变换进行分析研究。然而要强调的是拉普拉斯变换的基本介绍并非是它的全部价值所在。后面我们将很清楚，拉普拉斯变换是一个强大的数学工具，除了用于解决信号分析问题外，还可用来解决许多其他问题。尤其是如果使用拉普拉斯变换，则用一组微分等式表达的问题经常可以简化为一组非常简单的代数等式。由于机电系统的某类动力学问题在数学上用微分等式的形式来描述，因此在探索系统性能时，拉普拉斯变换方法具有极大的价值。

2.4.3 信号的零-极点描述

再次考虑函数

$$G(s) = 1/(s+\alpha) \tag{2-31}$$

正如在前面我们所看到的，它表示一个指数衰减的信号波形的拉普拉斯变换。我们将使 $G(s)$ 趋于无穷大和趋于零时的复频变量 s 的取值，分别称为 s 的极点和零点，这样就很方便地对相应信号进行绘图描述。在上述的简单例子中，当 $s=-\alpha$ 时，$G(s)\rightarrow\infty$，因此我们把 s 称为 $G(s)$ 的极点，而 $G(s)$ 没有零点。如果我们考虑常用的一个比较复杂的函数

$$G(s) = \frac{A(s-z_1)(s-z_2)(s-z_3)\cdots}{(s-p_1)(s-p_2)(s-p_3)\cdots} \tag{2-32}$$

式中，A 是一个常量；$s=z_1, z_2, z_3$ 等是 $G(s)$ 的零点；$s=p_1, p_2, p_3$ 等是极点。通过式(2-32)便可写出拉普拉斯变换式，因此可以根据一组极点和零点去定义信号。极点既可以是实数，也可以以复数共轭对的形式出现，零点也是一样。这样在 $s=\alpha+j\omega$ 的极点与在 $s=\alpha-j\omega$ 上的极点相匹配，这种情况总是适用于相应信号是时间的实函数的例子。

在图形上表示 $G(s)$ 的极点和零点是指在所谓的阿根图（复平面）上标出它们的位置。在这个坐标系里，复变量的实部沿着横坐标绘出，复变量的虚部沿着纵坐标绘出。$G(s)$ 函数的极点和零点都是普通的复数值，所以在复平面上绘出它们是很方便的，在此例子中，阿根图被广泛称为"平面"图。例如，假定有一个时间函数 $f(t)$，它的拉普拉斯变换式为

$$G(s) = \frac{4s(s-2)}{(s+1+j3)(s+1-j3)} \tag{2-33}$$

除了常数 4 之外，我们完全可以在复平面上通过绘出 $s=0$ 和 $s=2$ 时的零点和 $s=-1\pm j3$ 时的极点表示出函数 $G(s)$。

第3课　数据采样信号

3.1　简介

只有在某个特定的瞬间，数据采样信号才有具体的值，它适用于间歇测量和记录连续函数的场合。近年来，由于数字电子和计算技术的发展，此类信号变得越来越重要。我们不可能向数字计算机输入连续的数据，任何信号数据输入都必须被一系列数值所代替。在几乎所有情况下，这些数值代表了该连续信号在一连串等间隔点的采样值。

必须强调的是，这一连串采样值只有当连续采样点间隔足够小时才能充分替代该连续信号波形。

3.2　采用迪拉克函数的数学描述

迪拉克函数通常又称单位脉冲函数，它是一个持续时间极短且具有单位面积的脉冲。也就是说，它的持续时间和平均高度乘积为1，它精确的波形是不定的。该函数的物理意义可以用一个例子证明。假设我们用高尔夫球棒向高尔夫球传递一个机械冲量，其他条件是相同的，那么传递给球的动量和球滚动的距离由冲量的数值决定，而该冲量则由力与作用时间的乘积决定。或者，假设该力不是恒力，则冲量是由力-时间表中曲线下方面积决定的；因为迪拉克函数被定义为单位面积，所以可被用来表示一个机械冲量，并且在下面我们会用该方程描述一个作用于信号处理装置的突然电干扰。

单位脉冲函数用来描述数据采样信号，而该信号可被认为由一系列等距离的极短脉冲组成。我们可以很容易地把所有信号采样都看成在各自采样点的持续时间一定、高度与信号值成比例的脉冲。实际上，这个方法从数学上证明是有效的，只要脉冲持续时间与连续采样之间间隔相比是可忽略的。

为了更好地讨论数据采样信号的频谱，我们有必要对单位脉冲函数的频谱进行分析。我们用 $\delta(t)$ 代表在 $t=0$ 时产生的单位脉冲，则

$$G(j\omega) = \int_{-\infty}^{\infty} \delta(t) e^{-j\omega t} dt \tag{3-1}$$

我们用单位脉冲函数的所谓"移位算子"计算这个积分。把在 $t=a$ 时刻产生的用 $\delta(t-a)$ 表示的单位脉冲函数与函数 $f(t)$ 相乘。因为单位脉冲的面积为1，所以当该乘积从 $t=-\infty$ 到 ∞ 进行积分时，代表 $[f(t)\delta(t-a)]$ 的曲线下方面积等于 $t=a$ 时的 $f(t)$ 值，而结果正是该面积。因此，移位算子可表示为

$$\int_{-\infty}^{\infty} f(t)\delta(t-a)dt = f(a) \tag{3-2}$$

$t=0$ 时的单位脉冲的频谱 $G(j\omega)$ 可简化为 $e^{-j\omega t}$ 的值，而该值在 $t=0$ 时为1，所以有

$$G(j\omega) = 1 \tag{3-3}$$

这个结果告诉我们，所有的频率分量都可以被余弦分量所代表。当很多幅值相同而频率不同的余弦分量相加时，它们会在除 $t=0$ 点外互相抵消，在 $t=0$ 点它们会相互增强，因此当高频率分量越来越多时，便会形成一个以 $t=0$ 点为中心的极窄脉冲。

我们还可以饶有兴致地考虑一个发生在某一时刻 $t=T$ 的衰减单位脉冲的频谱。很明显，该脉

冲必定含有当它发生在 $t=0$ 时的相同频率分量，除非当中的每个频率分量都每隔 T 秒衰减一次，其频谱为

$$G(j\omega) = \int_{-\infty}^{\infty} \delta(t-T) e^{-j\omega t} dt = e^{-j\omega T} \tag{3-4}$$

$e^{-j\omega t}$ 项对应于任何 ω 值和 $-\omega T$ 相位偏移的单位幅值。如我们所知，对一个频率分量加一个 $-\omega T$ 弧度的相位偏移代表一个 T 秒的时间偏移。因此，$e^{-j\omega T}$ 表明所有的分量都由一个相等的幅值和一个与频率成比例的相位偏移组成。相同结果可由拉普拉斯变换得出：一个以 $t=0$ 为中心的单位脉冲的拉普拉斯变换为 1。由于移动一个时间函数 T 秒等价于将其拉普拉斯变换乘以 e^{-sT}，所以发生在 $t=T$ 的单位脉冲为 $1 \times e^{-sT}$，用 s 代替 $j\omega$ 可给出其频谱。

3.3 采样数据信号的频谱

3.3.1 离散傅里叶变换

采样数据信号可以用时域和频域表达式来描述。通过用 x 加权的单位脉冲函数来表达采样值 x，该信号可表示为

$$f(t) = x_0 \delta(t) + x_1 \delta(t-T) + x_2 \delta(t-2T) + x_3 \delta(t-3T) + \cdots \tag{3-5}$$

式中，T 为采样间隔。它的拉普拉斯变换式为

$$G(s) = x_0 + x_1 e^{-sT} + x_2 e^{-s(2T)} + x_3 e^{-s(3T)} + \cdots \tag{3-6}$$

它的傅里叶变换式为

$$G(j\omega) = x_0 + x_1 e^{-j\omega T} + x_2 e^{-j\omega(2T)} + x_3 e^{-j\omega(3T)} + \cdots \tag{3-7}$$

采样数据信号的傅里叶变换通常是指离散傅里叶变换（DFT）。这基于两个原因：其一，信号是离散的，这只有当信号采样点是离散的才成立；其二，通常我们用数字计算机求一个采样数据信号的频谱意味着它的傅里叶变换只能通过一系列 ω 的离散数据进行评估。这并不是说以上定义的方程 $G(j\omega)$ 是离散的，它是 ω 的连续方程；但我们只能在恰当的空间间隔上用计算机进行评估，并用所得的值代表连续方程。幸运的是，正如我们接下来所看到的，此频谱的离散表达式并不意味着我们漏掉任何关键环节。

尽管写出采集信号频谱的表达式很容易，但是想要对其可视化处理却不太容易。然而，此类频谱的一些主要特征可以参考以下一些方法。由于采样数据信号实际上由一系列加权单位脉冲组成，而每一个这样的脉冲包含着分布于很大范围频率的能量，所以可以预见，信号频谱也会相应地分布于一个"宽带"。由于诸如 $e^{-j\omega T}$、$e^{-j\omega(2T)}$、$e^{-j\omega(3T)}$ 这些项都是在频域内以 $2\pi/T$(rad/s) 周期重复的，它们的频谱一定也是周期的。最后，我们可以合理地认为这个波形的采样模型频谱不但反映采样脉冲的单位脉冲的频率特征，也反映了连续信号的分量。

如图 3.1 所示的两个典型连续信号和它们的采样模型的频谱证明了以上观点。例如，连续波形 $\cos \omega_0 t$ 在 $\omega \pm \omega_0$ 有两条谱线，该采样模型频谱从 ω 到 $2\pi T$ 会有无限延展的两条谱线，此处 T 为采样间隔。一个单矩形脉冲有一个连续的 $(\sin x)/x$ 形状的频谱，采样之后，这个频谱会以 $2\pi/T$(rad/s) 重复。在这两个采样信号频谱中，从 $\omega=-\pi/T$ 到 $\omega=\pi/T$ 的低频区域代表了连续信号波形的分量，而频谱的重复则是来源于采样过程本身。实际上，没有任何必要去计算频率高于 $\omega=\pi/T$ 采样数据信号频谱，因为它肯定是从 $\omega=-\pi/T$ 到 $\omega=\pi/T$ 的重复。

3.3.2 快速傅里叶变换

为了找到采样数据信号频谱，我们很有必要去计算下列形式函数的值，函数的表达式为

$$G(j\omega) = x_0 + x_1 e^{-j\omega T} + x_2 e^{-j\omega(2T)} + x_3 e^{-j\omega(3T)} + \cdots$$
$$= x_0 + x_1 \cos\omega T + x_2 \cos 2\omega T + x_3 \cos 3\omega T + \cdots -$$
$$j(x_1 \cos\omega T + x_2 \cos 2\omega T + x_3 \cos 3\omega T + \cdots) \tag{3-8}$$

该表达式的实部和虚部可以通过一系列恰当的 ω 值进行计算。假设采样间隔 T 是根据连续函数确定的，一个偶函数的采样数据信号有一个纯实数频谱，因此只要计算余弦项即可；相反，如果这个时间函数是奇函数，则只要计算正弦项即可。

在此，我们有必要讨论一下计算一个具有 N 个样本的信号的 DFT 需要多少次计算操作。正如我们所看到的，一个如此长度的信号被定义为 $N/2$ 谐波，并用一个余弦项和一个正弦项来表示。任何项的计算都要乘以每个信号样本值 $\cos n\omega T$ 或 $\sin n\omega T$，因此一个 N 个样本的信号频谱计算要涉及 N^2 个乘法。所以用于计算傅里叶变换的时间近似比例于样本数的平方。

然而，一个细致的研究指出，许多乘法运算都是重复的，而快速傅里叶变换(FFT)就是为了尽可能减少这些重复。离散傅里叶变换的这种形式是一种以增强计算效率为目的的方法或"算法"。结果显示，当用到 FFT 时，N 个样本的变换计算时间近似比例于 $N\log_2 N$，而当没有采取任何事先措施去减少重复计算时，N 个样本的变换计算时间近似比例于 N^2。快速傅里叶变换计算方法随着信号样本数目的增加而变得越来越有吸引力，而当该数目是 2 的整数次幂时该算法最有效。用于此种长度的信号，计算时间通常可以减少到其 1/50 或 1/100。

3.4　z 变换

3.4.1　简介

尽管描述采样数据信号的时域特性经常使用傅里叶变换或拉普拉斯变换，但还有另一种变换。z 变换不仅给这类信号的傅里叶变换提供一种非常有用的简化方式，而且也提供了一种通过用极点和零点定义信号的非常便利的方法。

采样数据信号的 z 变换可以简单地通过它的拉普拉斯变换作为出发点来定义。在 3.3.1 节中给出信号的拉普拉斯变换式为
$$G(s) = x_0 + x_1 e^{-sT} + x_2 e^{-s(2T)} + x_3 e^{-s(3T)} + \cdots \tag{3-9}$$
现在我们定义新的变量 $z = e^{sT}$，并且记信号的 z 变换为 $G(z)$，因此有
$$G(z) = x_0 + x_1 z^{-1} + x_2 z^{-2} + x_3 z^{-3} + \cdots \tag{3-10}$$
z 是一个新的一般含有实部和虚部的频率变量，即
$$z = e^{sT} = e^{\sigma T} e^{j\omega T} = e^{\sigma T}\cos\omega T + j e^{\sigma T}\sin\omega T \tag{3-11}$$
我们可以看到，z 存在与频率成正比的相位偏移，即有一个 T 秒的固定时间延迟。相反地，$e^{j\omega T}$ 项表示 T 秒的超前时间。这说明在一般描述方法中将 z 作为"移位算子"，表明超前时间等于抽样间隔 T。

上面的定义清楚地表明采样信号的 z 变换是 z^{-1} 的幂级数，而它不同项的系数等于它的采样值；用这种形式描述的变换，仅通过观察就可以得出它的时间函数。例如，假定有 z 变换为
$$G(z) = \frac{z}{z-a} = \frac{1}{1-az^{-1}} \tag{3-12}$$
我们可以用 z^{-1} 的幂级数重新表示为
$$G(z) = 1 + az^{-1} + a^2 z^{-2} + a^3 z^{-3} + a^4 z^{-4} + \cdots \tag{3-13}$$

3.4.2　z 平面的极点和零点

我们已经知道任何采样数据信号的频谱在形式上都是以 $2\pi/T(\text{rad/s})$ 重复的。既然一个信号

可以通过s平面上的极-零点来描述，这些点的位置和信号的频谱有直接的关系，那么一定可以想象到这种信号在s平面上的极-零点模式在形式上也是重复的。连续信号的采样形式和初始的s平面上的极-零点配置相同是确实有可能的，除非所有的极-零点在虚轴方向上不停地重复。

用s平面极-零点描述连续信号的优点之一是通过分析极点和零点距虚轴连续点的矢量长度和相位，很容易直观地得到其频谱。这样一来，采样数据信号分析技术变得没什么研究价值了，因为仅考虑几个有限的极点和零点就行了。这一点不足以强调z变换的优点，即通过几个极点和零点就可以表示一个采样数据信号。

为了更好地理解s平面和z平面上极-零点的关系，对于s取一定值时，研究复变量z产生的过程是很有用的。这个过程一般称为从s平面到z平面的映射。例如，s是一个纯虚数，那么

$$s = j\omega \quad \text{和} \quad z = e^{j\omega T} = \cos\omega t + \sin\omega T \tag{3-14}$$

如果我们现在考虑在z平面的阿根图上描出z的值，z的轨迹曲线随着ω的变化形成一个单位圆，这个轨迹从$\omega=0$时的实轴上开始，每隔$\omega=2\pi/T$时，重复它的轨迹。另一种是令$s=\sigma$，这里σ是实数，得出$z=e^{\sigma}$。如果σ是正数，那么z的值大于1；如果σ是负数，那么z是一个小于1的实正数。最后一种情况是假定$s=\sigma+j\omega$，得到$z=e^{\sigma}e^{j\omega T}$。这时$z$的值表示与正实轴成$\omega T$的弧度、模为$e^{\sigma}$的矢量。只要$\sigma$是负值，在$z$平面上单位圆内的点指定对应于$s$平面上左半平面上的点。因此，整个$s$平面上的左半平面就映射到$z$平面上的单位圆。

实际上，这并没有引入所有的情况。假定有一个采样数据信号，它唯一的z平面上的极点在如图3.2所示的右半部的B点。那么它的z变换式为$G(z)=1/(z-r_3)$，用$j\omega$替换s，得出它的频谱为

$$G(j\omega) = 1/(e^{j\omega T} - r_3) \tag{3-15}$$

分母可以看成从B点到单位圆上一点的矢量。每隔ω改变$2\pi/T$(rad/s)，绘出整个单位圆，在频谱上这个矢量的模和相位是重复的。因此，单个z平面上的极点(如B点)相当于s平面上极点的重复集，而它邻近的数在虚轴方向上相隔$2\pi/T$。s平面上左半平面的B点仅是这个无限集中的一点。这就解释了为什么使用z平面简化了采样数据信号的极-零点描述。

现在为了熟悉这种方法，我们将算出对应于一些z变换的时间函数。首先考虑的函数为

$$G_1(z) = (1 - z^{-8}) = (z^8 - 1)/z^8 \tag{3-16}$$

由根式$z^8=1$得出它有8个零点，以及它在$z=0$处有一个8次幂的极点。很明显，有2个零点是$z=1$和$z=-1$，其他6个零点分布在单位圆上，如图3.3所示。图3.3还表明了对应的时间函数和频谱，其中在$\omega=0$与$\omega=\pi/T$之间的3个零点对应于z平面上的A、B、C的3个零点。如果现在消除$z=1$时的零点，我们得出新的z变换形式为

$$G_2(z) = \frac{1}{z^8}\left[\left(z - \frac{1}{\sqrt{2}} + j\frac{1}{\sqrt{2}}\right)\left(z - \frac{1}{\sqrt{2}} - j\frac{1}{\sqrt{2}}\right)\left(z + \frac{1}{\sqrt{2}} + j\frac{1}{\sqrt{2}}\right)\cdot \right.$$
$$\left. \left(z + \frac{1}{\sqrt{2}} - j\frac{1}{\sqrt{2}}\right)(z+j)(z-j)(z+1)\right] \tag{3-17}$$

整理得

$$G(z) = z^{-1} + z^{-2} + z^{-3} + z^{-4} + z^{-5} + z^{-6} + z^{-7} + z^{-8} \tag{3-18}$$

对应的时间函数由一组8个单位高度的采样集合组成，如图3.4所示。它的频谱原则上不同于第一个例子，因为第一个例子具有有限个零频率，或者称为平均项。需要说明的是，消除$z=1$时的零点模式可以通过一个相应的极点来实现。因此，$G_2(z)$也可以写成

$$G_2(z) = (z^8 - 1)/[z^8(z-1)] \tag{3-19}$$

需要强调的是，一个给定的z变换经常可以表示成极点集和零点集，而不仅仅是零点集。

任何一个可以表示成 z 平面单位圆上零点的有限集的 z 变换都可以表示为一个有限持续时间的时间函数，这个时间函数除了与时间原点有一定的偏离之外，它或者是奇函数，或者是偶函数。如果我们回想一下单位圆上零点有限集等价于 s 平面内虚轴上的无限的重复集合，就很清楚这个结论了。任何一个从 s 平面上零点到虚轴上的一点的矢量总有一个 $\pm\pi/2$ 的相位角，因此所有这些零矢量的总的相位角总是 $\pi/2$ 的实数倍。这样频谱总是纯实数或纯虚数，它表明一个由余弦函数或正弦函数组成的时间函数。需要说明的是，图 3.3 所示的时间函数是一个奇函数，图 3.4 所示的时间函数为一个偶函数，这两个函数的差异只是在时间原点上有一定的偏移。

与用零极点表示信号稍有不同的信号表达式如图 3.5(a)所示，它是图 3.5(b)表示的无限衰减指数波形的截断形式。既然截断信号可以通过从图 3.5(b)所示的波形减去图 3.5(c)所示的波形来形成，那么可以观察到它的 z 变换可以写成

$$G_3(z) = \frac{z}{(z-a)}(1 - a^k z^{-k}) \tag{3-20}$$

因此，有 k 个零点平均分布在半径为 a 的圆上，除了 z=a 时的零点被取消，因为它碰巧是一个极点，图 3.6 所示的是当 k=12 时的情况。有趣的是，如果 a 和 k 被选定，以便可以忽略图 3.4 所示波形的所有项，那么信号的无限形式和截断形式效果上是等同的。既然无限形式可以通过 z=a 处单个的极点来表示，那么我们可以推出在半径为 a 的圆上零点的集合等价于在同样圆上的单个极点。

下面考虑的 z 变换为

$$G_4(z) = \frac{z(z-1)}{z^2 - 2\alpha z + (\alpha^2 + \beta^2)} \tag{3-21}$$

在 z=0 和 z=1 上有零点，在 z=α±jβ 处有一个复共轭极点对。拿一个数字来说明，假定 α=0.81 及 β=0.55，给出的极点位置如图 3.7 所示。这两个极点接近于单位圆，因此在 ωT=θ 上表现出很大的频率成分，或者 ω=θ/T(rad/s)。在 z=1 时的零点表示对应的时间函数有零平均值，因此它的所有采样值之和为零。为了进一步理解相应的信号，将 $G_4(z)$ 用 z^{-1} 的幂级数表示出来，这并不是很简单的，但是在图 3.7(b)中示出了这样做的结果。通过 z 平面上单位圆内的极点就可以引出具有无限项的幂级数，它是一个衰减振荡函数。因此，在平均采样时间上，采样频率是

$$f \approx \frac{1}{11T}\text{Hz} \quad 或 \quad \omega = 2\pi f = \frac{2\pi}{11T} = \frac{0.57}{T} \quad \text{rad/s} \tag{3-22}$$

这个值和我们想象的差不多，既然

$$\tan\theta = \frac{\beta}{\alpha} = \frac{0.55}{0.81} \tag{3-23}$$

因此 θ=34°，该值很接近于 0.57。这样在上面的例子中，z=0 时的零点仅表明 t=0 时是第一个采样有限值；如果不存在零点，那么第一个值就在 t=T 处。相反地，在原点处单个极点对应于 t=-T 时采样的第一个值。一般来说，在 G(z)的分子分母多项式中，z 的最高次幂相等时，第一个非零采样值出现在 t=0 时。

第 4 课 随 机 信 号

4.1 简介

至此，我们已经分析了具有确定波形的连续信号和采样信号。这类信号被描述成"确定性"信

号，由它们所得出的频谱详细表述了正弦波的幅值和相对相位，如果将这些正弦波叠加起来就会很精确地合成为原波形。相比较而言，在任何瞬间随机波形或信号的值是不确定的，也不可能根据它的过去很精确地预测它的未来状态。一种原因是我们对产生随机信号的物理过程理解不够；另一种原因是虽然理解产生随机信号的物理过程，但是预测信号的工作量太大而不值得去做。在这种情况下，通常计算出随机信号的一些平均属性以供信号分析时用。

对随机信号的第一反应是，在任何信号分析的科学理论中很少用到属性不明了的随机信号，但是事实刚好相反。例如，假如希望沿着一根电报专线传送信息。由于在接收端的人懂得的信息很少，所以传送一个清晰的、确定的信息几乎是不值得的。举一个简单的例子，传送一个连续的正弦波信号是无意义的，因为一旦接收者确定了正弦波的振幅、频率和相位特性，就没有进一步的信息传送了。但是，如果像在莫尔斯码信息中一样，信号在高电平和低电平之间转换，接收者不知道下一个信息是"点"还是"破折号"，这种未来非常随机和不确定的信号可以允许传送有用的信息。另一方面，确定莫尔斯码信息的一般属性是非常有可能的。因此，即使电报系统的设计者不知道在任何特殊时刻"点－破折号"传送的顺序，知道信息一般属性对他或许也是非常有用的。据此看来，随机信号理论在现代通信系统的发展中起到重要作用是不足为怪的。

信号的随机性表现在各方面，或许最普遍的随机性是在信号的振幅上，如图 4.1(a)所示。在这类例子中，即使考虑信号的过去状态，信号的未来值也不能被确定地预测出来。图 4.1(b)表明的是随机性的另一种普遍形式，这也是莫尔斯码信息用到的一般形式。这里的信号总是在两个定值之间变化，但是这两个定值的转换发生在随机时刻。某种确定事件发生的时间或在两种状态之间跃变的时间是随机信号，这种随机信号存在许多不同的研究领域中，例如，在排队论、核粒子物理学和神经心理学领域中，以及在电子通信领域中。图 4.1(c)给出一个前面涉及的随机信号，它也表示一个心电图或对心脏活动的电子记录。心跳具有某种不规则性，连续心电图的节拍具有随机性；另外，波形本身在形式上不会是完全重复的。

实际上，一个信号经常包含随机部分和确定部分。例如，图 4.1(c)所示的心电图或许经常被认为由一个严格重复的信号脉冲组成，只是振幅和节拍有小的随机性。这类随机性经常是由记录或测量误差引起的，或者是由于系统的某些点在实验员的控制之外而引起的。在电子电路中，随机干扰随处可见，这些干扰称为电子"噪声"。然而，这里介绍的随机信号的分析方法既可用于表示有用信号的波形，也可用于不期望的"噪声"。重要的是，这种随机信号的分析方法能够定量描述噪声波形，以便可以评估信号处理装置对波形的影响。后面我们应该看到，信号处理运算最重要的是，在不期望的干扰情况下，试图提取和改善信号波形。

正如前面所提及的，通常描述随机信号的方法是估计信号的某些一般特性。涉及的数学分支是统计学，它关心的是总体的定量属性，而不是单个元素的属性。这里我们认为总体是由一个随机信号波形的许许多多连续的值组成的。对随机信号分析有直接影响的另一个数学分支是概率论，它同统计学有很大的关系。概率论关心的问题是随机过程或现象产生各种结果的可能性，而统计学是用一般的方法来总结实际得出的结果。为了让我们在分析随机信号时，能熟练运用一些有用的一般方法，我们首先学习一下概率论中的一些基本概念。

4.2 概率论基础

4.2.1 事件的概率

假如重复地掷一个六面骰子。在掷的条件相同的情况下，随着实验次数的增加，我们观察到任何一面朝上的次数接近于总次数的 1/6。因此，要计算下一次实验时某一面朝上的可能性，我们可以计算出它等于 1/6。形式上，如果实验重复做上 N 次并且事件 A 发生了 n 次，则事件 A 的概率可以定义为

$$p(A) = \lim_{N \to \infty}(n/N) \tag{4-1}$$

尽管在许多严谨的数学处理中，对事件相对频率的概率的定义有相当大的困难，但它是客观上的要求。其中，主要的原因是任何实际的实验包含有限次的试验，没有令人信服的方法可以逼近极限。即使在简单的掷硬币的实验中，大量的投掷也许不能如预期的那样——正、反面的概率都等于 0.5。由于这个缘故，基于一组公理的概率定义的任意一个供选择的方法有时候可能被采纳。然而无论哪种定义被使用，毫无疑问，概率 1 表示必然发生，而概率 0 表示必然不发生。

定义事件（A 或 B）表示 A 发生或 B 发生。在大量的 N 次试验中，假定 A 发生了 n 次，B 发生了 m 次。如果 A 和 B 是互斥事件（即不同时发生），那么事件（A 或 B）发生了（$n+m$）次。由此有

$$p(A \text{ 或 } B) = \lim_{N \to \infty}\left(\frac{n+m}{N}\right) = \lim_{N \to \infty}\left(\frac{n}{N}\right) + \lim_{N \to \infty}\left(\frac{m}{N}\right)$$
$$= p(A) + p(B) \tag{4-2}$$

这是一个概率的基本加法定律，可以运用到任何互斥事件的反复试验的例子中。

4.2.2 联合概率和条件概率

现在我们引入两组可能发生的实验进行分析，那么，联合概率 $p(A \text{ 与 } B)$ 就等于第一组中发生结果 A 并且第二组中发生结果 B 的概率，如同时掷两枚骰子或从一副纸牌中抽出两张牌。假设在 N 次实验中，A 发生了 n 次，而 B 发生了 m 次。那么，m 就是事件 A 和事件 B 都发生的次数，则

$$p(A \text{ 与 } B) = \lim_{N \to \infty}\left(\frac{m}{N}\right) = \lim_{N \to \infty}\left(\frac{n}{N}\right)\left(\frac{m}{n}\right) \tag{4-3}$$

注意：当 $N \to \infty$ 时，(n/N) 的极限是 $p(A)$，(m/n) 的极限就是 A 已经发生的情况下 B 发生的概率。后一个概率称为"A 已发生下 B 的条件概率"，用 $p(B/A)$ 表示，则

$$p(A \text{ 与 } B) = p(A)p(B/A) \tag{4-4}$$

同理，也可以表示为

$$p(A \text{ 与 } B) = p(B)p(A/B) \tag{4-5}$$

则

$$p(A)p(B/A) = p(B)p(A/B) \tag{4-6}$$

或者

$$p(A/B) = p(A)p(B/A)/p(B) \tag{4-7}$$

这个结果称为贝叶斯定律，它把 B 已发生下 A 的条件概率和 A 已发生下 B 的条件概率联系起来。

显然，如果 A 和 B 相互完全不影响，那么 A 已发生下 B 的条件概率恰好就等于 B 的概率，因此有

$$p(B) = p(B/A) \tag{4-8}$$

并且

$$p(A \text{ 与 } B) = p(A)p(B) \tag{4-9}$$

这样的话，A 与 B 统计上就称为独立事件。

为了说明这个结论，我们假设有一个盒子，里边有 3 个红球和 5 个蓝球。第一个实验是从盒子中随机取出一个球，然后将它放回再取一个球，我们想知道第一次取到红球（结果 A）的概率和第二次取到蓝球（结果 B）的概率。这样的话，因为第一步取到的球被放回，第一步的结果决不会影响第二步的结果。因此，这个联合概率（先取到红球再取到蓝球）就是

$$p(A \text{ 与 } B) = p(A) \cdot p(B) = \frac{3}{8} \times \frac{5}{8} = \frac{15}{64}$$

现在，我们在取第二个球前不放回第一个球的情况下重复这项实验。在统计学上，实验中的两步不再是相互独立的。如果第一次取到红球，那么还剩下 5 个蓝球和 2 个红球，条件概率为

$$p(B/A) = \frac{5}{7}$$

下面通过两组可能结果的实验，也可以拓展到任意组，对联合概率和条件概率进行简略讨论。信号及信号分析的相关性如图 4.2(a)所示，图中显示了一个随机采样信号的一部分，信号的每个采样值都取 6 个可能电平中的一个。类似于有 6 个面的骰子，通过分析这个信号足够长的时间，我们可以计算下一个采样可能取到任何一个电平值的概率(注意，和均匀的骰子不同，6 个可能电平取得的概率可能各不相同)。因而任何一个采样值都看成一次实验的结果，这样的话，共有 6 种可能性。通过讨论是否信号的某个取值会影响到下面的取值，我们可以得到信号更详细的描述。例如，当取到的采样值为 3 时，由已经得到的简单概率，是否表明下一个采样值更可能为 2 或 4(或其他值)？这相当于问所取的特殊值是否一定是它前面的或后面的一个值，这个问题可以通过估算合适的条件概率来回答。这类事件将在后面讨论的相关函数中进行研究。

到此为止，我们已经考虑过有限次可能结果的实验，以及表现出有限次结果的随机信号的类似情形。如图 4.2(a)所示的信号例子中，对很长一段信号的分析让我们估计出 6 个可能的信号值出现的概率，据此绘出的图如图 4.2(b)所示，无论怎样，它与实际中具有连续范围振幅的信号比较吻合。由于信号在一个特定值处的概率变得非常小，所以我们不得不使用连续概率变量、概率密度函数去描述它。

4.2.3 概率密度函数

如图 4.3 所示的连续随机信号的振幅是一个无限集，因此出现某些特殊值(如 y 的概率)就变得相当小。在这个例子中，只要讨论信号在一个很小范围上[如在 y 和 $(y+\delta y)$ 之间]的取值的概率就是够了。简单来说，这个概率等于在这个范围上的时间占总时间的比例，或者可以通过在很长的时间记录(如 T_0 秒)所有时间区间 $\delta t_1, \delta t_2, \delta t_3 \cdots$ 之和除以 T_0 得到这个概率。记这个概率为 q，我们得出

$$q = \lim_{T_0 \to \infty} \left(\frac{\delta t_1 + \delta t_2 + \delta t_3 + \cdots}{T_0} \right) \tag{4-10}$$

很明显，q 依赖于区间 δy，由于 δy 趋于零，所以 q 也趋于零。然而当 $\delta y \to 0$，商值 $q/\delta y$ 趋于一个常数时，这个数称为概率密度 $p(y)$。$p(y)$ 给我们提供了一个计算概率的方法，它是独立于所选定的精确值 δy 的，落在 y 与 $(y+\delta y)$ 范围上的概率可以通过 $p(y)$ 乘以区间大小 δy 而得到，即

$$p(y) = q/\delta y \text{ 或 } q = p(y)\delta y \tag{4-11}$$

在 3.3.1 节中指出了非周期的时间函数的频谱 $G(j\omega)$ 应该当成一个密度函数来处理，因为在一些窄带频率上讨论频谱能量是最合适的。随机信号的概率密度函数也涉及类似的情况：讨论在一个连续信号上选定的特殊值 y 的概率是不合适的，只有讨论落在一个窄范围的 y 与 $(y+\delta y)$ 的概率才是合适的。

举一个简单的例子，假定有一个连续的随机信号 $f(t)$，其中，对于任意的 t 有 $-t<f(t)<t$，并且在整个区间上不偏向任何一个特殊的范围。换句话说，当 $-1<y<1$ 时，$p(y)$ 是一个具有常值的"矩形"或"偶"分布，如图 4.4 所示。$p(y)$ 的实际数值可以通过考虑一个小的数值范围[如 y_1 到 $(y_1+\delta y_1)$]得到。在这个范围上，信号的概率为

$$\frac{\delta y_1}{1-(-1)} = \frac{\delta y_1}{2} = p(y_1)\delta y_1 \tag{4-12}$$

$$p(y_1) = 0.5 \tag{4-13}$$

因此，在这个具体的例子中，当 $-1 < y < 1$ 时，概率密度是一个常数 0.5。对任何形式的函数 $p(y)$，信号在 $(-\infty, \infty)$ 范围上的概率一定为 1，即

$$\int_{-\infty}^{\infty} p(y)\mathrm{d}t = 1 \tag{4-14}$$

换句话说，任何概率密度函数必须是单位面积。

4.3 振幅分布和矩

我们已经看到两个概率的例子。它们描述了一个随机信号呈现一个特殊值，或者值落在一个狭窄区域范围内。这样的函数通常被当成"振幅分布"，并且像其他用来描述随机信号的长式子或平均的方法一样，也可以用来描述确定信号。例如，如图 4.5(a) 所示的正弦波，相对来说它可以接近 $\pm A$，但决不会超过 $\pm A$。振幅概率密度曲线如图 4.5(b) 所示。然而，总的来说，这类幅值概率函数并不用于描述确定信号，因为通常这类幅值概率函数本身简洁的（完整的）分析描述就足够了。

另一个要点是信号的幅值分布并不能详细给出它的结构和频谱信息，它也不是信号的唯一属性。图 4.6 表明一个随机信号和 3 个确定信号波形具有相同的幅值分布，但是它们有非常不同的结构和频谱。

有时信号的概率小于某一确定值 a，记这个概率为 $P(a)$，有

$$P(a) = \int_{-\infty}^{a} p(y)\mathrm{d}y \tag{4-15}$$

P 通常被当成"累积的"分布函数。相反，如果要求信号的概率大于 b，仅改变积分的上、下限即可，即从 b 积到 ∞。

随机信号一个重要的属性是中心矩，如果信号的幅值分布函数已知，中心矩便可得到。一阶中心矩通常作为函数的平均值或中间值，它和二阶矩或方差都是最重要的，尽管有时会出现更高阶的矩。首先考虑图 4.2 所示的常规随机信号的情况，我们可以假定 n 个离散的值 $y_1, y_2, y_3, \cdots, y_n$。如果我们取非常多次 $(N \to \infty)$ 的采样值，变量 y_1 就会出现 Np_1 次。这里 p_1 是它的概率，y_2 表示出现 Np_2 次，等等。采样值的均值可以通过把所有的采样值乘以它的总采样次数，然后再加起来而获得。因此，均值 $\bar{y}$ 为

$$\bar{y} = \sum_{m=1}^{n} p_m y_m \tag{4-16}$$

式中，大写西格马表示 n 项之和。这个均值又称期望值，通常写作 $E(y)$ 或 $<y>$。当随机信号是连续的，可以取无限多个值时，这个和可以用概率密度来表示。这种情况下，均值表示为

$$\bar{y} = \int_{-\infty}^{\infty} y p(y) \mathrm{d}y \tag{4-17}$$

二阶中心矩或方差通常用符号 σ^2 表示，其中 σ 称为标准方差。方差被定义为函数偏离它的均值的平方的平均值或期望值，因此它是信号波动的度量。根据类似如定义均值的理论，具有 n 个可能值的离散随机信号的二阶中心矩表示为

$$\sigma^2 = \overline{(y-\bar{y})^2} = \sum_{m=1}^{n} (y_m - \bar{y})^2 p_m \tag{4-18}$$

连续随机信号的二阶中心矩为

$$\sigma^2 = \overline{(y-\bar{y})^2} = \int (y_m - \bar{y})^2 p(y) \mathrm{d}y \tag{4-19}$$

我们必须把"中心矩"和函数的"矩"区分开来。前者是用来衡量均值的波动的,而简单的矩的计算并没有把均值考虑进去。具有 n 个可能值的离散随机变量的二阶矩简单表示为

$$\overline{y^2} = \sum_{m=1}^{n} p_m y_m^2 \tag{4-20}$$

很显然,中心矩和矩是紧密相关的,知道其中一个,就可以求出另一个。尤其是二阶矩与二阶中心矩和均值遵循下列关系:

$$\overline{y^2} = \sigma^2 + (\bar{y})^2 \tag{4-21}$$

式(4-21)解释了为什么二阶矩或方差是一个很重要的特征值。正如在 2.4.2 节所指出的那样,信号的均方值(即二阶矩)等于它的平均功率,习惯上可以把信号表示成通过 1Ω 电阻时的电压或电流。若均值的平方表示信号在零频率时(即直流时)的功率,二阶中心矩(σ^2)就应该表示其他所有频率段的功率,所以把它广泛地称为"交流功率"。

4.4 自相关函数和功率谱密度

4.4.1 随机信号频谱属性

我们以上所分析的概率函数并不能描述时域内随机信号的结构,也不能描述它的频谱特性。起初讨论随机信号频谱分析没有意义,因为它是不可预测的,并且用一个分析函数是定义不了的。然而,图 4.7 所示的两个随机信号充分证明了如果一个信号可以定义得很具体,那么对信号的频谱分析是十分有用的。例如,我们把这两个部分当成确定的信号,并且假定当 $-T_0 < t < T_0$ 时有具体的波形而在其他情况时为零,然后再来分析它们的频谱。如果只因为一个信号的频率远大于另一个信号,以及它也有实际的零频率,就很明显地看出它们的频谱是十分不同的。如果我们现在再取出随机信号上持续时间同为 T_0 的另一部分来分析它们的频谱,我们必定会发现由于波形上详细的区分而不同于第一种情况。但是,持续时间 T_0 越长,我们就对分辨出每一个波形属于哪一个母波形就更有信心,因此对一个给定的随机信号的不同部分进行频谱分析得出的结果将是一致的。换句话说,我们希望找到一个有用波谱段进行平均测量,即使波形的任何有限部分的频谱永远都不能期望与原随机信号完全匹配。广泛采用的平均测量就是所谓的功率谱,或者是它的时域内相关的函数——自相关函数。

4.4.2 自相关函数

信号波形的自相关函数(ACF)是在时域特性上对信号的平均测量。因此,当这个信号是随机的时,它可能具有特殊的用途。此外,我们应看到,ACF 在形式上不仅是一个令人好奇的函数,也是一个有价值的函数,并且它是分析随机信号频谱的关键。ACF 可定义为

$$r_{xx}(\tau) = \lim_{T_0 \to \infty} \frac{1}{T_0} \int_{-T_0/2}^{T_0/2} f(t) f(t+\tau) \mathrm{d}t \tag{4-22}$$

从式(4-22)可以看到,ACF 等于信号 $f(t)$ 和它的时移形式乘积的平均,它是施加了时移量的函数。式(4-22)用于描述持续时间无穷大的连续信号。如果式(4-22)用于描述持续时间为有限信号时,如一个单个脉冲,在一个长的时间间隔上,取任何值时,平均乘积将趋向于零。在这种情形下,常使用 ACF 的修正形式,通常称为"有限 ACF",它可被定义为

$$r'_{xx}(\tau) = \int_{-\infty}^{\infty} f_1(t) f_1(t+\tau) \mathrm{d}t \tag{4-23}$$

在采样数据信号时，当时移量等于抽样间隔 T 的倍数时，信号和它的时移形式的乘积仅有非零值，并且 ACF 可被定义为

$$r_{xx}(k) = \lim_{T_0 \to \infty} \frac{1}{(2N+1)} \sum_{m=-N}^{N} x_m x_{m+k} \tag{4-24}$$

这里的 x_m 和 x_{m+k} 代表每隔 kT 秒的两个抽样值，m 为整参数，它的取值是从 $-N$ 到 $+N$，这个和共有 $(2N+1)$ 项。在信号持续时间有限的情况下，应采用上述表达式的"有限"形式，即对信号有效长度相对应的一些项取平均值或单纯地求和。应该说明的是，有限或无限信号的自相关函数的精确定义在不同书本之间都有些不同的地方；然而重要的是，自相关函数的所有表达式是信号和它的时移形式的平均乘积。（在自相关函数里，广泛运用的下标"xx"或"11"表示一个信号和它自身的延时形式的乘积。这不同于非常接近的互相关函数。互相关函数是我们将在下面要用的，它是描述加在两个不同信号间时移的影响，下标通常用"12"或"xy"来表示。）

像前面所描述的一般方法一样，ACF 既可以应用于随机信号，也可以应用于确定信号，它的一些基本属性很容易得到论证。例如，假定有一个信号 $f(t)$ 由两个不同频率和相位角的余弦波组成。

这样，有

$$f(t) = A_1 \cos(\omega_1 t + \theta_1) + A_2 \cos(\omega_2 t + \theta_2) \tag{4-25}$$

得出 ACF 为

$$r_{11}(\tau) = \frac{A_1^2}{2} \cos \omega_1 \tau + \frac{A_2^2}{2} \cos \omega_2 \tau \tag{4-26}$$

因此，可以看出信号 $f(t)$ 中的每个频率成分构成自相关函数的一项，当原始信号在时间变量上时，它的每一项有相同的时移变量，每一项的振幅等于原始信号振幅平方的一半，如图 4.8 所示。在 ACF 中没有出现信号 $f(t)$ 相位角 θ_1 和 θ_2，因此它失去了与相对相位相关的所有信息。尽管上述 ACF 是通过仅由两个频率成分组成的函数 $f(t)$ 得出的，但是正弦函数和余弦函数的正交特性意味着它可以延伸到含有任意多频率成分的信号例子中去。不论相位是多大，原始信号每一个频率成分都引出 ACF 中的一个简单的余弦项。

既然任何 ACF 都是由余弦项组成的，那么它一定是 τ 的偶函数。换句话说，无论时间向前还是向后偏移，信号与它的时移形式的平均乘积是一样的，这个结论论证起来确实很简单。当 $\tau=0$ 时，所有余弦函数都达到正的峰值，因此相互加起来就得出 ACF 的最大可能值。在 τ 取其他值时，ACF 是否再次达到峰值取决于信号中各成分是否是谐相关的，但是在任何情况下都不会超过峰值。ACF 峰值表示为

$$r_{xx}(\tau)\big|_{\tau=0} = r_{xx}(0) = \lim_{T_0 \to \infty} \frac{1}{T_0} \int_{-T_0/2}^{T_0/2} [f(t)]^2 dt \tag{4-27}$$

简单地说，这是信号的均方值，或者称为平均功率。这样 $r_{xx}(0)$ 等于信号的二阶矩，正如我们所见的，也是从它的概率密度函数推导出来的。另外，根据 τ 的不同，绘出的 ACF 曲线给出了信号在时域上的信息，这是在概率密度函数中无法表现出来的。应该说明的是，在持续时间有限的信号的例子中，我们使用的是 ACF 的"有限"形式，关于 $\tau=0$ 时的 ACF 为

$$r_{xx}(0) = \int_{-\infty}^{\infty} [f(t)]^2 dt \tag{4-28}$$

式(4-28)表示的是信号的总能量，而不是在一个长时间范围上的平均功率。

4.4.3 功率谱密度函数

我们已经看到，信号波形中的 $A_1 \cos(\omega_1 t + \theta_1)$ 项是怎样推出它的自相关函数的 $(A_1^2/2) \cos \omega_1 \tau$ 项的。在 4.4.2 节中，已经表明振幅为 A_1 的正弦波的均方差或称为平均功率等于

$A_1^2/2$。因此，ACF 中的不同的余弦项只指出了原信号相应项的平均功率。

正如一个信号波形可以根据它的频谱来描述一样，一个自相关函数(关于时移变量 τ 的函数)在频域中有其相对应的部分。由上述的理论可知，其对应的部分在不同的频率成分下有不同的表示功率的频谱线，我们将其称为功率谱。

在图 4.9 中表明了一个典型的周期信号的频率成分和它的自相关函数的频率的关系。在一个具有连续频谱的非周期信号的例子中，它的 ACF 在频域中的对应部分也是连续的，我们将其称为功率谱密度。

对于在一个非常长的时间区间上平均功率趋近于零的持续时间有限的信号，必须对其做出单独的规定，在这类例子中，我们参考的是"能量谱"而不是功率谱。正如频率谱是 ACF 的常用形式在频域中的对应部分一样，能量谱与 ACF 的"无限"形式是等价的，它描述的是信号能量在频域上的分布。下面的讨论将主要是关于无限连续信号的功率谱，但是在有时间限制的信号中，我们可以假定它的能量谱本质上表现出类似的属性。功率谱和能量谱都将用符号 $P_{xx}(\omega)$ 来表示。

因此，我们看到 ACF 和功率谱或功率谱密度是在时域和频域的等量方法。换言之，通过傅里叶变换它们是相关的。它们的关系通常是由所谓的 Wiener-Khinchin 关系式来表示的。

$$P_{xx}(\tau) = \int_{-\infty}^{\infty} r_{xx}(\tau) e^{-j\omega\tau} d\tau \tag{4-29}$$

和

$$r_{xx}(\tau) = \frac{1}{2\pi} \int_{-\infty}^{\infty} P_{xx}(\omega) e^{-j\omega\tau} d\tau \tag{4-30}$$

此处 $P_{xx}(\omega)$ 是功率谱密度。如果我们使用的是延时变量 τ 而不是时间变量 t 的话，通过傅里叶变化和反变换可以把信号 $f(t)$ 和它的频谱 $G(j\omega)$ 相对应起来，按照傅里叶变换和反变换的形式，这些等式是完全相同的。

第 5 课 静 态 性 能

5.1 理想测量系统

理想测量系统的输出量与被测量是线性关系，在理想测量系统中不会引入像静态摩擦那样产生的误差，而且无论输入量如何，输出量总是输入量的真实反映。当然，这只是理论上的一种情况，理想测量值的作用只是和实际得到的测量值进行比较。一个测量系统(或仪器)是否是理想测量系统，通常以误差来界定，误差被定义为测量值与"真值"的差异。

这里给出的术语"真值"是指从仪器或测量系统中获得的值，目的是提供给系统作为输入量，而这些由专家认定的测量系统是精确可靠的。因此，在校准相对于净重测量仪的压力表时，下面读出的值将被作为"真值"。

检查测量和控制系统的性能有两种简单的方法。
(1) 当输入信号恒定不变时，稳定输出量和理想值比较后，得到系统的静态性能。
(2) 当输入信号变化时，输出量与理想值比较，给出系统的动态性能。

5.2 灵敏度

静态灵敏度被定义为输出变化量与相应的输入变化量的比值，即

$$k = \frac{\Delta \theta_o}{\Delta \theta_i} \tag{5-1}$$

式中，$\Delta\theta_o$ 为输出变化量；$\Delta\theta_i$ 为相应的输入变化量。

灵敏度有许多不同的单位，采用什么单位主要取决于仪器或测量系统。举例来说，铂电阻温度计根据温度的变化，其电阻值随之变化，因此其灵敏度的单位是 Ω/℃。

图 5.1(a)给出了输出量与输入量的线性关系，由此得出灵敏度等于标度图的斜率。非线性输入-输出关系如图 5.1(b)所示。灵敏度随着输出量的变化而变化。

记录仪和显示设备制造商倾向于提供灵敏度的相反值。举例来说，示波器的灵敏度是每厘米多少伏，而不是按定义得到的每伏多少厘米。

如果将具有静态灵敏度为 $K_1, K_2, K_3 \cdots$ 的系统中各个元件按顺序级联起来，如图 5.2所示，那么整个系统的灵敏度为

$$K = K_1 K_2 K_3 \cdots \tag{5-2}$$

假定在载荷影响下，$K_1, K_2, K_3 \cdots$ 的值没有任何变化。

经常发现系统元件的输入量和输出量是相同的物理量。例如，电压放大器中的增益，更多地称为电压增益而不是灵敏度；在使用杠杆系统的机械系统中，更频繁地使用放大倍数来描述位移的增加。术语"灵敏度""增益和放大倍数"的含义都相同。

例 5.1 一个测量系统由一个传感器、一个放大器和一个记录仪组成，各个仪器的灵敏度如下：

传感器灵敏度　0.2mV/℃

放大器增益　2.0V/mV

记录仪灵敏度　5.0 mm/V

利用式(5-2)，整个系统的灵敏度为

$$K = K_1 K_2 K_3 = 0.2\text{mV/℃} \times 2.0\text{V/mV} \times 5.0 \text{ mm/V} = 2.0\text{mm/℃}$$

5.3　准确度和精度

既然所有的测量都有误差，那么测量系统是精确的吗？这个问题毫无意义，因为答案总是否定的。"测量系统精度是多少"的答案更重要。

准确度通常引入误差来描述，在此有

$$\text{误差百分比} = \frac{\text{测量值-真值}}{\text{真值}} \times 100\%$$

但是，实际中，误差通常用仪器测量范围来表示，即

$$\text{误差百分比} = \frac{\text{测量值-真值}}{\text{最大刻度值}} \times 100\% \tag{5-3}$$

举例来说，如果 0 到 1bar 的压力表能够精确到 95%，那么误差将是 ±0.05。注意，如果该压力表用于测量范围的低端，那么 ±0.05bar 的误差与"真值"相比将导致更大的误差百分比。

例 5.2 当制造商标定压力表时，0 到 10bar 的压力表已经有 ±0.15bar 的误差。试计算：

(1)仪器的误差百分比。

(2)当测量得到的读数是 2.0bar 时，把可能的误差作为测量值的百分比。

(1)由式(5-3)得

$$\text{误差百分比} = \pm \frac{0.15\text{bar}}{10\text{bar}} \times 100\% = \pm 1.5\%$$

(2) 可能的误差＝±1.5bar，则

$$2.0 \text{bar 时的误差} = \pm \frac{0.15 \text{bar}}{2.0 \text{bar}} \times 100\% = \pm 7.5\%$$

因此，压力表越在低端测量越不可靠，实际应用中应该使用测量范围更大的压力表。

"精确度"这一术语有时常和准确度混淆，但是一个精确的测量仪可能不是一个准确的测量仪。当测量仪进行多次测量时，如果其输入量都是相同的，且测量结果都很近似，那么可以说这个测量仪是高精度的，过去把这种测量结果的近似性称为仪器重复性。

如果一块高质量的电压表被用来测量各种条件下恒定不变的电压（在仪器的准确度范围内），并且所有测量结果的读数是相同的，可以说这个读数是精确的。设想一下，实验做完后，当把电压表放在一边，可以注意到指针有偏移而且不指向零点。所有获得的读数可能是精确的，但不是准确的，图5.3以图解法解释了这两者的区别。

5.4 可能误差与概率误差

考虑一个涉及3个设备的测量系统，每个设备最大的可能误差分别是$\pm a\%$、$\pm b\%$和$\pm c\%$。3个设备同时有最大误差不太可能，所以用来表示整个系统误差更实际的方法是各个误差均方根，即

$$\text{整个系统的均方根} = \pm \sqrt{a^2 + b^2 + c^2} \tag{5-4}$$

例5.3 对于一个一般的测量系统，传感器、信号调节装置和记录仪上的误差分别是$\pm 2\%$、$\pm 3\%$和$\pm 4\%$，试计算系统的最大可能误差和概率误差或均方根误差。

$$\text{最大可能误差} = \pm(2+3+4)\% = \pm 9\%$$

由式(5-4)可得

$$\text{均方根误差} = \pm \sqrt{2^2 + 3^2 + 4^2}\% = \pm \sqrt{29}\% = \pm 5.4\%$$

因此，系统最大可能误差是$\pm 9\%$，而概率误差不超过$\pm 5.4\%$。

5.5 其他的静态性能术语

一致性 一个通用术语，用于表述一个仪表测试系统在许多不同情况下，对给定输入量显示同一读数的能力。

重复性 在给定使用条件下且短时间内恒定输入量的情况下，重复试验表现出的一致性。

稳定性 在与读数时间相比的长时间内、恒定输入量的情况下，试验结果表现出的一致性。

恒定性 在连续施加一个恒定输入量且测试条件在一定限制内允许变化（如温度变化等外部效应）时系统表现出的一致性。

量程 仪器或测量系统所能测量的整个范围。

范围 与设计好的输出信号工作范围相对应的输入信号范围。

公差 最大的误差。

线性度 输入量和输出量之间线性关系的最大偏差，即恒定灵敏度，以满量程的百分比表示。

分辨率 仪器所能检测到的最小输入变化，以满量程的百分比表示。

死区 由于摩擦或间隙效应，对系统不响应的最大输入变化量，以满量程的百分比表示。

滞后 对于相同的输入量，在输入量增加和减小时，读数的最大差值，以满量程的百分比表示。

第6课 动态性能

许多工业过程要求测量的参数是常量或缓慢变化量，如化学过程中的恒温、恒压，这类参数主

要反映的是测量系统的静态性能。但是，随着自动控制的发展，更多的重点放在了装置是否能够充分响应变化的信号。如果一个传感器对于一个输入参数的变化响应延迟，那么自动控制这个参数的可能变得很困难，甚至是不可能的。再举一个例子，例如，一个振动测量系统，它的参数是一个变化量，如果系统不能对振动频率做出响应，那么结果将毫无意义。

测量和控制系统的动态性能是非常重要的，并且由一些标准测试输入信号来详细说明。

(1) 阶跃输入信号：一个稳定值突然跃变到另一个值。阶跃输入信号能够反映系统应付变化及瞬态响应的能力。

(2) 斜坡输入信号：随着时间做线性变换，并做出线性响应。斜坡输入信号能够反映输入信号后的稳态误差。

(3) 正弦输入信号：给出系统的频率响应或谐波响应。正弦输入信号能够反映对频率 $f(\text{Hz})$ 或 $\omega(\text{rad/s})$ $(\omega=2\pi f)$ 作为循环输入量的响应能力。

以上3种输入信号如图6.1所示，在前面章节也曾遇到过。但是在这里，只考虑瞬态和频率响应，原因是做测试系统说明时需要这些。

所有的系统并不能完全准确地响应输入的变化，衡量一个系统响应程度的标准是它的动态特性。根据所应用的输入信号类型，用阶跃参数、瞬态参数或频率响应参数来表示。虽然输入信号种类不同，许多系统也能产生相同的响应曲线，这是由于系统动态特性是相似的，动力学或微分方程具有相同的形式。

6.1 零阶系统

在5.1节中提及的理想测量系统的特点是：无论输入量如何变化，它的输出量与输入量是成比例的。它的数学方程为

$$\theta_\text{o} = K\theta_\text{i} \tag{6-1}$$

式中，K 为系统灵敏度。

这就是零阶系统的数学方程，因为有相同的系数，式(6-1)还可以写成

$$\frac{\theta_\text{o}}{\theta_\text{i}} = K \tag{6-2}$$

该系统可以用结构图来表达（见图6.2）。

在实际测量系统中，最接近零阶测量系统的元件是电位器，电位器的输出电压与电刷（滑动触头）的位移成比例。

6.2 一阶系统

一个一阶系统的输入-输出动力表达式是一阶微分方程，即

$$a\frac{\mathrm{d}\theta_\text{o}}{\mathrm{d}t} + b\theta_\text{o} = c\theta_\text{i}$$

式中，θ_o 为输出变量；θ_i 为输入变量；a、b、c 为常数。

为了使得 θ_o 的系数为1，式(6-3)可以重新表示为

$$\frac{a}{b}\cdot\frac{\mathrm{d}\theta_\text{o}}{\mathrm{d}t} + \theta_\text{o} = \frac{c}{b}\cdot\theta_\text{i} \tag{6-3}$$

并能以标准形式表示为

$$\tau\frac{\mathrm{d}\theta_\text{o}}{\mathrm{d}t} + \theta_\text{o} = K\theta_\text{i} \tag{6-4}$$

式中，K 为静态灵敏度(单位由 θ_o/θ_i 获得)。

比较式(6-3)与式(6-4)，可得

$$\tau = \frac{a}{b}, \quad K = \frac{c}{b}$$

式(6-4)可用 D 算子形式表达，其中

$$D \equiv \frac{d}{dt}, \quad D^2 \equiv \frac{d^2}{dt^2} \text{ 等}$$

因此，$\tau D\theta_o + \theta_o = K\theta_i$，即 $(\tau D+1)\theta_o = K\theta_i$，则

$$\frac{\theta_o}{\theta_i} = \frac{K}{1+\tau D} \tag{6-5}$$

以 D 算子形式表示的比率 θ_o/θ_i 称为系统的传递算子。

对于一阶系统，式(6-5)表示传递算子的标准形式，其结构图如图 6.3 所示。需要指出的是，式(6-3)、式(6-4)、式(6-5)都可以表达相同的微分方程。

一阶系统的例子包括：
(1) 水银温度计，温度通过玻璃球以一阶微分方程形式传导给水银。
(2) 一个节气阀或波纹管系统的气压结构。
(3) 一个串联电阻-电容网络。

例 6.1 水银温度计的微分方程为

$$4\frac{d\theta_o}{dt} + 2\theta_o = 2 \times 10^{-3}\theta_i$$

式中，θ_o 为水银柱高度(m)；θ_i 为输入温度(℃)。测定水银温度计的时间常数和静态灵敏度。

根据式(6-4)所示标准形式，θ_o 的系数必须为 1，因此水银温度计的微分方程两边都除以 2 得

$$2\frac{d\theta_o}{dt} + \theta_o = 1 \times 10^{-3}\theta_i$$

将此式与式(6-4)相比较，即

$$\tau\frac{d\theta}{dt} + \theta_o = K\theta_i$$

可以得到

$$\tau = 2\text{s}$$

式中，$K = 10^{-3}\text{m/℃}(1\text{mm/℃})$。

式(6-4)和式(6-5)所表达的标准形式是很方便的，因为一阶系统无论对阶跃输入信号还是正弦输入信号，经常产生标准的响应。

6.2.1 阶跃响应

图 6.4(a)显示，最后的值呈指数上升，这是一阶系统的特征。动态误差是理想响应值和实际值之间的差距，两者之间的比较显示，误差随时间增加而减小。阶跃响应在图 6.4(b)得到更详细的反映，并且从图 6.4(b)中可以得到时间常数的定义：

如果保持初始速率，或者时间达到阶跃变化的 63.2%，时间常数就为达到最终值的时间。

值得注意的是，一阶系统不同于二阶系统，一阶系统在时间 $t=0$ 时的阶跃响应的初始斜率为零。

6.2.2 频率响应

通过已知振幅的正弦波获得响应，并且检查随着正弦波频率的变化输出响应的变化情况。

图 6.5 说明了系统不能如实地随输入信号的变化而变化，并且可以看出，输出信号滞后于输入信号。随着频率增加，输出信号更加滞后，并且信号振幅衰减。

输出信号振幅与输入信号振幅的比值称为振幅比。如果不考虑输入信号频率，任何时候振幅比都等于常量，如 $\theta_o = K\theta_i$。图 6.6 是一阶系统标准频率响应曲线，在 $K=1$ 或振幅比被认为是实际输出信号振幅与理想输出信号振幅的比值时，振幅比随着频率的变化而变化。频率轴通常表示为 $\omega\tau$，并且使曲线能够用于不同的时间常数。为了获得某一特定时间常数 τ 下 ω(rad/s) 的值，水平轴的单位要乘以因子 $1/\tau$。

例 6.2 如果一阶测试系统的时间常数为 0.01s，测定系统的频率保持在 10% 的输入信号频率的近似范围。

如图 6.6 所示，振幅比大于 0.09，即误差小于 10%，使 $\omega\tau \approx 0.5$。

∴ $\omega\tau = 0.5$ 定义了频率上限，即

$$\omega = \frac{0.5}{0.01}\text{rad/s} = 50\text{rad/s}$$

因此，必需的频率范围是 $0 \sim 50/2\pi$(Hz)，即 $0 \sim 8$(Hz)。

例 6.3 一阶系统的时间常数为 6ms，确定 $\omega\tau = 1$ 时的频率，并计算该频率下的误差百分比。

$$\omega = \frac{1}{\tau} = \frac{1}{6 \times 10^{-3}}\text{rad/s} = 166.7\text{rad/s}$$

∴ 频率 $f = \frac{166.7}{2\pi}\text{Hz} = 26.5\text{Hz}$

由图 6.6 可知，$\omega\tau = 1$ 时的振幅比近似为 0.7。

∴ 误差百分比 = 30%

6.3 二阶系统

二阶系统的输入-输出关系为

$$a\frac{d^2\theta_o}{dt^2} + b\frac{d\theta_o}{dt} + c\theta_o = e\theta_i \tag{6-6}$$

式中，a、b、c、e 为常量。

重新整理 $\frac{d^2\theta_o}{dt^2}$ 的系数，即

$$\frac{d^2\theta_o}{dt^2} + \frac{b}{a}\frac{d\theta_o}{dt} + \frac{c}{a}\theta_o = \frac{e}{a}\theta_i$$

以标准形式书写为

$$\frac{d^2\theta_o}{dt^2} + 2\xi\omega_n\frac{d\theta_o}{dt} + \omega_n^2\theta_o = K'\theta_i \tag{6-7}$$

式中，ω_n 为系统固有频率(rad/s)；ξ 为阻尼比；K' 为常量(如果静态下 $\theta_i = \theta_o$，则 $K' = \omega_n^2$)。

引入 D 算子，式(6-7)可变为

$$(D^2 + 2\xi\omega_n D + \omega_n^2)\theta_o = K'\theta_i$$

给出传递算子，即

$$\frac{\theta_o}{\theta_i} = \frac{K'}{D^2 + 2\xi\omega_n D + \omega_n^2} \tag{6-8}$$

或者除以 ω_n^2，得

$$\frac{\theta_o}{\theta_i} = \frac{K}{(1/\omega_n^2)D^2 + (2\xi/\omega_n)D + 1} \tag{6-9}$$

式中，K 为静态灵敏度。

传递算子可以用结构图表示，如图6.7所示。

ω_n是二阶系统响应速度的衡量尺度，ω_n越大表示系统能够更快地响应突变。ξ是系统阻尼的衡量尺度，等于实际阻尼与临界阻尼之比，其值决定阶跃响应的形状和频率。

(1) $\xi<1$　系统欠阻尼，阶跃响应下有振荡，而且频率响应中($\xi<0.707$时)有共振。此处的共振是指输出信号的振幅比理想输出信号的振幅更大。

(2) $\xi=1$　临界阻尼，阶跃响应下没有振荡或超调，而且频率响应中没有共振。这是从欠阻尼到过阻尼的转折点。

(3) $\xi>1$　系统过阻尼，响应处在迟滞状态，阶跃响应无超调，频率响应无共振。

最常见的二阶系统的例子是带阻尼的弹簧系统。大量的设备和机械装置都是这一类型，如紫外线镜反射式电流计、压电传感器、x-y绘图仪的开环控制系统等。

6.3.1 阶跃响应

二阶系统阶跃响应并不具有唯一的形式，根据ξ的不同值有许多形式。

图6.8显示一组ξ值下的阶跃响应，它反映了随着系统的阻尼减小，超调和振动随之增大。一组形状相同的曲线出现在以$\omega_n t$为时间坐标的图中，较大的ω_n值表示的是快速系统，较小的ω_n值表示的是慢速系统。图6.9显示了第一次超调和阻尼比的关系曲线，它可以用来估计欠阻尼阶跃响应的ξ值。"超调率"是指第一次超调值与最终值或稳态值的百分比。

例6.4　一个质量—弹簧阻尼系统，当输入一个阶跃外力时，第一次超调量为40%的稳定值。设从施加阶跃外力开始，第一次超调量的时间为0.8s，试求阻尼比ξ和ω_n。

如图6.9所示，$\xi=0.28$时超调量为40%，设$\xi=0.25$，则由图6.8可推出
$$\omega_n t = 3.2$$
$$\therefore \omega_n = \frac{3.2 \text{rad}}{0.8 \text{s}} = 4 \text{rad/s}$$

例6.4的解题思路是通过标准曲线图，提供了一个迅速、方便的由阶跃响应的超调百分比估计ξ的方法。

就阶跃响应而言，可以看出ξ越小响应越快但超调量增大，而ξ越大响应越迟缓但超调量为零。因此，最优环境是在可接受的响应速度和超调量之间进行协调。二阶测量系统通常规定$\xi \approx 0.7$。

6.3.2 频率响应

图6.10给出了一组典型的二阶系统频率响应曲线。振幅比轴以$K=1$为基准，绘制频率轴以适用于所有二阶系统曲线。在欠阻尼情况下，当输入频率接近于系统固有频率时，该系统产生共振。不规则的高输出响应仅当阻尼率小于0.7时产生。如果输入频率增大而超过固有频率，则由于系统不能响应更高的变化率，振幅率下降。

无论对测量系统还是自动控制系统，理想的频率响应都应对所有的频率有统一的振幅比。最符合这个理想状态的ξ介于$0.6 \sim 0.7$之间，此时振幅比是个常量，误差在$-3\% \sim 3\%$之内，频率范围在无阻尼固有频率f_n的$-60\% \sim 60\%$之间。

对于阶跃响应和频率响应，ξ的最优值介于$0.6 \sim 0.7$之间，通常取0.64，测量系统的制造商必须确保当前阻尼值是正确的。这里所谓的"最优"是指对阶跃信号和正弦信号最佳的响应，但是特殊设计的过阻尼设备对于变化的信号仅给出平均值，在这种情况下，ξ的值越大，效果越好。

例6.5　在一个二阶系统频率响应测试中，已知最大振幅比为1.36，当频率为216Hz时发生共振，试求系统ξ和ω_n。

由图 6.10 可得

$$\xi \approx 0.4$$

则振幅比为 1.4，此共振发生在 $\omega/\omega_n = 0.8$，
又

$$\omega = 2\pi f = 2\pi \times 216\text{Hz} = 1357\text{rad/s}$$

$$\omega_n = \frac{\omega}{0.8} = \frac{1357\text{rad/s}}{0.8} = 1696\text{rad/s}$$

注意，当频率低于无阻尼固有频率时系统发生共振，但当阻尼为 0 时是个例外。

6.4 阶跃响应指标

描述系统的阶跃响应应用 3 个指标：响应时间、上升时间、稳定时间。

响应时间(t_{res})：系统输出量从 0 上升至系统稳态值的 100% 的时间。响应时间仅适用于欠阻尼系统。

上升时间(t_r)：系统输出量从最终稳态值的 10% 上升至 90% 的时间。

稳定时间(t_s)：系统输出量到达并稳定在稳态值的一定误差内的时间，误差典型值为 2% 和 5%。这些参数如图 6.11 所示，阶跃信号存在振荡。在图 6.11 中，t_s 是指误差为 5% 的稳定时间，此时系统输出量达到并维持在稳定值的 95%～105% 的范围内。

6.5 频率响应指标

有许多方法标定测量系统、仪器、控制系统的频率响应，而这容易导致混淆。可以在装置之间进行一个划分：工作于一个频率范围的装置称为交流装置；工作于直流的装置(即频率低至 0)称为直流装置。

6.5.1 交流装置

对于此类型的装置，增益或振幅比通常为一给定频率范围内的常数，但在低频或高频时，振幅比下降，如图 6.12 所示，用带宽表示频率范围，其值等于 $(f_2 - f_1)$Hz。

带宽是振幅比维持在 -3dB 内的频率范围。

因此，一个示波器中的交流放大器具有典型的 8～10Hz 频带，也就是说，在 8Hz 与 10Hz 之间，一个恒定振幅的输入信号的跟踪频率是中间频率的 70%，即 1kHz。因此，从示波器读到的频率值比实际值低 30%。

6.5.2 直流装置

在这方面，直流装置并不是指仅响应稳态或直流信号的装置，而是指可以同时响应交流和直流信号的装置。对于这种设备的典型频率响应如图 6.13 所示，例如，它可用于笔式记录仪。在此，工作范围就是曲线上部的频率，这一部分的增益或频率比下降到误差带范围外。图 6.13 所示误差为 ±3%，而其他常见误差为 ±5%，即前面提到的 -3dB 或 ±1dB(等同于 ±10%)。

第 7 课 传感器和电阻式传感器的基础知识

传感器是一种把被测量转换为光的、机械的或更平常的电信号的装置。能量转换的过程被称为换能。

按照转换原理和测量形式对传感器进行分类，可把用来测量位移的电阻式传感器归为电阻式位移传感器，其他的传感器可分为压力波纹管、压力膜和压力阀等类型。

7.1 传感器

除特例，大多数的传感器都由敏感元件、转换器或控制元件组成，如图7.1所示。

如振动膜、波纹管、应力管和应力环、低音管和悬臂都是敏感元件，它们对压力和力做出响应，把物理量转变成位移。然后，位移可以改变电参数，如电压、电阻、电容或感应系数。机械式和电子式元件合并形成机电式传感设备或传感器。这样的组合可用来输入能量信号。热的、光的、磁的和化学的元件相互结合形成热电式、光电式、电磁式和电化学式传感器。

7.2 传感器灵敏度

通过校正测量系统获得的被测物理量和传感器输出信号的关系称为传感器灵敏度 K_1，即

$$K_1 = \frac{输出信号增量}{测量增量}$$

实际中，传感器的灵敏度是已知的，通过测量输出信号，输入量为

$$输入量 = \frac{输出信号增量}{K_1}$$

如图7.2所示，当弹簧受到10kN的作用力时，弹簧伸长0.05m，试求弹簧伸长0.075m时的作用力大小。

灵敏度为

$$K_1 = \frac{x}{F} = \frac{0.05}{10\text{kN}}$$

所以，弹簧伸长0.075m时所需的作用力为 $0.075\text{m} \times \frac{10\text{kN}}{0.05\text{m}} = 15\text{kN}$。

7.3 理想传感器的特性

高质量的传感器应该具有以下特性。
(1) 高保真性：传感器输出波形应该真实可靠地再现被测量，并且失真很小。
(2) 对被测量的干扰最小：任何时候，传感器的出现要尽可能地不改变被测量。
(3) 尺寸：传感器必须能尺寸合适地放在所需的地方。
(4) 被测量和传感器信号之间应该有一个线性关系。
(5) 传感器对外部影响的灵敏度应该小，例如，压力传感器经常受到外部振动和温度的影响。
(6) 传感器的固有频率应该避开被测量的频率和谐波。

7.4 电传感器

电传感器具有许多理想特性。电传感器不仅具有远程测量和显示的功能，还具有很高的灵敏度。
电传感器可分为以下两大类。
(1) 变参数型，包括：
　　a) 电阻式；
　　b) 电容式；
　　c) 自感应式；
　　d) 互感应式。

这些传感器的工作依靠外部电压。

(2) 自激型，包括：

a) 电磁式；

b) 热电式；

c) 光栅式；

d) 压电式。

这些传感器根据测量的输入量产生输出电压，而且这一过程是可逆的。例如，在一般情况下，压电式传感器可根据晶体材料的变形产生一个输出电压；但是，如果在材料上施加一个可变电压，该传感器可以通过变形或与变电压同频率的振动来体现可逆效应。

7.5 电阻式传感器

电阻式传感器可以分为以下两大类。

(1) 那些表现为大电阻变化的物理量可通过分压方式进行测量，电位器就属于此类。

(2) 那些表现为小电阻变化的物理量可通过桥电路方式进行测量，这一类包括应变仪和电阻温度计。

7.5.1 电位器

绕线式电位器由许多绕在非导体骨架的电阻丝及滑行在线圈上的触头组成。电位器的结构原理如图 7.3(a)和(b)所示，触头能够转动、直线运动或这两运动合成的螺旋式运动。

如果测量设备的电阻比电位器的电阻大，那么激励电压既可以是交流电压也可以是直流电压，且输出电压与输入的运动量成正比。

这样的电位器存在着分辨率和电子噪声的问题。分辨率是指传感器能检测到的最小输入增量，分辨率大小取决于线圈与滑动触头围成的面积。因此，输出电压为触头从一端移到另一端时的一系列阶跃信号，如图 7.4(a)所示。

电子噪声(即有害信号)产生的原因可能是接触电阻的振动、触头摩擦形成的机械磨损或从敏感元件传出的触头振动。另外，测得的运动量可以通过惯性和电位器中移动元件的摩擦获得较大的机械载荷。触头表面的磨损将电位器的寿命限制为一定的转数。这个转数通常指的是生产商在说明书中提及的"寿命转数"，其典型值为 $20×10^6$ 转。

如图 7.4(b)所示的空载电位器电路的输出电压 V_o 可由下面的公式推导得出。

设电阻为

$$R_1 = \frac{x_i}{x_T} R_T$$

式中，x_i 为输入位移(m)；x_T 为最大可能位移(m)；R_T 为电位器的总电阻(Ω)。那么，输出电压为

$$V_o = V\frac{R_1}{R_1 + (R_T - R_1)} = V\frac{R_1}{R_T} = V\frac{x_i}{x_T} \cdot \frac{R_T}{R_T} = V\frac{x_i}{x_T} \tag{7-1}$$

式(7-1)表明，对于空载电位器，其输出电压和输入位移呈直线关系。

通过提高激励电压 V 来获得电位器式电阻传感器高的灵敏度。但是，V 的最大值由电位器线圈金属丝的功率损耗 P 决定，即

$$V = \sqrt{PR_T} \tag{7-2}$$

例 7.1 一个电位器式电阻传感器的总线圈电阻为 10kΩ，最大位移为 4mm。如果最大功率损耗不超过 40mW，激励电压为最大，输入位移为 1.2mm，那么就可以确定输出电压。

运用式(7-2),激励电压为

$$V=\sqrt{PR_T}=\sqrt{0.04\text{W}\times10000\Omega}=20\text{V}$$

由式(7-1)可得

$$V_o=V\frac{x_i}{x_T}=20\text{V}\times\frac{1.2\text{mm}}{4\text{mm}}=6\text{V}$$

载荷电位器

当在电位器两端连接一个测量设备如仪表,以及一个电阻 R_L,电流流入仪表,这样就可在电位器上产生负载效应,从而产生一个与不同于线性的输入-输出关系图,如图7.5所示。

在负载条件下,电路特性为

$$V_o=V\left[\frac{x_T}{x_i}+\frac{R_T}{R_L}\left(1-\frac{x_i}{x_T}\right)\right]^{-1} \tag{7-3}$$

这个电路的特性是非线性的,而且随着 R_T/R_L 比值的增加,非线性也增加。

例 7.2 当负载电阻为电位器电阻值的两倍时,在触头运动到满量程的50%时,计算电阻式电位器的误差。

运用式(7-1)和式(7-3),空载输出电压为

$$V_o=\frac{V}{2}=0.5\text{V}$$

运用式(7-3),载荷输出电压为

$$V_o=V\left[\frac{1}{2+1/2(1-0.5)}\right]=\frac{V}{2.25}=0.44\text{V}$$

因此,有

$$误差=\frac{0.44\text{V}-0.5\text{V}}{0.5\text{V}}\times100\%=-1.2\%$$

注意,负号表示读数太小。

7.5.2 电阻应变仪

电阻应变仪是由机械应变产生电阻变化的传感器。电阻应变仪可以分为耦合应变仪或非耦合应变仪。

1. 耦合应变仪

运用黏合剂可将电阻应变仪与被检测的结构或部件的表面黏合或黏牢。

耦合应变仪分为:

a) 黏合在绝缘纸背后的金属细丝仪;

b) 在环氧树脂上粘贴导电箔片的光栅;

c) 在环氧树脂上粘贴铜或镍的半导体丝。

耦合应变仪既可作为单个元件,仅在一个方向测量应力,又可作为几个元件的组合体(如芯状组合体),可在几个方向同时测量应力。

2. 非耦合应变仪

典型的非耦合应变仪如图7.6所示,它表明细电阻丝在悬臂弹簧偏差作用下,通过改变电阻丝张力来改变电阻丝的阻值。在实际应用中,通常在力、负载、压力传感器上使用非耦合应变仪。

7.5.3 电阻温度传感器

电阻温度传感器的材料有以下两大类。

(1) 金属，如铂、铜、钨、镍的阻值会随着温度的升高而增大，即有一个正温度电阻系数。

(2) 半导体，如用锰、钴、铬或镍的氧化物制成的电热调节器，其阻值变化与温度变化存在一个非线性关系，即通常有一个负温度电阻系数。

1. 金属电阻温度传感器

在窄温度变化范围内，此类传感器的电阻为

$$R_1 = R_0[1+\alpha(\theta_1-\theta_0)] \tag{7-4}$$

式中，α 为阻抗系数；R_0 为 $\theta_0=0℃$ 时的电阻。

国际上，通用的温度刻度以铂电阻温度计为准，其温度范围为 -259.35~630.5℃。

铂电阻温度计的典型特征曲线如图7.7所示。

例7.3 如果铂电阻温度计的电阻在 0℃ 时为 100Ω，那么在 60℃ 时电阻为多少（$\alpha=0.003\ 92℃^{-1}$）？

由式(7-4)可得

$$R_1 = R_0[1+\alpha(\theta_1-\theta_0)] = 100\Omega \times [1+0.003\ 92 \times 60] = 123.5\Omega$$

2. 电热调节器（半导体）电阻温度传感器

电热调节器为感温电阻器，其阻值变化与温度变化呈非线性关系。通常此类传感器有一个负温度系数，如图7.8所示。对于小的温度增量，电热调节器阻值变化与之大体呈线性关系，但是如果存在大的温差，测量电路要运用特定线性化技术，使其阻值随温度变化呈线性关系。

电热调节器通常被制成附有玻璃质瓷釉的半导体圆盘形状。由于电热调节器的长度可以小到 1mm，所以响应时间非常快。

例7.4 如图7.9所示电路中的仪表读数为满量程的一半时，请运用图7.8中类型1的电热调节器的特征曲线确定测量温度。

总电阻为

$$R = \frac{V}{I} = \frac{10\text{V}}{0.5 \times 10^{-3}\text{A}} = 20\text{k}\Omega$$

∴ 电热调节器电阻 = 20kΩ - 5kΩ = 15kΩ（忽略仪表阻抗）

因此，特征温度 ≈ 20℃。

7.5.4 光敏元件

光敏元件如图7.10所示，采用光敏半导体材料制成。当照射在半导体上的光强度增大时，金属电极间的阻抗就会降低。光敏元件常用的光敏半导体材料有硫化镉、硫化铅和铜锗化合物。

光敏元件的频率的有效范围由所用材料决定。硫化镉主要适用于可见光，硫化铅在红外线区有峰值响应，所以光敏元件最适合于光故障检测及温度测量。

7.5.5 放射性光元件（可变导体或可逆电阻）

当光照射到放射性光元件（见图7.10）的阴极时，电子就会获取足够能量到达阴极。放射性光元件的阴极会吸收这些电子，产生一个通过电阻 R_L 的电流，从而形成一个输出电压 v_o。

产生的光电压为

$$v_o = I_p R_L \tag{7-5}$$

式中，I_p 为光发射电流；$I_p = K_t \Phi$，K_t 为灵敏度（μA/lm），Φ 为输入照度（lm）。

尽管放射性光元件的输出电压能够表示照明的强度，但放射性光元件却更多地应用于计算或调节，这里照射到阴极的光可被中断。

例 7.5 一个放射性光元件与一个 5kΩ 电阻串联。如果此元件灵敏度为 30μA/lm，当其输出电压为 2V 时，试计算输入照度。

由式(7-5)可得
$$v_o = I_p R_L = K_t \Phi R_L$$

∴ 照度 $\Phi = \dfrac{v_o}{K_t R_L} = \dfrac{2V}{3 \times 10^{-6} A/lm \times 5 \times 10^3 \Omega} = 13.3 \text{ lm}$

第 8 课 电容式、电感式传感器及其他器件

8.1 电容式传感器

平行极板电容器的电容为
$$C = \varepsilon_0 \varepsilon_r \frac{A}{d} \tag{8-1}$$

式中，ε_0 为空气的介电常数(8.854×10^{-12} F/m)；ε_r 为极板间介质的相对介电常数；A 为极板的遮盖面积或有效面积(m^2)；d 为极板间的距离(m)。

平行极板电容器的电容随着 ε_r、A 或 d 的变化而变化。电容式传感器的一些例子如图 8.1 所示。

图 8.2 中的特征曲线表明，在空间的一段范围内，A 和 ε_r 的变化与平行极板电容器的电容变化呈线性关系。

通过对式(8-1)微分，我们能得出平行极板电容器的灵敏度(单位是 F/s)为
$$\frac{dC}{dd} = -\frac{\varepsilon_0 \varepsilon_r A}{d^2} \tag{8-2}$$

因此，对于 d 微小的变化，平行极板电容器的灵敏度是很高的。

不像电位器，变极距型电容式传感器有无限大的分辨率，这最适合测量微小的位移增量。

例 8.1 平行极板空气间隔电容器的有效面积为 $6.4 \times 10^{-4} m^2$，极板间的距离为 1mm，如果空气的相对介电常数是 1.000 6，求平行极板空气间隔电容器的位移灵敏度。

对式(8-1)求导得
$$\frac{dC}{dd} = -\frac{\varepsilon_0 \varepsilon_r A}{d^2} = \frac{-8.854 \times 10^{-12} F/m \times 1.000\ 6 \times 6.4 \times 10^{-4} m^2}{(1 \times 10^{-3} m)^2} = -56.6 \times 10^{-10} F/m$$

注意，负号表示随着 d 的增大，电容在减小。

8.2 电感式传感器

缠绕在电感式传感器磁路上的线圈电感为
$$L = \frac{\mu_0 \mu_r N^2 A}{l} \text{(H)} \tag{8-3}$$

式中，μ_0 为真空磁导率；μ_r 为相对磁导率；N 为线圈匝数；l 为磁路长度(m)；A 为磁路的截面积(m^2)。

式(8-3)也可写为
$$L = \frac{N^2}{S} \tag{8-4}$$

式中，S 为电感电路的总磁阻。

电感式传感器的电感可通过改变其电感电路的阻抗来调节。电感式传感器的一些例子如图8.3所示。电感式传感器的一些特征曲线如图8.4所示。

例8.2 分别求在a)相对磁导率变化、b)磁路长度变化条件下，单线圈电感传感器的灵敏度。

a) 对式(8-3)中的μ_r进行微分得

$$\frac{dL}{d\mu_r} = \frac{\mu_0 N^2 A}{l}$$

b) 对式(8-3)中的l进行微分得

$$\frac{dL}{dl} = -\frac{\mu_0 \mu_r N^2 A}{l^2}$$

电容式传感器和电感式传感器的测量技术如下。
(1) 用差分式电容或电感作为交流电桥。
(2) 用交流电位计电路进行动态测量。
(3) 用直流电路为电容器提供正比于容值变化的电压。
(4) 采用调频法，C或L随着振荡电路频率的变化而改变。

电容式传感器和电感式传感器的一些重要特性如下。
(1) 分辨率无限大。
(2) 精确到满量程的$\pm 0.1\%$。
(3) 位移范围从25×10^{-6}m到10^{-3}m。
(4) 上升时间小于$50 \mu s$。

电容式传感器和电感式传感器的典型被测量是位移、压力、振动量、声音和液位。

8.3 线性调压器

典型的调压器如图8.5所示，它由一个主线圈、两个副线圈和可移动的铁芯构成。

高频励磁电压V_p作用于主线圈，两个副线圈分别产生电压V_{s1}和V_{s2}。副线圈的电压幅值取决于主、副线圈的电磁耦合及铁芯偏移量x。

由于副线圈串行反向连接，变压器铁芯的偏移量x使V_{s1}增大的同时使V_{s2}减小。在理想条件下，电压V_{s1}和V_{s2}的相位差应该是180°，从而中心位置的输出电压为零。但是，由于一般两电压的相位差不是确切的180°，因为就会产生一个小的无效输出电压，如图8.6所示。

线性调压器的一些重要特征：
a) 无限分辨率；
b) 线性度优于0.5%；
c) 励磁频率为50～20kHz；
d) 无效电压小于满量程输出电压的1%；
e) 最大偏移频率为励磁频率的10%；
f) 位移范围是2×10^{-4}～0.5m；
g) 运动部件无磨损；
h) 调幅输出，即输出电压为其幅值随输入位移量变化且频率恒定的波形。

典型的被测量为可被转化为位移量的量，如压力、加速度、振动、作用力及液位。

8.4 压电式传感器

当在某晶体材料上施加一个力时，就会产生负极电荷，这就是压电效应(压电源于希腊语的"压")。

压电式传感器由天然晶体(如石英、酒石酸钾钠晶体)、合成晶体(如硫化锂)或极化陶瓷(如钛酸钡)构成。由于这些材料能够产生一个与施加力成比例的输出电荷,所以它们不仅最适宜测量力本身,还可测量驱动力变量(如压力、负载和加速度)。

压电材料是很好的电绝缘体。因此,与其连接的金属板共同构成平行极板电容器,如图 8.7(a)所示。当在这个电容器的压电材料上施加一个力时,在压电效应的作用下可对这个电容器充电,如图 8.7(b)所示的等效电路。遗憾的是,任何与这个电容器连接的电子测量仪常常会放电,即压电式传感器的稳态响应很不理想。我们可通过带有充电放大器的高输入阻抗($10^{12} \sim 10^{14}$ Ω)的测量放大器来克服此缺陷,但这样会使得测量系统费用增高。

例 8.3 一个压电式传感器灵敏度为 80pC/bar。如果其电容为 1nF,其输入电压为 1.4bar,请确定其输出电压。

电荷为

$$q = 灵敏度 \times 压力 = 80 \frac{\text{pC}}{\text{bar}} \times 1.4 \text{bar} = 80 \times 1.4 \text{pC}$$

输出电压为

$$V = \frac{q}{C} = \frac{112 \times 10^{-12} \text{C}}{1 \times 10^{-9} \text{F}} = 112 \text{mV}$$

8.5 电磁式传感器

电磁式传感器是利用线圈在磁场中移动而产生电流的原理制成的。电磁式传感器的输出电压如下所示。

1. 对于变磁通的线圈而言

输出电压为

$$v_o = -N \frac{\text{d}\Phi}{\text{d}t}$$

式中,N 为线圈匝数;$\frac{\text{d}\Phi}{\text{d}t}$ 为磁通变化率(Wb/s)。

2. 对于在磁场中运动的单个导体而言

输出电压为

$$v_o = Blv$$

式中,B 为磁通强度(T);l 为导体长度(m);v 为垂直于磁通方向的导体移动速度(m/s)。

常用的速度传感器的构造原理如图 8.8 所示。

电磁式传感器的一些重要特性如下。
(1)输出电压与输入运动速度成正比。
(2)通常质量较大,因此它们固有频率低。
(3)具有高的功率输出。
(4)厂家技术说明中提供 10Hz~1kHz 的频率响应范围。

8.6 热电式传感器

将两块不同金属或合金首尾相连便形成一个热电偶(见图 8.9),且其两端温度不同,则在两导体间产生热电势,并在回路中有电流经过,这就是热电效应。热电势的大小取决于连接处的温度差和所用的材料。热电效应就是众所周知的塞贝克效应,它被广泛应用于温度测量与控制系统中。

热电偶的主要问题是:易被腐蚀、氧化或被放置处的空气污染等。可以通过选用与空气或液体不发生反应的防护外壳来克服热电偶的这些问题。

尽管热电偶能够提供直流输出电压,但这个电压通常是毫伏级,通常要将这个电压放大。

热电偶的优点包括:

(1)由于热电偶的体积小,可测局部点的温度。

(2)耐用,温度范围为-250~2600℃。

8.7 光电管(自发电式)

光辐射在光电管的两块不同材料连接处并产生一个电压,这就是光电效应。典型光电管的构造如图8.10(a)所示,其中标明了金属夹层、半导体材料及透明层。通过透明层的光会产生一个电压,此电压为光强度的对数函数。这样的装置灵敏度极高,频率响应效果好,且由于电压与光强度呈对数关系,所以该装置非常适合在一个大的光强度范围内产生光电效应,光电管的特性如图8.10(b)所示。

8.8 机械式传感器及敏感元件

许多换能系统由两类连续或层叠式的传感器组成。在电子传感器中,输入电子传感器的信号由敏感元件提供。敏感元件本身常为机械式传感器,此传感器把被测量转化为位移或力,然后再通过电子传感器把位移或力转变为电参数。

一些较普遍的机械式传感器分为以下3类。

8.8.1 力—位移传感器

1. 弹簧

图7.2中的弹簧是最简单的机械式传感器。

在图7-2中,为保持弹簧受力平衡,有

$$F = \lambda x$$

式中,λ为弹簧的刚度(N/m)。

$$x = \frac{F}{\lambda} \quad 或 \quad \lambda = \frac{F}{x}$$

由于灵敏度 $K_1 = \frac{x}{F}$,所以灵敏度 $K_1 = \frac{1}{\lambda}$,也就是说,弹簧的刚度越大,其灵敏度越小。

2. 悬臂梁

给图8.11中的悬臂梁上施加一个作用力F,就会产生一个偏移y,那么F与y的关系为

$$偏移 = 常数 \times 作用力$$

即

$$y = kF$$

其中,k取决于悬臂梁的材料和尺寸大小。

8.8.2 压力—位移传感器

1. 振膜

压力可用图8.12所示的钢圈振膜测得。如果振膜位移小于振膜厚度t的1/3,振膜位移就与压力差(P_1-P_2)成正比。压力差(P_1-P_2)与振膜位移(偏移)间的关系为

$$偏移 = 常数 \times 压力差$$

即
$$x = k(P_1 - P_2)$$
其中，k 取决于振膜的材料和尺寸。

振膜通常为一个弹簧钢圈薄片，它与一个电子传感器一起可制成一个高灵敏度的小型传感器。

2. 低音管

低音管这类传感器（见图 8.13）在市场上常用于压力表中。低音管的主要特点是测量范围大。如果横截面的长轴 a 比短轴 b 大得多，那么输入压力 p 与管末端偏差 φ_0 有以下关系：
$$\varphi_0 = 常数 \times p$$

3. 波纹管

波纹管主要是一个带气垫的板簧，如图 8.14 所示，常被用于气动设备中。

在图 8.14 中，为了保持波纹管受力平衡，作用在波纹管底部的力为
$$pA = \lambda x, \quad x = \frac{A}{\lambda} p$$
式中，A 为波纹管横截面的面积（m²）；p 为输入的压力（N/m²）；λ 为波纹管的硬度（N/m）。

8.8.3 位移—压力传感器

典型压力舌簧系统的原理如图 8.15(a) 所示。

对位移—压力传感器施加一个固定压力 p_s，其舌簧的运动引起空气流动，从而改变了控制压力 p_c。位移—压力传感器的特征曲线为非线性的，如图 8.15(b) 所示，但当位移—压力传感器用于气动设备时，就可通过反馈装置使这个传感器的特征曲线产生一个窄的线性区，通常这个线性区可以扩展。由于位移—压力传感器中的空气流动速度很小，所以可通过气动放大器加大控制压力 p_c。

第 9 课　模　拟　仪　器

据科学调查，电子电路设计、电路计算和电子维修基本上都涉及精密测量、选择性分析和数学推导过程。其中最重要的是测量，它被定义为一个过程。通过这个过程，物理量和模拟量被转化为用来分析我们周围事物的数据。

把真实物理量转化为我们可以处理的数据是通过使用测试仪来完成的。测试仪有不同的测试范围和不同的描述类型，每种类型的测试仪都要执行特定的测量。

在所有的测试仪中，最基本、最古老的仪器是动圈表或模拟表。曾有一段时间，模拟表是唯一可靠的电子测量方法。近年来，随着数字技术的发展，使得能设计出消除由模拟表的机械元件而产生误差的仪器成为可能。然而，有争议的是动圈表不如现代数字试验测试仪测量精确。值得怀疑的一点是，由比较精确的测试仪获得的结果是否一定比那些由廉价仪器获得的结果更有意义呢？结果是，模拟表作为一种可靠的测试方法已经巩固了它在实验室领域的重要地位。

9.1　仪表基础

所有的动圈表需要 3 种力来实现正常工作。这 3 种力是偏转力、支配力、阻尼力。偏转力通常由通过流经线圈的电流形成的磁场产生，这个磁场和一个控制磁场相互作用产生一个阻抗力。当动圈表中机械指针和其中的一个磁场共同作用时，运动就产生了。

最早的动圈表是奥斯特实验测试仪,它被称为检流计。事实上,它是一个置于待测电流通过的电线下面的罗盘针。在没有电流通过时,罗盘针为南北指向,有电流通过时,罗盘针按照和电流成比例的关系向着东西方向偏转。在此仪表中,地球磁场是支配力,流过电线的电流产生的磁场是偏转力。通过此仪表中的指针枢轴内产生的摩擦来防止指针来回摆动(有时又称自动瞄准)。

后来,开尔文通过在罗盘上绕上弯的金属线改善了这一装置的灵敏度,并在低电流条件下,强化了该装置的偏转力强度。

如今的大部分仪表体现出 D'Arsonval 设计。D'Arsonval 运动由处于永久磁铁的磁场中的一个运动线圈或回转线圈形成。永久磁铁要比线圈大得多,且形状如图 9.1 和图 9.2 所示。

图 9-2 中所示的扭矩为

$$T = NBILW$$

式中,N 为线圈匝数;B 为磁场强度;I 为电流;L 为线圈的垂直长度;$W/2$ 为悬置点的距离。

扭矩会使线圈旋转,直到线圈在原始位置转过旋转角 ϕ 时达到平衡。该位置上,由悬垂物产生的力矩和由电流和磁场产生的力矩完全相等,此时旋转角为

$$\phi = \frac{NBLW}{k} I$$

式中,k 为一个常数。它由悬垂物或复位弹簧的厚度、宽度及弹簧的性能决定。

9.2 电流表

悬挂式 D'Arsonval 型仪表可以测量 $1\mu A$ 或 $1 \times 10^{-6} A$ 的电流。要测量更大的电流,必须校正并增大该仪表的量程。这可以通过对该仪表使用一个外部电流分流器来完成,电流分流器起的作用与该仪表并联一个电阻的作用是相同的,这样就可以形成一个电流分配器。

电流分配器规则是:任何两个并联支路中的电流的比率与该支路中电阻的比率成反比,即 $\dfrac{I_1}{I_2} = \dfrac{R_2}{R_1}$。用仪表内部电阻 R_m 代替 R_1,分流电阻 R_s 代替 R_2,该式变为

$$\frac{I_m}{I_s} = \frac{R_s}{R_m}$$

如果 $I_s = I_{total} - I_m$,那么分流电阻器的电阻为

$$\frac{I_m}{I_{total} - I_m} = \frac{R_s}{R_m} \Rightarrow R_s = \frac{I_m R_m}{I_{total} - I_m}$$

分流电阻器可以提高该仪表的电流范围,从而能够显示 I_{total}。例如,为了使一个内阻为 100Ω 的 1mA 电流表能测量 100mA 的电流(见图 9.4),分流电阻器的电阻为

$$R_s = \frac{0.001 \times 100}{0.1 - 0.001} = \frac{0.1}{0.099} = 1.01\Omega$$

应用该原理,只要使用更合适的分流电阻器就可以改变量程,任何微安表、毫安表或安培表都可以测量一个更大数量级的电流。事实上,这就是商业仪表如何只用一个表头就可以测量很大量程的电流的原因。图 9.5 所示的就是 7 种量程商业电流表的原理图。注意,旋转开关将被接入不同的分流电阻器中以选择该仪表的量程。

9.3 电流表测量误差

电流表通常串联到待测电路中。为了避免在测量中影响电流值,电流表的阻值必须远远低于电路的阻值。在理想状况下,电流表的内阻应该为零,但是这样的电流表是不存在的。电流表的接

入对电路的影响如图 9.6 所示。

电流表用于测量由 10V 电源供电和 10Ω 负载连成的电路的电流。由欧姆定律可知,在理想条件下,流过该电路的电流为

$$I = \frac{E}{R} = \frac{10}{10} = 1\text{A}$$

然而,如果电流表内阻为 1Ω,该表就不会显示为 1A,而会显示 0.91A 的电流值。如图 9.7 所示,当电路接入电流表时,计算值与测量值的差别是很明显的。

运用欧姆定律,我们能推导出精确的电流表测量值。当电流 $I=10\text{V}/11\Omega$ 时,实际电流仅为 0.909A。电路中接入测量仪器所带来的误差为

$$误差 = \frac{计算值 - 测量值}{计算值} \times 100\%$$
$$= \frac{1\text{A} - 0.909\text{A}}{1\text{A}} \times 100\%$$
$$= 9.1\%$$

如果电流表的阻值减小到 0.1Ω,那么测量值将是 $I=10\text{V}/10.1\Omega=0.99\text{A}$,而电流仪表引起的误差接近 1%。

9.4 电压表

微安电流表不但可以测量电流,也可以测量电压,因为微安电流表的偏差与其指针所示的电流成正比,也与通过其线圈的电压成正比。因此,该表的刻度也可被标定成指示电压而不是电流。

大部分微安电流表的线圈阻值非常小,大约为 1000Ω 或更小,因此其线圈电压非常小。例如,一个 50μA、2000Ω 的微安电流表的满量程电压偏差为

$$E = IR = 5 \times 10^{-5} \times 2 \times 10^3 = 1 \times 10^{-1} \text{(V)}$$

即 0.1V。遗憾的是,测量小于或等于 0.1V 的电压的情况很少见。

如图 9.8 所示,通过给微安电流表串联一个电阻,就可以轻易地增加待测电压的范围。这个电阻限制了通过微安电流表的电流,以便微安电流表能够测量比原来只使用内部电阻时更高的电压。这些电阻无论安装在微安电流表内部还是微安电流表外部,都称为倍程器。倍程器的阻值可以由满偏电流和待测电压范围决定。既然流经微安电流表的电流和待测电压成正比,根据欧姆定律可以计算该电阻为

$$R_s = \frac{E}{I_m} - R_m$$

式中,R_m 为微安电流表线圈的电阻;R_s 为倍程器电阻;I_m 为微安电流表的满量程偏差电流。

一个内部电阻 R_m 为 100Ω,满量程为 100μA 的微安电流表,为了表示量程为 0~1V 的电压,倍程器电阻为

$$R_s = \frac{1}{1 \times 10^{-4}} - 100 = 10000 - 100 = 9.9\text{(k}\Omega\text{)}$$

当与微安电流表串联一个 9.9kΩ 的电阻,且把它的测量范围调到伏特档时,它就成为了一个 0~1V 的电压表。

使用一块微安电流表和许多倍程器及旋转开关就可以构成多量程电压表,它有两种配置电路。在图 9.9(a)中,通过旋转开关选择串联到电表中的倍程器电阻。在图 9.9(b)中,倍程器电阻是串联的,每个电阻之间的连接点被连接到旋转开关的一端上。

以上两种电路中,图 9.9(b)是最经济实用的,因为它只有一个电阻(R_a)是非标准值。非标准

值是用来补偿微安电流表的内阻的。一旦微安电流表的内阻接近标准阻值，其他所有的阻值将减小到标准阻值的范围内，这样就降低了微安电流表的制造成本。对于图 9.9(a)中的配置，需要每个电阻的阻值必须都是非标准阻值，因为每个电阻的阻值都单独受到微安电流表内阻的影响。

9.5 电压表测量误差

与电流表不同的是，电压表经常并联在电路中。由于电压表需要一定的电流（无论多小）去工作，因此其电路也受到影响，这就是电压表的负载效应。该效应会降低电源电压，从而影响测量结果。

该影响可以通过图 9.10 的例子进行说明。在该例子中，使用一个电压表来测量两个串联电阻电压。在图 9.10 中，不接电压表的 A 点到 B 点电势为 1V，这是因为 R_1 和 R_2 的阻值相等，所以它们把电源电压进行了均分。

然而，当电路中接入电压表时，电路的相应结构就改变了。如图 9.11 所示，电压表可以用一个并联在 R_2 的固定电阻表示。

假设电压表的有效电阻为 5000Ω，也就是 R_s 和 R_m 的阻值。此时 A 与 B 点间的等效电阻为

$$R_L = \frac{R_2(R_s + R_m)}{R_2 + (R_s + R_m)} = \frac{5000 \times 5000}{5000 + 5000} = 2500(\Omega)$$

应用欧姆定律和新的电流值，可得

$$I = \frac{2}{7500} = 2.66 \times 10^{-4}(A)$$

计算流经 A 与 B 间的电压为

$$V_{A-B} = (2.66 \times 10^{-4}) \times 2500 = 0.665(V)$$

如我们所见，电压表并不显示测试电路的正确电压值，而是修改了这个值，即给出修正测量值。它会显示 0.66V，而不是正确的 1V。

电压表的分流系数称为灵敏度，可表示为

$$灵敏度 = \frac{R_s + R_m}{满量程电压值}$$

电压表灵敏度越高，分的电流越少，负载效应越小。上述例子中，电压表灵敏度为

$$灵敏度 = \frac{5000}{1} = 5000(\Omega/V)$$

理想的电压表有一个无限大的灵敏度，在这样的电压表中，电流参数不会影响测量结果。实验室或测量模拟仪器的灵敏度为 20 000Ω/V。

减少图 9.9 中电压表影响测量结果的一个方法是：用 10V 的满偏电压表代替 1V 的电压表，这时它的内阻为

$$R_{in} = 满量程电压值 \times 灵敏度 = 10 \times 5000 = 50(k\Omega)$$

这个内阻是之前电压表内阻的 10 倍，相对于这个大内阻，电路的负载较小，对测量结果影响也较小。当这个内阻代替图 9.11 中相应的阻值时，表的读数为 0.95V，只有 5% 的误差。

遗憾的是，模拟表在其量程的低端是不精确的。由于绕组的旋转有一定的摩擦力，所以指针的偏转前必须施加一定的力矩。随着流经绕组电流的增加，反作用力也增大，所以摩擦力变得不是那么明显。由此可知，一个机械表在低于 1/3 量程时的读数没有半量程或全量程时准确。

9.6 欧姆表和电阻的测量

欧姆表可以用来测量电阻。如欧姆定律所定义的那样，电阻是电流与电压之比。因此，如果我

们知道流经一个仪器的电流及其电压降(两个数值都可以用仪表测量得到),我们就可用欧姆定律计算它的阻值。数学上,这个电阻表示为

$$R = \frac{E}{I}$$

这种测量电阻的电流-电压法是一个间接的方法,因为它是一个计算值,即这两个仪表的读数没有一个是实际的阻值。

通过把电压或电流中的一个设为常量,并测量另外一个变量,我们可对仪表标定刻度,使其直接得出阻值。现在广泛应用的欧姆表有两种:串联欧姆表和分流欧姆表。

9.7 串联欧姆表

串联欧姆表提供了一个常压电源,并通过变化的电流值来测量它的阻值。串联欧姆表的基本电路结构如图 9.12 所示。

电源 E 与可变电阻 R_1、固定电表电阻 R_m、1 个 0~1mA 量程的电表表头串联。两个测试点 P_1 和 P_2 代表探针连接被测的未知电阻 R_x。由于电表在满量程时测得的电流为 1mA,所以 R_1 和 R_m 为

$$\frac{E}{R_1 + R_m} = 1\text{mA}$$

当表笔两端被短接时,电表就能达到满偏电流。这种情况表明,R_x 为 0Ω,从而电表的刻度就可相应定出。

当一个大于 0 的电阻连接 P_1 和 P_2 两个测试点时,电表电流相应地成比例减少。电表测得的电流为

$$I_m = \frac{E}{R_1 + R_x + R_m}$$

当 R_x 等于 $R_1 + R_m$ 时,可得

$$I_m = \frac{E}{R_1 + R_m + (R_1 + R_m)}$$
$$= \frac{E}{2R_1 + 2R_m}$$
$$2I_m = \frac{E}{R_1 + R_m}$$

也就是说,电表表针应指向其 1/2 量程处。这一点的值可由相应合适的阻值得出。例如,当一个电表内阻为 100Ω 时,这一点的值应是 100。通过使用较大阻值的电阻进行校准,就可以构成一个能反映未知电阻的阻值刻度表。该刻度表如图 9.13 所示。

可以看到刻度是非线性的,在高阻区刻度变得很密集。随着 R_x 的增加,电表内阻成为次要因素,并对电流和电表影响较小。随着 R_x 接近无穷大,电表电流减少到接近 0。实际上,它通过梯度量来替代阻值,这样就造成了密集的刻度情况。

9.8 分流欧姆表

当要测量一个低阻值电阻时,分流欧姆表优于串联欧姆表。分流欧姆表依靠其内阻(R_m)来推导出电阻测量值。分流欧姆表电路如图 9.14 所示。

工作中,分流欧姆表中的电流先由 R_1 分流调整,以使分流欧姆表表头指示出一个满偏电流。该读数对应一个无穷大的阻值,并指示出刻度。

当表笔之间接入一个未知电阻时，R_x的作用就如一个电流分流计，它分到部分电流，同时按比例减少分流欧姆表表头的读数。当$R_x=R_m$时，未知电阻分得的电流就等于分流欧姆表表头显示的电流，并且其表针指示量程的1/2处。如分流欧姆表内阻为100Ω，那么R_x等于100Ω。该值显示在刻度盘上，且校对过程继续进行。结束后，刻度指示如图9.13(b)所示。

应当注意的是，分流欧姆表是从左到右校准的；而串联欧姆表则相反，是从右到左校准的。该电路的天生缺陷在于：电池总是接在电路上，只有在分流欧姆表不用时，电池才不消耗，而其他时候它的电池消耗特别快。

9.9 欧姆表的精度

由于使用了模拟表而产生的电阻测量误差会以多种形式出现。欧姆表刻度的非线性使得被测量在无穷大区域的测量很不准确。在该区域，欧姆表指针位置的稍微不准确就会带来极大的误差。

另外一个问题是，由工厂生产欧姆表本身的精度造成的。欧姆表的生产精度是基于一个全量程直流电压的。而该误差是由欧姆表固有的机械条件限制而来的，称为误差弧。欧姆表指针回到原点之前，必须克服线圈轴的摩擦损失或游丝拉伸产生的滞后效应。根据欧姆表指针的运动方向，欧姆表指针可以落在误差弧的任何一边或它们之间任何地方。

假设产品误差为满量程的1%，这意味着电压读数在量程中间时，误差为满量程的1%或指示电压的2%。

因为欧姆表刻度的非线性，误差弧在测量电阻时，可导致比测量电压或电流时更大的误差。直流电压的误差弧和欧姆表刻度精确性之间关系如图9.15所示。当串联欧姆表刻度被证明之后，同样道理可用于证明分流欧姆表刻度。

当我们根据欧姆表刻度的非线性来考虑误差弧时，很显然，当欧姆表指针指示量程一半时精度最高。无论串联欧姆表还是分流欧姆表，这种情况出现在当R_x等于P_1和P_2之间的有效内阻时。

9.10 伏特—欧姆—毫安表

现实中很少见到模拟欧姆表作为一个仪器单独存在。一个欧姆表通常是一个伏特—欧姆—毫安表(VOM)或万用表的一部分。VOM是一个集直流/交流电压计、直流/交流电流计和多档位欧姆表于一身的多功能仪表。有一个功能选择开关供选择任何一项功能，一个量程开关为改变量程提供方便。

实际上，VOM是许多仪表的集合体，共同使用一个表头来读数。每个功能的电路结构都是基本相同的，电流表、欧姆表电路如上所述。欧姆表中使用的精确电阻通常用来组合电路，所以只要两个选择开关来配置欧姆表：功能开关和量程开关。

使用VOM时，只要简单地选择所需的功能和量程，就可以开始测量了。一个典型的VOM如图9.16所示。

使用VOM功能开关可以很方便地选择测量正直流电压、电流，负直流电压、电流或交流电压、电流。其量程开关位于VOM中央的表头下方，可选以下量程：电压(2.5～500V)、电流(1～500mA)、电阻(1Ω～100kΩ)。通过利用VOM面板上不同表笔插口来使用一些附加的功能。

标有COMMON(-)和(+)的插口适用于所有电压、电流和电阻的测量。当量程开关打在10A处时，+10A和-10A插口用来测量直流电流。这些插口允许换挡，并通过电流分流进行适当的量程转换。当直接换挡没有电流分流时，则要用+50 A/250 mV插口，把量程开关打到50V，并且用COMMON作为回路。该模式下最大输入电压为250mV。

测量超过500V的直流或交流电压时，把VOM表笔从"+"输入插口拔出，插入1000V插口。

263

当量程开关打到 500V 位置时，可有效地使 VOM 的量程加倍。对于 250mV～2.5V 之间的电压，可插入+1V 插口，把量程开关打到 2.5V，从 0～10 的刻度进行读数。

多年来，具有许多优点的 VOM 已经不仅仅是一个万用表了。对一些人来说，它已成为一种必备的工具。对于一个工程领域的技术人员，如果必须携带专业仪表去工作（如果可能的话），他会感到非常不便。

9.11 电子伏特表

由动圈表和多量程电阻组成的伏特表还有许多显著的缺陷。如上所示，最严重的是它有一个固有的低值内阻，这导致电路中的电流分流，从而产生测量误差。它也不能测量非常低的电压。当使用一个可以扩大和缓冲输入信号的仪器时，它的局限性可以得到克服。这种仪器的例子有电子伏特表和电子万用表。

电子万用表最初称为真空管电压表（VTVM），这是因为在设计中使用了真空管。随着晶体管和集成电路的出现，真空管被更稳定的半导体元件所代替，于是该仪器就简单地称为电子伏特表（见图 9.17）。

9.12 晶体管电子伏特表

晶体管电子伏特表是把一个测量装置安放于差动放大器中。差动放大器既对输入信号进行放大，又对输入信号进行隔离。如图 9.18 所示，可做如下分析。

晶体管 Q_1、Q_2 和电阻 R_2、R_5、R_4 构成了一个差动的发射极耦合的放大器。当 Q_2 的基极电压为零且信号电压被加到 Q_1 的基极上时，两基极的电压差被放大并在集电极得以反映，所以该放大器由此而得名。

差动放大器需要正、负极电源供电才能正常工作，这两个电源的公共端是虚地的。在 Q_1 和 Q_2 的基极接地（0V）时，通过发射极电阻 R_4 的电压降为

$$V_{R4} = 0 - V_{BE} - (-V_{EE})$$

或者当 V_{EE} 小于 0.7V 时，通过 R_4 的电流为

$$I_{E1} + I_{E2} = \frac{V_{R4}}{R_4}$$

当 $E_{in}=0$ 时，则 $I_{E1}=I_{E2}$。若晶体管参数匹配良好，则 $I_{C1}=I_{C2}$，$V_{R2}=V_{R5}$。

这种状况使得 Q_1 的集电极电压等于 Q_2 的集电极电压，若是在两电极间装一个仪表或其他的电压监测装置，表头显示没有电压差，而且其指针保持静止。

由于 R_4 的电压差必须保持，以满足 Q_2 的需求电压 V_{BE}，且加到 Q_1 基极的正电压将导致通过 Q_1 的电流增加、通过 Q_2 的电流减少。相反地，这将在 R_2 上产生一个较大的电压降，而在 R_5 上产生一个较小的电压降，Q_1 和 Q_2 的集电极上分别产生较低和较高的电压值。

由此产生的结果是在集电极间产生一个差动电压，这一点在仪表上表现为：在 Q_2 的集电极间有一个正电势而在 Q_1 的集电极间有一个负电势，测得的电压和输入电压 E_{in} 成正比，从而得到一个代表信号电压在 E_{in} 时的仪表读数。

由电阻串制成的电压分配器将电压限制在差动放大器的输入值范围内，并设定仪器的范围。如图 9.19 所示，包含电阻 R_a 和 R_e 的电压分配器，在电压输入晶体管之前已经将待测电压分压了。当为了使测试电压正确而设定选择开关时，会将输入晶体管的电压限制到小于 1V。

在所测试的电路中，尽管设定了电压的范围，但电路的衰减器配置一个恒定输入电阻。灵敏度这个术语对于电压表不再有意义。这个术语应用于电压表时代表输入阻抗。

然而，当使用如上所述的双极性晶体管的衰减器时，晶体管的输入阻值对于仪表的输入阻抗会产生影响。由于基极的电阻和分压器并联，以及放大器负载效应，在衰减中就会产生一些误差。假设图9.18中R_4为10kΩ，在Q_1基极输入电阻为20～40kΩ，所取值取决于晶体管的特性。当并联负载放置于分配器电阻之间时，它降低了总电阻值，减小量可以由单一并联电阻法则来求得。为了便于讨论，假设基极输入电阻为40kΩ，当选择开关置于50V时，这个40kΩ的电阻与R_e并联。这将产生的有效电阻为

$$R_{e'} = \frac{R_{ib}R_e}{R_{ib}+R_e}$$

$$= \frac{40\text{k}\Omega \times 100\text{k}\Omega}{40\text{k}\Omega + 100\text{k}\Omega}$$

$$= 28.6\text{k}\Omega$$

对于R_e来说，这与原来期望的100kΩ有较大的差距，而且基本上是不能接受的。为了解决该问题，必须增加差动放大器的输入阻抗。

9.13 场效应管伏特表

一种可以完成该任务的元件为场效应晶体管(Field Eflect Transistor，FET)。当FET连接到差动放大器的输入端时，电压表的输入电阻的阻值会显著增加。由于FET门电流比50nA小，所以它有一个2×10^{10}Ω的输入阻抗。而MOSFET则有更大的输入阻抗。图9.20说明了由此产生的电流。在这种配置中，源端电压将按相同的量随着FET门端电压的改变而改变，此电压称为电源跟随电压。

在图9.18中，尽管已经为晶体管Q_1和Q_2的h_{FE}设定了一个定值，但在差动阶段h_{FE}还是可以在两个晶体管之间充分变化而产生不平衡，即使两个晶体管近似相同。因此，归零控制(h_{FE})可用来平衡流经Q_1和Q_2的电流，从而使操作规范。把仪表的表笔短接，调整之后仪表显示0V。

9.14 运算放大器电子伏特表

近期，集成电路的发展导致了运算放大器的产生。运算放大器在电子伏特表中的应用性能比较稳定。集成电路运算放大器的输入电阻（如同它们的名字一样）是非常高的。那些具有双极性晶体管输入的地方有典型的输入电流，其值为0.2A。而FET输入运算放大器理想值是0Ω，它足以提供驱动偏差仪所需的电流。

运算放大器电子伏特表根据应用形成3种电路配置。它们分别是电压跟随器、放大器和电压—电流转化器。

运算放大器式电压跟随器类似于一个晶体管射极输出放大器电路，它唯一的目的是把高阻输入信号转变为低阻输出信号。如图9.21所示，运算放大器的增益通过运算放大器的输出端连到反相输入端来统一确定。通过设定放大器的增益，电压跟随器在运算放大器的输入端重新产生输入信号，使基极—发射极电压不下降，其输出电压总是等于输入电压。

仪表被连接到串联有电阻网络的运算放大器输出端上。电阻R_m是仪表的内阻并且是仪表本身的固有特征。R_s是用来校正仪表读数的串联限流电阻。衰减电路为放大器提供输入信号。

当仪表被用来测量比仪表本身所能显示的更小电压时，则运算放大器电路既是缓冲器也是放大器，其应用电路类似于图9.22所示的同相放大电路。

设计放大器电路是非常容易的。反馈网络由R_4和R_3组成，通过调节从运算放大器输出端到反相输入端的电路，可以设置放大倍数。电路的增益为

$$A_v = \frac{R_3 + R_4}{R_3}$$

选择比运算放大器输入偏流 I_B 大许多的电压分配器电流 I，以便 I_B 对反馈电压几乎没有影响。整个网络($R_3 + R_4$)的电阻为

$$R_3 + R_4 = \frac{V_{\text{out}}}{I}$$

通过 R_3 的电压总等于 E_{in}，即

$$R_3 = \frac{E_{\text{in}}}{I}$$

除了由于仪表偏差或电阻的不精确引起的误差外，还有由于运算放大器的内部增益引入的误差。但是，这两种电路中的误差和正常的仪表偏差相比是可以忽略的。

在图 9.22 所示的运算放大器电路配置中，电压—电流转换器是可调的。在电压—电流转换器中，仪表被电阻 R_4 代替，R_3 做部分调整。通过运算放大器配置，R_3 的电压降总等于输入电压 E_{in}。如果输入信号增加或减少，V_{R3} 也随之增加或减少。因此通过仪表的电流为

$$I = \frac{E_{\text{in}}}{V_{R3}}$$

换句话说，仪表读数是输入电压的直接作用。R_3 的一部分可用来校准仪表的刻度。对运算放大器非转换输入端输入一个电压并调整 R_3 直到仪表指针指向所用的电压，从而完成了对仪表的校准。对于正常的工作电路，通过 $R_m + R_3$ 的电压必须总是小于供电电压 V_{CC}。

9.15 电流测量

电子万用表主要是电压测量设备。因此当使用电压表测量电流时，电流必须转化为电压。如图 9.23 所示的电路是用来测量电流的。

该电路起了电阻分流器的作用，通过电阻分流器产生一个电压。当电压被测量并显示在有刻度的模拟电表上时，根据欧姆定律，它代表的是电流值，即

$$I = \frac{E_R}{V_m}$$

式中，E_R 是通过电流分流器的电压；V_m 是仪表读数。

对于量程从 1A 开始的仪表，附加的电阻器被串联到测试电路中直到总电阻值为 450Ω。注意，选择适合的电阻值，以便设置的每个量程分流电压决不会超过 0.455V。

遗憾的是，大部分的电子万用表中的电流会给测试电路引入电阻值。该电阻值要比实验室的电流分流器显示的值大得多，该分流器在相同的电流下会产生 50mV 的电压降。对于实验室研究来说，由电子万用表引起的误差是不能接受的，但一个功能多合一的仪表对于现场维护和工作真是太方便了。

9.16 电阻测量

通过给未知电阻提供一个固定电流，然后用电压表测量流经该未知电阻两端的电压来确定其电压降，从而测量出电阻。电压表测量电阻时所需的器材是一个电池、几个标准电阻和合适的开关。电压表测量电阻电压通常有 3 种方法。

1. 频率响应

影响交流信号测量过程的一个重要参数就是频率。不同类型的仪表有不同的内阻、电容和电感。后两个参数与仪表的总阻抗相关，仪表的频率响应和带宽是影响交流信号测量值的主要因素。

在直流信号测量过程中,电压表并联到测试电路中,而电流表串联到测试电路中。交流仪表电路如图9.24所示。

当频率变化时,每个元件的某种因素都起作用。在低频时,输入极间耦合电容器的容抗即交流仪表的输入阻抗必须比电容器的电抗大100倍多。交流仪表的电容应该通过测试表笔被隔离,使其对被测量的影响最小。

由于示波器只是作为一般测量仪器,我们已习惯于在屏幕上观察非正弦交流波形。该波形似乎仅仅由我们所见到的以某一重复频率出现的波形组成。然而根据傅里叶分析,我们知道这与事实差别很大。许多波形尤其是诸如方波、锯齿波的谐波部分远远超过基频的扩展,而且是波形能量谱的复杂成分。

2. 交流电子伏特表

在交流电压的测量中出现了在直流电压的测量中不曾出现的特殊问题。该问题源于交流电压除幅值外还有附加参数影响测量结果。让我们分析最常见的交流电压波形——正弦波。

3. 正弦波分析

如图9.25所示,标出了一个正弦波形的参数。正弦波的电压值从零到最大值并回到零。这个交替过程称为一个循环。一个循环持续的时间称为周期(T),1s内周期循环重复次数称为频率。

该正弦波电压瞬时值为

$$e = E_m \sin(\omega t)$$

式中,e 为瞬时值(V);E_m 为最大幅值(V);ω 为角速度(rad/s);t 为时间(s)。

以下表达式可用来进一步研究正弦波的属性。

峰-峰值(E_{pp}):它是最大值(E_m)的两倍,即

$$E_{pp} = 2E_m$$

平均值(E_{avg}):任何电压的平均值在数学上都被定义为电压曲线下的面积除以周期。在正弦波中平均值为零,因为正的部分刚好等于负的部分。因此,总面积为零。

均方值(E_{rms}):交流的均方根值,又称有效值,它被定义为是获得与一个直流电压相同加热效果的交流电压值。例如,需要14.14V直流,则 E_{rms} 可表示为

$$\frac{E_m}{E_{rms}} = \frac{14.14}{10} = 1.414 = \sqrt{2}$$

因此,有

$$E_m = \sqrt{2}\, E_{rms}$$

$$E_{rms} = \frac{E_m}{\sqrt{2}} = 0.77 E_m$$

迄今为止,E_{rms} 是上述3个值中最重要的一个。因为它是唯一一个为直流信号和交流信号的效果提供直接和精确的比较参数,而不管其波形如何。

第10课　数字仪器

虽然模拟仪器在实验室里已经占据了一定的位置,但是仍然受到有限的机械精度的限制。模拟测量方法有很大的主观性,受到视觉和心理感知误差的影响,同一实验对不同的观察者就会得到不同的读数。大多数的误差来自视差,即由于观察仪器指针的角度不垂直而造成的测量误差,即使

我们已经解释了这种生理特性，还是想逃避自身责任。如果测量值有轻微的偏差，我们总是倾向于补偿误差，将记录值向感觉的真值靠拢。更糟的是，事实上两次测量中仪器指针本身不会一直指向相同的位置，这是由于指针是移动的。

另一方面，数字指示器是真正客观的读出器，对所有的用户可以得到相同的指示。直接的数字读数可以减少人为误差和烦琐的测量过程。"欺骗"很难发生，因为数字就在你面前，像"白纸黑字"那样存在着（很显然）。数字仪器同模拟仪器相比，其相关外围技术包括自动控制、波形测量和更高的抑制共模干扰特性等，因而其具有更多的优越性。

10.1 数字显示器

第一台数字显示器是机械结构的，由一定数目的被伺服控制发动机驱动的齿轮组成。随后出现了各种形式的小型白炽灯结构，这种技术得到了惊人的发展，直到现在流行的发光二极管（LED）和液晶显示器（LCD）。

白炽灯显示器的熔丝是一根脆弱的细金属丝。即使工作在很低的电流状态下，白炽灯的寿命仍然是以小时来计算的。除了寿命短，白炽灯的热反应时间还对显示器的滞延时间带来了问题。尽管现在白炽灯已经很少用了，但是它所建立的七段显示标准却一直沿用至今。

等离子显示器是通过少量的氩氖离子或充电气体来照亮。通常用少量的汞来控制电离电压从而改变光谱。这种显示器通常被制成很多类型，有些类型还有外部电极。等离子显示器输出光很好，而且等离子设备的开关特性允许方便地进行多路扩展。柏洛兹制作了一种数码管，它是一种耐用的气体照亮显示器。这种数码管（氖发光管）具有9个内部阴极，每个阴极为一个数字的形状。所有的阴极是垂直于视线层叠排列的。当一个阴极被激活时，它所表示的数字形状就会发光。在某种程度上，将9个数字放在不同的面上是不太保险的，但是可以接受。等离子显示器的缺点是需要很高的电离电压来点亮它。额定电压为150~200V，这就给现在的低压晶体管电路提出了接口问题。虽然等离子显示器不占有统治地位，但是它还是有一定的应用空间的（见图10.1）。

真空荧光灯是真空电子管技术的一个产品。荧光显示器是在阳极上覆盖了一层和荧光灯里的物质相似的发光物质磷。常见的为蓝绿色，其他颜色也有。当阴极发射体发射出电子轰击荧光屏时就会发出荧光。在大多数情况下，数字采用七段结构显示。虽然这种显示器继承了真空电子管的缺点——脆弱性、寿命短和需要高电压驱动，但是它具有很好的美观性，因此它仍然有很多应用，如自动化计算机显示。

场致发光显示是利用磷发光原理来代替晶体管结构。这种发光系统类似于一个电容。将涂敷有磷的介质叠层在两个透明的导体面之间，当有交流电压驱动时，电容上电流的变化导致磷发光。

随着半导体材料和加工技术的不断发展，发光二极管成为数字显示广为应用的设备。它具有亮度高、成本划算、兼容使用晶体管的电压和电流标准、耐用等优点。最初发光二极管只有红色，现在可选的颜色范围很广，包括红色、橙色、黄色、绿色和蓝色。

液晶显示器不发光，通过改变自身晶体材料特性来反射光。视觉上，液晶显示器在发光二极管不能用的情况下起作用，周围光越强它越亮。而发光二极管在周围光很强的时候就失效了，几乎看不到它发出的光。另一方面，液晶显示器在低光线下也不能像发光二极管一样具有可读性。由于液晶显示器不是发光器，因此它的功耗很低。正因如此，液晶显示器被普遍用于便携的电池供电的设备中。

10.2 电子数字计数器

所有数字仪器共同的部分是电子数字计数器。电子数字计数器经常被认为是一个统计输入事件的装置，是所有数字测试和测量应用的基本结构。

一个简化的用于频率测量的电子数字计数器结构如图10.2所示。一个简单的频率计数器主要由7个部分组成：输入转换电路、时基振荡器、时基分配器、主控门触发器、主控门、计数寄存器、读数屏（显示屏）。主控门是电子数字计数器工作的核心。

主控门就是一个逻辑与电路。当输入转换电路和主控门触发器同时为逻辑真时，主控门打开一定时间，这个时间由时基分配器决定。当主控门打开时，经过转换的输入脉冲通过主控门进入计数寄存器，并在此进行统计，然后通过读数屏输出。在计数阶段的最后，主控门关闭，然后计数器复位，开始下一个采样阶段。

10.3 输入信号转换

首先，输入信号经过过滤排除所不希望的噪声，并且使信号电平与内部数字处理电路相兼容。由于频率计数器必须提供一个宽范围的输入信号，使之能够适用于不同电压、噪声成分和直流偏移的情况，所以输入信号在充分变换前必须经过许多转换电路（见图10.3）。

如果需要的话，所要测量的输入信号首先经过一个衰减器以减少信号的振幅。衰减器是一个电压RC网络，主要提供1、10、100的输入量程开关选择。某些设备中，RC网络可以用一个分压计替代，以对输入信号进行连续可变的衰减。

一个包含大量直流成分的交流信号会使这个电流信号改变，从而不在仪器电流可用范围之内。因此，与衰减器耦合的输入信号既可以是直流的，也可以是交流的，这取决于输入信号的输入特性。随着交流耦合，输入信号经过一个电容，这个电容可以阻止直流电压通过，并为包含较大偏移电压的输入信号建立一个参考地。虽然交流耦合作用很小，但是如果输入信号的周期大于输入信号的脉宽时，仪器显示的便是输入信号脉冲的积分值，即显示的是输入信号的平均值。同理，当测量频率可变的信号时，就应该避免交流耦合，这种情况通常出现在开关电源和电力控制器上（见图10.4）。

为了防止仪器在输入信号超过衰减器的输入信号设置范围时发生意外过载而损坏，通常使用二极管 D_1、D_2 来限制输入电压。因此，输入信号通过一个阻抗匹配网络传输到一个宽带放大器。

放大器的输入信号一般是高阻抗（1MΩ）。输入信号的高阻抗可以减少输入线路采集到的噪声量，使仪器具有更好的灵敏度。50Ω的输入信号可以使仪器具有20mV的灵敏度，而典型的高阻抗输入信号可以使仪器具有250mV的灵敏度。当输入信号存在交流耦合时，低阻抗的输入信号可以减少脉冲积分的影响。当选择输入阻抗时，必须考虑仪器的负载能力。

触发器电平设置了一个参考电压 V_T，在这个电压之上的输入信号可以触发频率计数器，如果 V_T 设置太高，如图10.5(b)所示，输入信号超过全部滞后范围，不会有输入脉冲被计数，通过降低 V_T，如图10.5(a)所示，输入信号通过全部滞后范围，每一个输入脉冲都会被计数，对于负输入电压，V_T 的设置如图10.5(c)所示。

AGC电路可以调整放大器的灵敏度以适应不同数量级的输入信号。每个输入信号都会出现波形上的毛刺，AGC能够使输入信号平滑输出，从而为施密特触发器提供一个恒定的输出电压。在AGC响应信号和可计数的最小频率输入信号之间存在着一定的互换关系，即50Hz输入信号对应高频调幅的AGC响应信号。AGC电路趋向于响应输入信号的峰值电压，从而忽略输入信号的波谷电压。在一个高频调幅输入信号轨迹中，响应时间可能不足以快到捕捉调制的低电平，因此导致频率计数器的错误计数。

放大器的输出信号是施密特触发器的输入信号。施密特触发器为输入信号触发提供必需的迟滞作用，以调整所有信号，并满足缓慢的上升时间、缓慢的下降时间，或者两者都满足（如在低频正弦波中），从而提供一个具有边沿的矩形输出信号。该信号作为主控门的一个输入信号。

10.4 时基振荡器

当主控门的输入信号是逻辑1时，施密特触发器的输出信号才传给计数寄存器。每个施密特触发器从时基分配器得到一个负信号时，其状态便改变一次。因此，设定时基分配器每秒提供一个负信号，施密特触发器输出信号在1、0之间每秒交替变化一次。因此，主控门也在1、0之间每秒交替变化一次。换句话说，当主控门控制施密特触发器的输出信号传给计数寄存器时，这个输出信号要在计数寄存器中储存1s。施密特触发器定时由时基振荡器驱动的时基分配器控制。

时基振荡器是一个精确的晶体振荡器，用以驱动时基分配器。时基分配器会测量时基振荡器(RTXO)、温度补偿晶体振荡器(TCXO)和过控晶体振荡器的输出信号。

温度是所有振荡器的"天敌"。一个振荡器的频率稳定性直接依赖于电路组件的温度稳定性。对于一个晶体振荡器，这个组件就是晶体本身。可以通过调温装置的控制进行温度补偿或选择适当的晶体类型等方法来尽量减小温度对振荡器的影响。表10.1总结了这些方法的优点。

对所有振荡器来说，比例控制的恒温调节炉控制的温度是最稳定的。恒温调节炉能够将炉温控制在$\pm 0.0001°C$的误差范围内。当炉温低到与晶体的转变温度相一致时，恒温调节炉的温度系数是非常低的。一个炉子要获得最大的设定精确度一般需要24h，而一个炉子通常能够在20min内达到7×10^{-9}的精确度。

当不用炉子时，晶体可以通过一组电抗来进行温度补偿，此处电抗的温度系数抵消了晶体的温度系数。另外一个可选的方法是使用能随温度变化的的变容二极管，如图10.6所示。感应器的值通过下面公式选择：

$$\omega_s < \sqrt{\frac{1}{L_1 C_1}} < \omega_0$$

式中，ω_0为振荡频率期望值；ω_s为干扰频率。尽管此方法并未给予如恒温调节炉那样的低温度系数，但由于其低功耗及无显著升温时间，还是很有使用价值的。

10.5 时基分配器

主控门触发器时钟频率通过在时基分配器中划分时基振荡器的频率获得。时基振荡器包括一个一组十进制的计数器，每个计数器都将输入频率除以10。

一个十进制计数器触发器的输出频率刚好是输入触发频率的1/10。也就是说，输出信号周期是输入信号周期的10倍。被1MHz的时基振荡器驱动的第一个十进制计数器的输出信号周期是$10\mu s$，则第二个十进制计数器(由第一个计数器的输出触发)的输出周期是$100\mu s$，第三个十进制计数器的输出信号周期是1ms，依次类推。如果一共有6个十进制计数器，则可获得的时间范围是$10\mu s \sim 1s$。

10.6 计数寄存器

计数寄存器是计算和记录数字脉冲的十进制计数器的阵列。每一个显示的数字都有它自己的十进制计数器，这个十进制计数器由锁存电路和BCD七段解码器组成(见图10.7)。

十进制计数器是脉冲的累加器和加法器。在电路中，脉冲被计数，并将计数的数值转化为十进制格式。十进制计数器串联起来可以完成加法处理。当第一个十进制计数器输出一个脉冲给第二个十进制计数器(以十为单位)时，第一个十进制计数器就翻转一次。当第二个十进制计数器从第一个十进制计数器中累加到10个脉冲时，第二个十进制计数器就翻转一次，并输出一个脉冲给第三个十进制计数器(以百为单位)，依次类推。显示阿拉伯数字的数目决定数字计数器的分辨率。

在计算处理中,锁存电路用于隔离十进制计数器和读数显示器。在计数期间,显示器与十进制计数器输出端是断开的。在计数的最后期间,锁存电路的信号把计数结果装载到计数电路的最终条件中。BCD七段解码器把十进制计数器输出的二进制数通过七段数码管显示(见图10.8)。

10.6.1 数字频率计数器

数字计数器自从第一次出现以来已有很长时间。甚至原来它被作为最有用的测量仪器出现在实验室的示波器中。数字计数器最原始的应用就是频率计数器。

当通过精确的时间间隔控制主控门时,计数器处在频率模式。频率模式与上面描述的累加模式相似,但它们门控的操作方式是不同的。

对于频率测量,数字计数器的计数时间间隔精确到几秒。举例来说:当主控门被打开1s,如果45 500个脉冲通过计数器寄存器,那么输入信号的频率为45.5kHz。依靠读出的阿拉伯数字,可以直接显示或以10的幂表示频率值。

为了测量超过计数寄存器的最大计数值的频率,主控门被打开的时间间隔将减小,这个变化有个比例因子。例如,在计数寄存器中0.1s内累积相似的脉冲数(45 500),这就代表着输入信号的频率为455kHz。在计数寄存器中10μs内累积25 000个脉冲,就代表着2.5MHz,依次类推。当时基在各个范围之间不停地转换时,比例因子能通过数字读出器的小数点的位置自动设置。

某些仪器具有自动范围修正系统,该系统能够自动选择时间范围,以取代人工选择合适的时间范围,该系统称为自动测距仪。自动测距仪通常由电路组成,该电路产生的电压与输入频率大致成比例。根据实际电压值,几个晶体管开关之一被开启用于选择正确的时基和数字读出器的小数点位置。

数字频率计数器可以作为频率仪使用,它除了具有电子计数功能外,其他的功能也正在开发。数字频率计数器能指示输入信号的期间、比较比例模式中的两个信号、表示一个波形上两点间的时间或做倒数计算。操作模式的选择仅仅是基本计数器结构的重新配置,如图10.2所示。

10.6.2 周期模式

信号周期是信号频率的倒数。周期测量是指测定一个信号完成一个振荡周期所需的时间。在周期模式下,计数寄存器计算来自时基振荡器的脉冲,输入信号用于控制主控触发器的使能端。电路配置如图10.9所示。时基脉冲的数值直接与输入信号周期成比例。

例如,测得的信号是100kHz(10μs),时基基准振荡器的频率是1MHz(1μs)。主控门被打开10μs,在这段时间内,来自时基振荡器的10个脉冲通过主控门。然后这10个脉冲通过计数寄存器处理和标定以提供10μs的读数。

10.6.3 比例模式

第三种操作模式是比例模式。在比例模式中,两个输入信号的比值被显示在计数寄存器的显示器上。如图10.10所示,通过对来自门电路和允许低频信号控制主控门的时基振荡器的信号进行分频,可以测量频率比值。

实际上,低频信号决定了主控门打开多长时间,以此决定来自上一级的更高频率信号通过主控门的时间有多长。因此,主控门输出信号是两个信号的比值。如果低频信号仅仅允许一个高频信号经过,那么比例是1∶1。如果低频信号允许10个高频信号通过,那么比例是10∶1。当参考频率低于被参考的信号(即输入信号A的频率小于输入信号B的频率)时,计数寄存器将不工作。因为主控门没有足够长的时间保证信号脉冲传递给计数寄存器。

10.6.4 时间间隔模式和相位测量法

在时间间隔模式下，允许计数寄存器对任何两时间点之间发生的事件数进行计数。实际上，这时的计数寄存器等同于电子秒表，电子秒表使用开和停输入信号控制主控门的开和关。其等效电路如图10.11所示。在计数期间，计数寄存器计算来自时基振荡器的脉冲。在大部分仪器中，时基分配器的选择可以改变时基的频率。

时间间隔模式的特殊应用是相位测量。通过时间间隔测量法，频率计数器能够测量两个频率相同信号在波形的同一点之间的相位差，并且计算从开始测量时的相位。

如图10.12所示的时序图，T_1和T_2是参考信号与未知信号的相位相同的一点。T_ϕ是未知信号与参考信号的相位差。为了测量T_ϕ，频率计数器采用开/停时间间隔模式，频率计数器从T_1开始，到T_2结束。

时间测量的相位间隔可能比以前讨论的其他模式具有更精确的门控信号。如图10.13所示，主控门有3个输入信号：一个来自主控门的触发器，另两个来自信号调节装置状态(开和停)。

当开始信号触发主控门时，从时基振荡器到计数寄存器，主控门触发器提供一个精确的振荡频率。类似地，停止信号输入触发器，停止频率计数器的计数过程。然后，两个信号间的相位角通过下式计算：

$$F = (T_\phi/T) \times 360(°)$$

开始和停止点通过输入信号的触发电平来选择。

10.7 数字电压表

数字电压表是数字测试仪中最常用的仪器。除了读数以数字显示而不是以机械式测量仪的指针指示外，它基本上与前面讨论的模拟电压表相似。

虽然有许多种方法用于把模拟量转化为数字量，但是基本的操作原理是相同的。即电压未知；使用上面描述的数字计数电路对脉冲进行计数，计数值以七段式数码管显示。模/数转化通常有5种方法，它们是：

电压—频率型
单积分式（斜坡式）
双积分式
连续近似电位器法
连续平衡电位器法

第11课 基于计算机的测试仪器

各类公司迅速运用基于PC的测试仪器来实现计算机辅助测试(CAT)，这种快捷的方法可以使工程师在程序开发和结构搭建上节约大量的时间。这样的仪器能够降低劳动成本，提高生产率，减少读取和处理中的人为失误，并通过使用软件来完成许多传统上由硬件搭建的结构所实现的功能。

这类仪器的总体设计是靠计算机平台实现的，如IBM PC或IBM PC AT。众所周知，PC仪器分为内置和外置两种。

11.1 内置 PC 仪器

内置 PC 仪器做在一个或多个计算机适配板卡上。这些板卡在物理性能上与视频卡和 I/O 卡一致,计算机依赖这些板卡才能正常工作。要将该仪器装进计算机,只要将相应板卡插进母板的扩展槽,而计算机系统性能不受影响。通过计算机键盘由软件来控制内置 PC 仪器的使用。这里没有用户可触摸到的手柄、旋钮开关或指示器。内置 PC 仪器的全部操作通过计算机与其接口来实现。

R. C. 电子公司的 COMPUTERSCOPE-IND IS-16 分析示波器就是一种内置 PC 仪器。IS-16 的数据采集模块由一块 16 通道的模/数转换板卡、外置仪器接口盒和相应软件组成。

工作过程中,IS-16 为 16 个独立输入通道提供 1MHz、分辨率为 12 位、输入电压范围为 $-10 \sim 10V$ 的信号。全自动按键指令(程序员称为热键)供用户操作内置 PC 仪器的各种功能,包括通道选择、触发控制(内部、外部、$+/-$电平或上升沿)、采样率、从 1KB 到 64KB 的内存缓存。事实上,热键是定义在计算机键盘上的键,其作用犹如示波面板上的旋钮和拨动开关。

除了模仿一般示波器所具有的面板功能外,IS-16 拥有一个环形缓存,可存储触发前任何间隔长度的数据。软件指令允许时间轴的拉伸和缩小(向左、向右移动),垂直方向增益调整,以及波形存储与再现。

正是后一个特征,波形存储和再现,使得基于 PC 的 IS-16 示波器区别于普通示波器。输入的测量数据可存储在计算机文件中或临时计算机内存(RAM)中作为备份或进一步处理。由于动态输入实际上变成静态处理,使用户有能力以各种方式修正现存文件内容,因此后续的信号分析可有效地执行。整个操作以软件形式完成,以下是输入信号的采集过程(见图 11.1)。

计算机通过程序将输入信号的逐字备份并创建为计算机数据文件,采用二进制字串来实时地表达瞬态事件和数值。一旦数据被保存下来,整个测量过程就变成将文件记录保存到计算机来记录每个数值。同样地,音乐被永久地刻录在磁盘或磁带中,并根据需要可反复播放。每个数据通道可采用这种方式分别进行存储和处理。

已存储的数据现在可通过软件来处理,而不是由内置 PC 仪器完成。表 11.1 给出了原始数据被处理的过程。当人们认识到所有这些测量可这样进行时,其意义是重大的。对于传统测量来讲,难以再现的瞬态事件,只要计算机用适当的软件记录下数据,就可在任何时候进行处理。

此外,可通过内置 PC 仪器的一个参数求得另一个参数。例如,开始时,可调整内置 PC 仪器的噪声和放大因子,产生一个数据文件,通过对它求平均、积分或差分,来求得所需的参数,而不必改变测试中的测试仪器或设备的物理装置。

11.2 外置 PC 仪器

上述性能不局限于内置 PC 仪器,相似的情形也表现在采用外部连接的 PC 仪器上。

在外置 PC 仪器中,测试仪器装在计算机外面的柜子里。装在柜子里的既可以是输入/输出接插件、选择开关,也可以是一两个 LED。还有些接插件缆线从外置 PC 仪器中引出,并连接到计算机上。

外置 PC 仪器与计算机以电缆的形式连接,电缆用于传送电压、数据及两台设备之间的控制信号。如果没有这种连接,外置 PC 仪器就只是一台孤立的设备,无异于本书前面分析的仪器,然而,当接到计算机时,它变成一个虚拟工作站。

计算机内有一个接口卡,它将外部测量仪器的信号转变为满足计算机总线要求的带有电平和定时请求的数字脉冲(见图 11.2)。

测试音频精度的多功能音频分析器是一个外置 PC 仪器。多功能音频分析器被设计为能与计算机接口连接的音频测试工作站,它能对音频放大器、磁带系统及相应的音频部件执行超过36种的标准性能测试。

多功能音频分析器包括一个超低失真正弦波发生器(频带中心达0.0008%)、一个电平探测器、一个测频放大器、一个谐波失真分析仪和一个计算机接口卡。

多功能音频分析器没有控制和显示部件,只有电源开关。其仪器面板上的音频连接器将多功能音频分析器和测试设备连接起来。所有与用户相关的控制功能,如测试类型的选择、电平调整、频率输出和显示,都通过键盘和IBM PC兼容的个人计算机来完成。

像COMPUTERSCOPE-IND IS-16一样,多功能音频分析器所有测试功能由PC键盘上的热键来控制。因为大多数音频测试是被成批测量的,而不是被单独测量的,所以在发出请求时,只要触摸一下控制键,就能通过程序去执行一系列测量。

通常,音频测试采用扫描频率方式,其频率范围大于被测设备的频率。表11.2列出了一些可测项。

测试的类型(如频率响应)可从标准测试单中选择。在这些测试项目中,扫描范围、扫描率和扫描幅度根据SMPTE、DIN、CCIR、EIA、IHF、NAB及其他确定的音频标准来预设。每种测试模式记录在多功能音频分析器应用包的软件中,便于计算机通过屏幕菜单进行存取。

用户也可在扫描前,通过启用计算机控制屏来改变任何参数。控制面板实质上是机械控制面板的可视化版本,这样就不用通过转动手柄来改变范围。如果要改变频率,只要将光标移到表示当前频率的位置,输入新数值即可。传统的硬件面板仪器受限于测量单位,为此,设计者留出面板空间和测量空间,而对于任何特定阻抗或函数,多功能音频分析器的面板值可在"V""dBV""dBu""W""dBm"或百分比等范围内改变。如果需要的话,新参数可存储到计算机文件中以备将来使用。

一旦测量开始,多功能音频分析器的软件就开始测试。在测试过程中,测量数据被显示在计算机的视频监视器上。根据测试单元的响应,测量结果表示在 X-Y 坐标图上。通过将正确的数值输入展开的菜单中,就会看到多线图。

测量数据也可存储在计算机文件中。如同前面描述的PC示波器,这些文件可重新读取以重新进行已执行的测试或修正测试结果。与COMPUTER-SCOPE不同的是,多功能音频分析器进行的是实时检测。在测试单元连接到测试设备上执行时,由于采用了直接测试,程序化的测量比由计算数值导出的测量更准确,这并不是说新的测量和结论不能从多功能音频分析器创建的数据中得出。因此,虚拟仪器具有优于其他测试设备的特性。

11.3 模/数转换

任何虚拟仪器性能的关键是模/数转换器(ADC)。ADC将模拟输入电压值转换为二进制数。例如,2.00V输入电压可表示为0011001001。其转换的精度完全依赖于二进制数的位数。这个数值可精确到10位。

实际上,有两种模/数转换方法用于PC虚拟仪器:逐次逼近式转换器和闪存式转换器。

11.3.1 逐次逼近式转换器

直到单片电路出现,逐次逼近(SAR)式转换器因为其低的转换率被认为是ADC家族中的落伍者。随着当今技术的发展,超快比较器、快速电流开关和ECL逻辑电路都很好地应用于逐次逼近式转换器中。

逐次逼近技术的一个显著特点是它几乎不需要硬件。因此,可以将这种逐次逼近式转换器集成到成本很小的12位或更高位的单片电路。

逐次逼近式转换器的框图如图11.3所示,采用ADC形成一个反馈闭环电路。ADC通常是一个电阻网络,此网络采用梯阵精密电阻生成与数字编码相应的输出电压。当恒定电流通过电阻阵列时,其输出电压正比于总电阻,总电阻依赖于以二进制数方式插入电路的电阻数量和阻值。

在实际操作中,移位寄存器在逼近寄存器的 MSB 锁存器中置一个 1,其他所有的锁存器预置为 0,数/模转换器(DAC)的模拟输出电压与模拟输入电压相比较,判定哪个值更高。如果模拟输入电压小于模拟输出电压,那么,MSB 锁存器清零,下一个逼近寄存器中最有效位置为 1。所有这些发生在一个时钟周期内。

在接下来的时钟周期中,新的 DAC 输出电压与模拟输入电压比较,判定哪个比这个数字化的输出电压大。此时,控制逻辑在当前锁存器中保留一个值,有序地在下一个锁存器中置一个 1。下一个时钟周期,这个新的二进制数(0110000…)与模拟输入电压比较,做出继续插入新的 1 还是不插入的决定。

在下一个时钟周期中,这一过程不断重复进行,伴随着寄存器有效位的一点点减少,直到用完。因此,DAC 输出值是模拟输入电压的渐进精确近似,N 个时钟周期获得 N 位分辨率。转换结束时,二进制数作为输入值被读进计算机且另一个转换开始。当转化率达到 15MHz 时,可采用现在的设计技术,大多数用于虚拟仪器的逐次逼近式转换器的工作范围是 1~5MHz(见图 11.4)。

11.3.2 闪存式转换器

逐次逼近式转换器虽然价格低廉,但它不能应用于视频或数字采样电路。对于这种应用,可以采用另一类 ADC,称为闪存式转换器。

闪存式转换器在单个时钟周期内执行转换并输出数据。图 11.5 说明时钟输入信号对上升沿和下降沿很敏感,所有有效操作都对应于这些边缘。

模/数转换通过一个并行电压比较器阵列来完成。采用图 11.6 中的 128 个并行电压比较器能获得 7 位分辨率。

一方面,128 个并行电压比较器的参考电压由电压均分的分压器得到,分压器横跨输入信号的正、负端。另一方面,所有 128 个采样输入信号连接起来,并接到模拟输入端。

当模拟输入电压从一个值变到另一个值时,并行电压比较器根据模拟输入信号的大小决定闪存式转换器的打开或关闭。这些输出信号由一个逻辑电路(称为 127-7 的解码器)来分析,转化为代表闪存式转换器开/闭的二进制数。在时钟脉冲下降沿,此值锁在输出寄存器中,并输入计算机后进行处理的过程。

11.4 计算机接口

我们已介绍了很多 PC 仪器的数字化数据流被输入计算机后进行处理的过程,但很少介绍其转换方法。在内置 PC 仪器情形下,采用计算机内部数据总线,数据流被直接传送到计算机。计算机与外部的唯一接口是测试仪器的输入/输出接口。

在外置 PC 仪器的设计中,必须做必要的准备,以把外部机盒与内部计算机总线连接起来。在多功能音频分析器的音频测试站中,这种连接是通过插在扩展槽中的数字接口卡来实现的。这种接口卡将数字输入脉冲转换为计算机总线可辨识的信号。许多外置 PC 仪器采取这种做法,即采用各种专用接口结构。

在某些情形下,外置 PC 仪器制造商试图使其接口标准化,并可与更广泛的计算机兼容,而不局限于 IBM PC 家族。其方法之一是采用 RS-232 接口,这种接口在各种计算机和许多其他型号的设备中都可见到。

典型的 RS-232 接口的虚拟仪器是 Maron 公司的 SAV 10 系列电压表。SAV 10 系列电压表是多通道 ASCII 电压表,它监视四路模拟电压输入,将模拟量转化为数字量,并且将其作为 ASCII 解码信息以一种与标准数据显示终端和RS-232接口兼容的形式传输。单个SAV10系列电压表不需要主机的控制信息，并独立于操作计算机系统运行。事实上，SAV 10系列电压表是一种卫星数据采集设备，它将原始数据输入计算机并进行后续存储或处理。

11.5 通用接口总线

随着电子测量的应用在数量和复杂性上的发展，越来越多的仪器用户认识到传统的设计和测试过程的不足。现代技术人员和工程师有了新需求，这些需求只有将单个测试仪器集成为交互的自动化系统才能达到。随着数百万台个人计算机被工程师和科学工作者使用，仪器测量和自动测试将不可避免地走进以计算为平台的王国（见图11.7）。

达到这一目标的第一步是20世纪70年代早期由Hewlett-Packard通过Hewlett-Packard接口总线（HP-IB）的引入完成的。HP-IB本质上是一个通信连接，这使得仪器之间可通过标准总线相互通信。HP-IB之前，将程序化测试仪器与计算机平台或局域网（LAN）连接起来是一项主要任务。仪器制造商间的不同标准产生了专有接口，这把用户绑定到一个制造商，拒绝使用非这个供应商的测试仪器。HP-IB标准果断地定义了接口，使计算机控制的测试系统集成一体成为可能。

1975年，HP-IB标准被IEEE协会接纳为IEEE-488标准。IEEE-488标准于1978年被修订为IEEE-488-1978，并正式命名为通用接口总线或GPIB。

IEEE-488接口总线是一种异步的、并行的计算机LSI-11总线（它是基于DEC的极为流行的PDP-11微处理机系统）。GPIB系统的关键操作是控制器。控制器相当于接线员，它通过网络接收指导和发送信息。GPIB通过IEEE-488接口总线，利用发送器和接收器来交换数据。发送器通过控制器将数据传送给接收器（见图11.8）。

发送器（通常只作为讲话设备）只能通过总线发送数据，不能接收数据或命令。发送设备相当于一个电压表，它只发送测量信息。接收器对于IEEE-488接口总线测试系统通常不理想，因为控制器不能将它打开或关闭。此类设备如电压表，必须由触发器或前面板上的按键来激发。一个只包含发送设备的系统是一种极简单的只包含两个部件的装置，如电压表和数字打印机。在这种条件下，随着电压表数据的不断产生，数字打印机将周期性地打印出电压数值。然而，当用于GPIB系统的发送仪器有多个设备时，发送设备将不占用总线，并妨碍由控制器发出的命令或来自其他设备的总线数据。

一个接收器或接收设备只能通过总线接收数据，但不能发送数据。虽然接收设备在IEEE-488接口总线测试特性中不具有共性，它们对一些非测量设备是很实用的。接收设备相当于一个电源，它的功能是通过总线接收程序化的信息：电压和电流，而不是发送信息。然而，即使在这种情况下，提供电源让话筒在掉电情况下工作是有益的。报警信号由电源通过GPIB发送，由控制器使电源停止工作，以免测试单元被破坏。

IEEE-488接口总线测试系统的首选仪器是收发器，它可通过总线同时发送和接收信息。收发器具有十分简捷的规划、解决竞争和维护IEEE-488接口总线测试系统的能力。大多数新的GPIB仪器是各种不同的收发器。当需要更严格的控制时，通过软件指令或硬件开关，这些仪器通常可被设定为只收或只发工作状态。

第12课 工 业 总 线

12.1 什么是工业总线

什么是工业总线？传统上，中央计算机同现场设备通过工业总线进行通信。中央计算机是大型机或小型机，现场设备可能是一个智能设备，如流量表、温度传感器、CNC单元或机器人这样的复杂设备。随着有计算能力设备成本的下降，计算机之间也可以通过工业总线进行相互通信，以协调工业生产。

如同人类语言一样，人们用许多方式来设计计算机同设备之间的通信，也同人类相似，大部分通信与任何别的系统之间的通信是不兼容的。不兼容可分成两类：物理层和协议层。

Modbus和Data Highway是两种使用RS-232和RS-422/485标准的流行工业总线。Modbus由Modicon公司为它的PLC系列及984系列控制器而开发。Modbus可以通过4线模式配置为RS-232或RS-485标准的工业总线。Data Highway是工业总线的名称。RS-485接口可以用在PLC-2、PLC-3和PLC-5等型号的控制器上。参考控制器的随机手册以确定它支持何种总线。使用RS-232和RS-422/485标准的工业总线及与多种工业总线都兼容的产品都列在控制器的随机手册上。B&B产品只在物理层支持这些总线，并且主要作为中继器、线性延伸器和隔离器。B&B产品也用于定制设计服务，来解决由工业总线产生的特殊问题。

可以以电话系统为例来定义物理层和协议层。任何一种语言都可以通过电话线进行传输。只要对话双方彼此能懂得这种语言，通信就可以实现。电话线不关心传送信号的内容，它只负责在物理上实现将信号从一端传送到另一端。这种将信息从一点传送到另一点的通道就是物理层。而对话双方关心的是电话线上的传送内容是什么。如果说话者使用西班牙语，而听者只熟悉英语，信息交流是不能实现的。尽管物理层在工作，但语言或协议如果不正确，信息交流依然无法进行。工业界已经开发了各种不同的物理层和协议层的通信标准。如果将它们全部列出，将会占用大量的篇幅。所以，我们这里仅讨论符合RS-233和RS-422/485标准的物理层的工业总线。RS-232和RS-422/485标准之间的最大区别在于信息的传送方式不同。

12.2 数据线的隔离原理

当要保护数据线免受瞬间的电干扰时，一般首先想到的就是振荡抑制。振荡抑制的概念是一种直觉，并且市场上有大量不同的振荡抑制设备以供选择（见图12.1）。振荡抑制设备可以用来防止从计算机到应答设备及串行端口为RS-232、RS-422或RS-485的设备受到任何干扰。不幸的是，在绝大多数的串行通信系统中，使用振荡抑制设备并不是最佳选择。大多数的雷电和感应振荡都会引起通信系统中各点之间对地的电动势差异。系统覆盖的物理区域越大，各地点对地电动势的差异就越有可能存在。

通过水的类比可有助于解释这种现象。我们不要想象管子中的水，而是更大，如海洋中的浪。问任何人"海洋的海拔是多少？"大家都会回答是零——因此我们通常将它称为海平面。尽管平均海拔是零，我们知道潮汐和海浪都能够引起水面实际高度短暂出现很大变化，这和接地非常相似。大量电流进入大地的效果就像水波从原点向四周漫开一样。两点之间的对地电压差距很大，直到这种能量消失。

换个类比角度，在大浪中保护船只最好的方法是什么？我们可以将船只捆在固定的码头上，强制船只保持在某一个高度。这种方法能够对付较小的浪，但这种解决方法明显有局限性。尽管这种比较很粗略，但基本描述了典型的振荡抑制设备要完成的工作。将局部设备上的能量振荡抑制在一个安全的水平，就需要振荡抑制设备必须能够完全吸收或旁路瞬间的能量。

如果不将船只系在固定的码头上，而是让码头浮动。船只就可以随着巨浪升降（直到碰到浮动码头的柱子的一端）。

不是与自然抗争，而是顺应自然，这就是我们数据线隔离的基本思想。

隔离不是一个新概念，它早已应用在电话和以太网设备上了。对于异步数据传送的应用，如许多采用 RS-232、RS-422 和 RS-485 标准的系统，光学隔离器是最常用的隔离元件。通过隔离，两个不同的接地（最好当成参照电压）处于隔离元件的两边，而不让任何电流穿过隔离元件（见图 12.2）。光学隔离器是用 LED 和感光的晶体管完成光电隔离的，光只在这两个元件之间穿过。

光学隔离器的另外一个优点就是它不受安装质量的影响。用于保护数据线的典型振荡抑制设备是利用专门的二极管将多余的能量传输给大地。安装者必须提供一个相当低的接地阻抗来处理这些频率可能是数十万兆赫、电流可能是数千安的能量。小的接地阻抗，如 1.8m 长的电线，可以引起数百伏的电压降——这个电压足以损坏大多数设备。而对于光学隔离器，则无须附加接地，它对安装质量的要求不高。

光学隔离器并不是解决电干扰的完美方法。隔离电路要附加隔离电源。这样的间隔电源可能是内置或外置的 DC-DC 转换器。简单的振荡抑制设备无须提供电源。

光学隔离器的隔离电压也是很有限的，通常是 500～4000V。在某种情况下，振荡抑制设备和隔离设备的组合应用是一个有效的方法。

当为系统选择保护数据线时，考虑全部选项是很重要的。振荡抑制设备和光学隔离器都有优点和缺点，而对于大多数系统来说，使用光学隔离器是较为有效的方法。如果不能确定使用哪种方法，就选择使用光学隔离器。

第 13 课　可编程序控制器

可编程序控制器（PLC）又称可编程逻辑控制器，是计算机家族的一员。它们应用于商业和工业领域。PLC 通过监视输入信号，根据程序得到输出结果，以实现自动控制。下面就介绍基本的 PLC 功能和配置信息。

13.1　PLC 的基本操作

PLC 包括输入模块或输入点、一个中央处理单元（CPU），以及输出模块或输出点（见图 13.1）。输入模块可以接收来自不同现场设备（传感器）的各种数字或模拟信号，然后将它们转换为 CPU 可以使用的逻辑信号。CPU 根据存储器中的程序做出判断，然后发出控制指令。输出模块将来自 CPU 的控制指令转换为现场控制的各个设备（执行机构）所能使用的数字或逻辑信号。编程设备用于输入期望的程序，由这些程序来决定特定的 PLC 做什么。通过操作界面可以显示过程信息及输入新的控制参数。在这个简单的例子里（见图 13.2），按钮（传感器）连接到 PLC 输入模块，由 PLC 的输出模块发出控制指令给电动机的启动器（执行机构），然后由电动机的启动器控制电动机启动和停止。

13.2　硬接线控制

使用 PLC 之前，许多控制任务是通过接触器和继电器来完成的，这就是常说的硬接线控制。对于硬接线控制，必须先设计电路图（见图 13.3），确定元件，并且进行安装及建立接线表，再连接执行特定任务的元件。如果接线中有一个错误，就必须重新连接来纠正。如果改变功能或扩充系统，就要更换大量元件，并且线路也要重新连接。

13.3　PLC 的优势

对于同样或更加复杂的硬接线控制任务，可以用 PLC 来做。现场设备同继电器触点之间的连线由 PLC 中的程序来替代,虽然依然要通过一些硬接线来连接现场设备,但这种硬接线连接已很少了。修改 PLC 中的应用软件或纠正错误是比较方便的。在 PLC 中建立或修改程序比起修改或重新连接电路要容易得多。

下面列出的只是 PLC 的一小部分优点。
(1) PLC 比通过硬接线控制装置的物理尺寸小。
(2) 使用 PLC 会使控制过程的修改更简单且更快速。
(3) PLC 有综合诊断和忽略功能。
(4) PLC 的应用软件能被立即生成文件。
(5) PLC 的应用程序能够被更快地复制且费用更低。

13.4 西门子 PLC

西门子制造了几种属于 SIMATIC S7 产品系列的 PLC。它们是 S7-200 PLC、S7-300 PLC 和 S7-400 PLC。

1. S7-200 PLC

S7-200 PLC 由于尺寸小,被称为微型 PLC。S7-200 是块状设计结构,即它的电源和 I/O 点都在一块板上。S7-200 PLC 可以用在小型的、独立的应用设备中,如电梯、汽车冲洗器或搅拌机。它也能用到更复杂的工业应用中,如装瓶机和打包机。

2. S7-300 PLC 和 S7-400 PLC

S7-300 PLC 和 S7-400 PLC 被用到更复杂的应用中,它们支持更多数量的 I/O 点。这两种 PLC 都是模块化和可扩展的。由分离模块组成的电源和 I/O 点都与 CPU 连接。根据应用的复杂性及以后的扩展需要来选择 S7-300 PLC 或 S7-400 PLC。西门子公司可以提供任何型号的西门子 PLC 的附加信息。

13.5 CPU

CPU 是包括系统存储的微处理器系统,是 PLC 的决策单元。CPU 监控输入信号,并根据保存在程序存储器中的指令做出决策。然后 CPU 执行继电器、计数器、计时器、数据比较和顺序操作等功能。

13.6 编程设备

程序是在一个编程设备(PG)中建立的,然后被传送到 PLC 中。可以使用专用西门子 SIMATIC S7 编程设备来建立 S7-200 PLC 的程序,如已经安装了 STEP 7 Micro/WIN 软件,就可以使用 PG 720 或 PG 740。安装了 STEP 7 Micro/WIN 软件的个人计算机(PC)也可以作为 S7-200 PLC 的编程设备。

13.7 软件

必须要有一个软件程序告诉 PLC 应执行哪一条指令,这个软件程序就是编程软件。一个编程软件只针对一个 PLC 或一个 PLC 系列,如 S7 系列,不能应用到其他 PLC 上。基于 Windows 平台的 S7-200 PLC 的编程软件称为 STEP 7-Micro/WIN32。PG 720 和 PG 740 使用前必须先安装 Micro/WIN32 软件。在个人计算机上安装 STEP 7-Micro/WIN32 的方法同安装其他计算机软件的方法类似。

13.8　连接电缆 PPI(点对点接口)

连接电缆用来将编程设备的数据传送到 PLC。只有当两个设备使用相同的语言或协议时才能够进行通信。西门子编程设备同 S7-200 PLC 之间的通信采用 PPI 协议。对于 PG 720 或 PG 740 的编程设备，需要一个合适的连接电缆。S7-200 PLC 使用 9 针的 D 型连接器。这是一个直插串行接口设备，与西门子编程设备(MPI 接口)兼容，D 型连接器接口是一个标准接口。

当个人计算机作为编程设备时，需要一个专用电缆，也就是 PC/PPI 电缆。这根电缆能够连接 PLC 的串行口与个人计算机的 RS-232 串行口并进行通信。PC/PPI 上的 DIP 开关用来设置 PLC 同个人计算机之间信息传输的合适速度(比特率)。

第 14 课　遥　感

遥感是一种远距离获得陆地、水面或某个物体的信息的过程，它不需要传感器和被测物体之间的直接接触。遥感这一名词通常是指用装在航天飞机或卫星上的设备进行数据采集。遥感系统经常用于对地球资源和环境的测量、描绘和监控等。它们也用于对其他行星的探索。

遥感设备有许多不同的种类。一些遥感系统使用照相机拍照的方式来记录可见光的反射能量。另外一些遥感系统记录可见光以外的电磁能，如红外线和微波。多频带扫描仪可以对可见光和红外频谱进行成像。

14.1　传感器

我们最熟悉的电磁能就是可见光，它是电磁波谱中人眼能够感觉的那部分。当照相机的胶卷曝光的时候，它可以记录电磁能。50 多年来，用空中摄影机获取的照片图像一直用于城市规划、森林管理、地形测量与绘图、土壤保护、军事监察和其他很多应用中。

红外传感器和微波传感器能够记录不可见的电磁能。一个物体的热量可以通过它所发出的红外能量来测量。红外传感器可以得到反映一个地区温度变化的图像，对于传统的照相，这通常是非常困难甚至根本不可能的。热红外传感器可用于水温的测量、已损坏的地下管道定位、地热和地质结构图的绘制。

微波传感器如雷达，可以向物体传送电磁能并且记录这些物体如何反射能量。微波传感器工作在波长很长的电磁波段，它能穿透云层，当云层覆盖使别的传感器不能成像时，它却能成像，它的这个特点是非常有用的。通过用雷达扫描一个地区，并用计算机处理收集到的数据，科学家们可以建立雷达地图。表面被厚厚的云层所包围的金星就是用这种办法绘制出它的表面外貌。雷达成像还被用于绘制地质图、估算土壤湿度、确定海上冰的状况以辅助航海。

多波段扫描仪可以提供不同电磁频谱范围的电子数据。科学家常用计算机来提高图像的质量或借以进行自动信息收集和绘图。利用计算机，科学家还能把工作在不同频率上的扫描仪得到的几个图像进行组合。

14.2　卫星

卫星对遥感系统的发展起到非常重要的作用，这一点已被证实。1972 年，美国发射了地球资源一号卫星，这是第一次为遥感系统专门设计的一套卫星。如今，地球资源五号卫星在每 16 天就可以给出地球大部分表面的图像。每一幅地球资源卫星图像都覆盖超过 31 000 km²（11 970 平方英里）。用地球资源卫星专业绘图仪绘制的图像可以看见 900 m² 的物体，它是一种多频带扫描仪。地球资源卫星数据有很多用途，如绘制地图、管理森林、估计农作物产量、监控放牧情况、分析水体质量及保护野生动植物。

从 1990 到 1996 年，有大约 50 个遥感卫星被发射到赤道上方。1986 年，法国的人造定位及跟踪卫星提供了 100 m² 物体的图像及立体图像，这对于绘制地形图是非常有用的。欧洲航天局及日本、俄罗斯、印度等国家也已经发射了地球观测卫星。

气象卫星如美国海洋大气局所用的卫星，可以提供供海洋和陆地天气预报用的图像。气象卫星上的远程传感器可以跟踪云层的运动，记录大气的温度变化。

14.3　展望

遥感技术正在迅速发展。一些卫星携带的装置可以对与汽车一样大小的物体成像，不断进步的技术可以提供更好的办法来解决现在乃至将来出现的问题。计算机辅助图像分析技术推动了遥感技术许多新的应用。20 世纪 90 年代后期，美国宇航局计划发射地球观测系统卫星，这个计划是美国地球行星任务的关键环节，这个计划会发射一系列卫星去研究行星的环境变化。

第 15 课　多传感器数据融合

15.1　简介

多传感器数据融合（MDF）是在融合多个传感器数据的过程中涌现的一项技术，以便通过测量和检测，得到对环境更加准确的估计。MDF 的应用领域广泛，包括如自动目标检测与跟踪、战场监视等军用方面，以及如环境监视与监控、复杂机械监控、医疗诊断、智能建筑、食品质量检验及精细农业民用方面。数据融合技术是多门学科的综合，包括信号处理、模式识别、统计估计、人工智能和控制理论。计算机的迅速发展、微机电系统传感器的不断产生及数据融合技术的成熟都为数据融合的日常应用提供了一个基础。

15.2　技术背景

多传感器数据融合提供了一种提高单个传感器性能的途径。一般来说，从单个传感器很难得到高质量的测量数据。如果利用多个传感器进行同一个测量，再将这些传感器的测量数据以某种方式组合，那么合成结果将很可能超过每个单独测量的精度。

总之，正如其名称所指，MDF 是一项将多个传感器的数据通过一个中央数据处理器进行组合，以提供全面而准确的信息为目的的技术。这项技术在适应环境变化和可预见的冲击时具有强大的潜能，比传统单个数据源更稳定（即使该单个数据源高度可靠）。因而，MDF 使得创造一个协同处理的过程成为可能，在这个过程中，单个数据的合并创造出一个其总值大于部分之和的联合数据源。

尽管这个概念不是新的，但 MDF 技术仍然处在它的初期。从 20 世纪 80 年代末期开始持续到现在，这项技术经历了迅速的发展。美国国防部（DoD）对这项技术开展了许多早期研究，并在军事监视和陆地战斗管理系统中探索了它的有用性。数据融合技术在商业项目（如机器人学和通用图像处理)和非军事政府项目(如天气监视和美国航空航天局使命)中的应用也在迅速增长。在目前的情况下,该技术可以融合多种类型的传感器信号,包括雷达、红外线、声呐和视觉信息。

在测量与仪器的数据分析与处理中,模式识别技术是必需的。模式识别用于开发数据融合算法。人工神经网络是根据对人脑机制的研究而开发的理念,应用该理念进行统计模式识别要远胜过其他常规方法。Linn 和 Hall 研究过 50 个以上的数据融合系统。这些系统中仅 3 个系统使用了人工神经网络方法。这么低的使用数据表明可能低估了人工神经网络在数据融合领域的重要性。

人工神经网络已被广泛用于解决复杂问题,如模式识别、快速信息处理和自适应。一个神经网络的构造和执行就是人脑结构和活动的模拟简化版。生物神经组织与生俱有的海量处理能力启发了人们对其结构本身的研究,将其作为设计和组织人工计算结构的模型。在测量与仪器中,综合神经网络模式识别的 MDF 方案是一种前景广阔、可完成高质量数据分析与处理的结构。

15.3 方法

有以下 3 种可供选择的方法用于 MDF。
(1)将传感器的输出数据直接融合。
(2)用特征矢量表示传感器的输出数据,然后将特征矢量融合。
(3)处理每个传感器的输出数据,得到高级推理或决策,再对其进行融合。

利用不同的融合方法,使得上述每种方法都受到一些应用类型的推动和影响。它们不能被推广到所有数据融合方法中。下面将集中介绍基于特征的数据融合方法。

可以从每个传感器的输出数据中定量提取出特征矢量。提取出的特征矢量可以组成的矩阵如下:

$$\boldsymbol{F} = [f_1, f_2, \cdots, f_n]^{\mathrm{T}} \quad (n \geq 2) \tag{15-1}$$

式中,f_i 是从第 i 个传感器得到的特征矢量;n 是传感器的个数。

对于每个输入的特征矢量,具有代表性的模式识别技术是神经网络,这些特征矢量要被区分并归入适当的类别。n 个神经网络的特征矢量在数据融合时是必须进行合并的。

假定 $O_1^i, O_2^i, \cdots, O_m^i (i=1,2,\cdots,n)$ 是神经网络对第 i 个传感器的 m 个输出数据,合并后的输出数据应该是如下形式:

$$O_j^M = c_1 O_j^1 + c_2 O_j^2 + \cdots + c_n O_j^n \quad (j=1,2,\cdots,n) \tag{15-2}$$

式中,$c_1, c_2, \cdots, c_n$ 是由赋值或最优化得到的合并系数。

神经网络 MDF 方案的框图如图 15.1 所示,该框图应该按如下步骤运行。
(1)从 n 个传感器的每个输出数据中提取特征。
(2)通过一个神经网络分类器对特征矢量分类。
(3)神经网络分类器将所有 n 个传感器的输出数据合成 m 个输出数据。
(4)对产生的每个融合结果进行后续处理。

15.4 应用

我们已经开发或将要开发许多 MDF 在测量与仪器中进行高质量的数据分析与处理方面的应用。这些应用的领域在工程文献中尚鲜有报道。

1977 年,Gros 写了一本书,那是第一本将多传感器融合和数据融合专用于 NDT(非破坏性测试)或 NDI(非破坏性检验)的书。这本书为希望探索 NDT/NDI 数据融合的读者提供了详细的专题研究和实用指南。从那以后,各个地区进行了 NDT/NDI 数据融合的研究和应用。然而,一个统一的、能解决多种应用问题的数据融合系统模型却是很难设计的,甚至是不可能。一般意义下的各种数据融合模型对每个研究和应用的专业领域是必要的。

NDI超声波和涡流图像方法是在老化飞机面板的腐蚀检测中研究出来的。为了增加单传感器的检测精度,一种适于同种应用的NDI数据融合方法被开发出来。

我们开发出了另一种集成了人工神经网络分类器的NDI数据融合方法。超声波和涡流图像数据通过神经网络分类器来识别类似飞机面板上的腐蚀点。在腐蚀检测中,为了评估NDI成像的总体性能,将两个来自不同图像传感器的分类后的腐蚀数据进行融合很重要。通过MDF,我们希望产生一幅完整的画面,而且具有更准确的腐蚀检测精度。

带神经网络分类器的NDI成像数据融合方案的框图如图15.2所示。该方案可以按如下步骤进行。

(1)以一个指定大小的窗口:2×2,3×3,…,12×12,扫描超声波和涡流图像,提取统计特征。

对每幅图像特征提取之后,产生两组新数据:特征矢量和符号矩阵。符号矩阵用来记录图像中每个像素代表的意义。

0:表示图像中该像素没有腐蚀。

1:表示图像中该像素可能有一定程度的腐蚀,但该像素不会包含在神经网络分类中,这是由于其周围有太多的未腐蚀像素,因而窗口特征提取跳过了它。

2:表示图像中该像素表现出一定程度的腐蚀,而且它将包含在神经网络分类中。

(2)神经网络分类器产生分类后的超声图像和涡流图像。

(3)合并神经网络分类器对不同传感器的特征矢量,包括合并后的符号矩阵。

符号矩阵有助于寻找神经网络分类器对不同传感器的输出数据之间的关系。合并后的符号矩阵形式如下。

0:在超声和涡流符号矩阵中,同一像素处都是0。

1:在超声和涡流符号矩阵中,同一像素处或是0或是1。

2:在超声和涡流符号矩阵中,同一像素处至少一个是2。

(4)产生一个分类后的融合图像。

这个方案执行了二次合并功能:特征符号矩阵的合并,以及神经网络分类器对不同传感器的输出数据的合并。融合后的图像可能胜过来自两个不同图像传感器的每幅图像,因为融合结果综合了两个不同传感器所获取的信息。

这个方案的结果是:一种是分类后的超声图像,另一种是分类后的涡流图像和一个融合过的分类图像。为了评价数据融合模型的性能,这3幅图像与同一个标本基准X-射线腐蚀图像(见图15.3)相比较,图15.4是腐蚀性飞机面板标本的一个超声波图像。图15.5是标本的涡流图像。图15.6是以4×4窗口大小提取特征且经神经网络分类后的超声波图像,它与X-射线数据匹配率达50.61%。图15.7是以4×4窗口大小提取特征且经神经网络分类后的涡流图像,它与X-射线数据匹配率达57.70%。图15.8是图15.6和图15.7融合分类后的图像,匹配率高达65.49%。

表15.1总结了不同窗口大小下神经网络分类器数据融合与X-光数据的匹配率。结果表明,数据融合稳定地提高了超声图像和涡流图像数据各自的神经网络腐蚀检测精度。

第16课　磁耦合共振式无线电能传输技术

20世纪初,在电网出现之前,尼古拉·特斯拉(Nikola Tesla)便致力于无线电力传输的构想。然而,典型实例(如特斯拉线圈)也包含了难以实现的电力系统。过去十年,自主电子设备(笔记本计算机、手机、机器人、PDA等)的使用量激增。因此,人们对无线电能传输的兴趣重新燃起了。辐射式电能传输虽然非常适合传输信息,但对于电能传输应用来说,存在许多困难:如果辐射是全方位的，电能传输效率很低，单向辐射需要不间断的视线和复杂的跟踪机制。最近的一篇理论论文详细分析了利用共振物体通过其非辐射场尾端耦合进行中距离电能转移的可行性。从直观上看，同一共振频率的两个共振物体往往能有效地交换能量，而在共振物体的外部耗散的能量相对较少。在耦合共振系统中(如声学、电磁、磁、核)，通常有一个"强耦合"的工作状态。如果我们能在给定的系统中在该状态下工作，电能传递将是非常有效的。以这种方式实现的中距离电能传输几乎可以是全方位和高效的，无论周围空间的几何结构如何，对环境物体的干扰和损耗都很低。

无论共振的物理性质如何，上述考虑都适用。在这里，我们集中讨论一个特殊的物理现象：磁共振。磁共振特别适用于日常应用，因为大多数常用材料不与磁场相互作用，因此与环境物体的相互作用被进一步抑制。通过在兆赫频率下探测非辐射(近场)磁共振感应，我们能够识别两个耦合磁共振系统中的强耦合区域。乍一看，这种电能传递类似于通常的磁感应；然而，请注意通常的非共振感应对于中距离应用非常低效。

16.1 基本概述

有效的中距离电能传输发生在参数空间的特定区域，描述了相互强耦合的共振物体。利用耦合理论来描述这个物理系统，我们得到以下一组线性方程：

$$\dot{a}_m(t) = (i\omega_m - \Gamma_m)a_m(t) + \sum_{n \neq m} ik_{mn}a_n(t) + F_m(t) \tag{16-1}$$

其中，指数表示不同的共振物体。定义变量 $a_m(t)$ 使对象 m 包含的能量为 $|a_m(t)|^2$，ω_m 是该孤立对象的共振角频率，Γ_m 是其固有衰减率(如由于吸收和辐射损失)。在这一模型中，一个参数 ω_0 和 Γ_0 非耦合、无驱动振荡器按 $\exp(i\omega_0 t - \Gamma_0 t)$ 随时间演化。$k_{mn} = k_{nm}$ 是以下标来区分的共振物体之间的耦合系数，$F_m(t)$ 是驱动项。我们将处理限制在发射端和接收端表示的两个对象的情况下，这样发射端(由下标 S 标识)以恒定频率从外部驱动，并且两个对象具有耦合系数 k。电能传输是借助于负载(下标 W)通过设备(下标 D)来实现的，该负载通过电路电阻连接，并贡献额外的 Γ_W 给衰减率为 Γ_D 的设备对象。因此，装置的总衰减率为 $\Gamma_D' = \Gamma_D + \Gamma_W$。提取的功率由负载中的耗散功率决定，即 $2\Gamma_W|a_D(t)|^2$。对于加载 Γ_W，可以通过最大限度地提高转移效率 η 来解决阻抗匹配问题，如式(16-1)。我们发现，当电源和设备共振时，该方案的效果最好，在这种情况下，效率为

$$\eta = \frac{\Gamma_W|a_D|^2}{\Gamma_S|a_S|^2 + (\Gamma_S + \Gamma_W)|a_D|^2} = \frac{\frac{\Gamma_W}{\Gamma_D}\frac{k^2}{\Gamma_S\Gamma_D}}{\left[\left(1+\frac{\Gamma_W}{\Gamma_D}\right)\frac{k^2}{\Gamma_S\Gamma_D}\right] + \left[\left(1+\frac{\Gamma_W}{\Gamma_D}\right)^2\right]} \tag{16-2}$$

当 $\Gamma_W/\Gamma_D = [1+(k^2/\Gamma_S\Gamma_D)]^{1/2}$ 时效率最大。显而易见，高效能量传递的关键是让 $k^2/\Gamma_S\Gamma_D > 1$。这通常被称为强耦合机制。共振在这种功率传递机制中起着至关重要的作用，因为相对于感应耦合，非共振物体效率提高了约 ω^2/Γ_D^2 ($\sim 10^6$ 典型参数)。

16.2 谐振线圈的理论模型

我们的实验实现方案包括两个谐振线圈。一个线圈(发射线圈)感应耦合到振荡电路；另一个(接收线圈)感应耦合到电阻负载(见图 16.1)。谐振线圈依靠分布电感和分布电容的相互作用来实现谐振。线圈由总长度为 l、横截面半径为 a 的导电丝制成，并将它绕成 n 圈、半径为 r、高度为 h 的螺旋线。据我们所知，文献中没有关于有限螺旋线的精确解，即使在无限长线圈的情况下，解也依赖于不充分的假设。然而，我们发现，下面描述的简单准静态模型与实验结果(误差 5%)是一致的。

A 是一个半径为 25cm 的单铜环路，是驱动电路的一部分，它输出频率为 9.9 MHz 的正弦波。

S 和 D 分别是本文提到的源线圈和器件线圈。B 是连接到负载（灯泡）上的一圈电线。不同的 ks 表示箭头所示对象之间的直接耦合。调整线圈 D 和回路 A 之间的角度，以确保它们的直接耦合为零。线圈 S 和 D 同轴对齐。B 和 A、B 和 S 之间的直接耦合可以忽略不计。

从观察到线圈两端的电流必须为零开始，我们做出了有根据的猜测，即线圈的谐振模式很好地近似于沿着导线长度的正弦电流分布。我们对最简单模式感兴趣，所以如果我们用 s 表示沿着导体长度的参数化坐标，使其从 $-l/2$ 到 $+l/2$，那么与时间相关的电流分布形式为 $\lambda_0 \cos(\pi s/l) \exp(i\omega t)$。根据电荷连续性方程，线性电荷密度分布的形式为 $\lambda_0 \sin(\pi s/l) \exp(i\omega t)$，因此线圈的一半（垂直于轴相切方向）包含振荡总电荷（振幅 $q_0 = \lambda_0 l/\pi$），其大小相等，但与另一半的电荷符号相反。

当线圈共振时，电流和电荷密度分布是 $\pi/2$ 互补的，这意味着当一个实部为零时，另一个实部最大。同样地，线圈中所含的能量在某个时间点完全取决于电流，而在其他时间点则完全取决于电荷。利用电磁理论，我们可以定义每个线圈的有效电感 L 和有效电容 C 如下：

$$L = \frac{\mu_0}{4\pi |I_0|^2} \iint dr dr' \frac{J(r)J(r')}{|r-r'|} \tag{16-3}$$

$$\frac{1}{C} = \frac{1}{4\pi\varepsilon_0 |q_0|^2} \iint dr dr' \frac{\rho(r)\rho(r')}{|r-r'|} \tag{16-4}$$

其中，空间电流 $J(r)$ 和电荷密度 $\rho(r)$ 分别由沿着隔离线圈的电流和电荷密度，以及物体的几何体获得。根据定义，L 和 C 具有储能性质，线圈中包含的能量 U 为

$$U = \frac{1}{2} L |I_0|^2 = \frac{1}{2C} |q_0|^2 \tag{16-5}$$

考虑到这种关系和连续性方程，得到的共振频率 $f_0 = 1/[2\pi(LC)^{1/2}]$。我们现在可以通过定义 $a(t) = [(L/2)^{1/2}]I_0(t)$，将此线圈视为耦合理论中的标准振荡器。

通过研究电流分布的正弦曲线，即峰值电流平方的平均值为 $|I_0|^2/2$，我们可以估计耗散的功率。对于由电导率为 S 的材料制成的 n 匝线圈，我们相应地修改了欧姆（R_o）和辐射（R_r）电阻的标准公式：

$$R_o = \sqrt{\frac{\mu_0 \omega}{2\sigma}} \frac{l}{4\pi a} \tag{16-6}$$

$$R_r = \sqrt{\frac{\mu_0}{\varepsilon_0}} \left[\frac{\pi}{12} n^2 \left(\frac{\omega r}{c}\right)^2 + \frac{2}{3\pi^3} \left(\frac{\omega h}{c}\right)^2 \right] \tag{16-7}$$

式(16-7)中的第一项是磁偶极子辐射项（假设 $r \ll 2\pi c/\omega$，其中 c 是光速）；第二项是线圈的电偶极子，小于实验参数的第一项。因此，线圈的耦合理论衰减常数为 $\Gamma = (R_o + R_r)/2L$，其品质因数为 $Q = \omega/2\Gamma$。

通过考量从电源传输到设备线圈的功率，我们得到耦合系数 k_{DS}，假设在稳态解中电流和电荷密度随时间变化 $\exp(i\omega t)$：

$$P_{DS} = \int dr E_S(r) J_D(r) = -\int dr [\dot{A}_S(r) + \nabla \phi_S(r)] J_D(r)$$
$$= -\frac{1}{4\pi} \int dr dr' \left[\mu_0 \frac{J_S(r')}{|r'-r|} + \frac{\rho_S(r')}{\varepsilon_0} \frac{r'-r}{|r'-r|^3} \right] J_D(r') \equiv -i\omega M I_S I_D \tag{16-8}$$

式中，M 为有效互感；ϕ 为标量势；A 为矢量势；下标 S 表示电场是源引起的。然后，我们从标准耦合理论的论点得出结论，$k_{DS} = k_{SD} = k = \omega M/[2(L_S L_D)^{1/2}]$。当线圈中心距 D 远大于其特征尺寸时，k 满足双一双耦合的 D^{-3} 特征尺寸。k 和 Γ 都是频率的函数，k/Γ 和效率在特定的 f 值下最大化，对于相关的典型参数，f 值在 1 到 50MHz 的范围内。因此，为给定线圈尺寸选择合适的频率，正如我们在本实验演示中所做的，对优化电能传输起着重要作用。

16.3 实验验证

用于电能传输实验验证的两个相同螺旋线圈的参数分别为 $h=20$ cm、$a=3$ mm、$r=30$ cm 和 $n=5.25$。两个线圈都是铜制的。螺旋线圈各匝之间的间距不均匀,我们通过将 $10\%h$(2 cm)来概括其一致性的不确定度。给定这些尺寸的预期共振频率 $f_0=(10.56\pm0.3)$ MHz,与 9.90 MHz 下测量的共振频率相差约 5%。

回路的理论 Q 值约为 2500(假设 $\sigma=5.9\times10^7$ m/ohm),但测量值为 $Q=950\pm50$。我们认为,这种差异主要是由于铜线表面导电性差的氧化铜层的影响,在这种频率下,电流受到短集肤深度(~20 mm)的限制。因此,在随后的所有计算中,我们使用实验观测到的 Q 和推导出的 $\Gamma_S=\Gamma_D=\Gamma=\omega/2Q$。

通过将两个谐振线圈(微调,稍微调整 h 达到相同的谐振频率)设置一个 D 间距并测量两个谐振模式频率的差值,我们找到耦合系数 k。根据耦合理论,这个差值应该 $\Delta\omega=2[(k^2-\Gamma^2)^{1/2}]$。在目前的工作中,我们关注的是两个线圈是同轴对齐这种情况,尽管对于其他方向也获得了类似的结果。

16.4 效率的测量

最大理论效率仅取决于参数 $k/[(L_SL_D)^{1/2}]=k/\Gamma$,此值大于 1 即 $D=2.4$ m(8 倍线圈半径)。因此,在整个探测距离范围内,我们都在强耦合状态下工作。

作为驱动电路,我们使用一个标准的考毕兹振荡器,其电感元件由一个半径为 25 cm 的单圈铜线组成;该线圈感应耦合到源线圈,并驱动整个无线电能传输设备。负载由校准的灯泡组成,并连接到其自身的绝缘导线回路上,该绝缘导线放置在装置线圈附近并感应耦合到该线圈上。通过改变灯泡和设备线圈之间的距离,我们可以调整参数 Γ_W/Γ,使其与理论上由 $[1+(k^2/\Gamma^2)]^{1/2}$ 得出的最佳值相匹配。(连接到灯泡的回路为 Γ_W 增加了一个小的无功分量,这可以通过略微重新调整线圈来补偿。)我们通过调节进入考毕兹振荡器的功率来测量所提取的电能,直到负载处的灯泡在其额定亮度下发光。

通过用电流探针测量每个自谐振线圈中各点的电流(电流探针不会显著降低线圈的 Q 值),我们来确定源线圈和负载之间发生转移的效率。这给出了一个测量电流参数 I_S 和 I_D 的方法。然后,我们依据 $P_{S,D}=\Gamma L|I_{S,D}|^2$ 计算每个线圈的功耗,并依据 $\eta=P_W/(P_S+P_D+P_W)$ 求得效率。为了保证双目标耦合模式理论模型能够很好地描述实验装置,我们对器件线圈进行了定位,使其与连接到考毕兹振荡器的铜线圈回路的直接耦合为零。使用这个装置,我们可以传输几十瓦的电能,从 2 m 以外的地方完全点亮一个 60 W 的灯泡。

作为交叉检查,我们还测量了从壁装电源插座到驱动电路的总功率。然而,无线传输本身的效率很难用这种方式来估计,因为考毕兹振荡器本身的效率并不精确,尽管预计离 100% 差得很远。然而,从驱动电路中提取的功率与进入驱动电路的功率之比对效率给出了一个基本估值。例如,当在 2 m 的距离内将 60 W 的功率转移到负载时,流入驱动电路的功率为 400 W。这就产生了因移动造成的 15% 效率下降,这是合理的,因为在该距离处无线功率转移的预期效率为 $40\%\sim50\%$,并且考毕兹振荡器的效率较低。

参 考 文 献

[1] 田坦,姜弢. 电子电信工程专业英语[M]. 哈尔滨:哈尔滨工业大学出版社,2001.

[2] LEDN W. COUCH II. 数字与模拟通信系统[M]. 5版. 北京:清华大学出版社, Prentice Hall 公司. 1997.

[3] 尚雅层,赵朝兰. Electromechanical Engineering English Readings[M]. 西安:西安工业学院出版社,1998.

[4] 陈逢时. English for Electronic Testing Engineers[M]. 西安:西安电子科技大学出版社,1994.

[5] 翟俊祥,杨向明. 信息与控制专业英语[M]. 西安:西安交通大学出版社,2000.

[6] T. J. Byers, Electronic Test Equipment-Principles and Applications,1985.

[7] IEEE Transaction on Tnstrumentation and Measurement.

[8] IEEE Signal Processing Magazine.

[9] "PHILPS PM6680 Timer/Counter"Operators Manual.

[10] HEWLETT-PACKARD 8175A Digital Signal Generator. Operating and Programming Guide.

[11] 王宏文. 自动化专业英语教程[M]. 2版. 北京:机械工业出版社,2008.

[12] 祝晓东,张强华,古绪满. 电气工程专业英语实用教程[M]. 2版. 北京:清华大学出版社,2006.

[13] Kurs A, Karalis A, Moffatt R, et al. Wireless power transfer via strongly coupled magnetic resonances [J]. Science,2007,317(5834):83-86.

[14] Karalis A, Joannopoulos J D, Soljacic M. Efficient wireless non—radiative mid—range energy transfer [J]. Annals of Physics,2008,323(1):34-48.

[15] Liao Chenlin, Li Junfeng, Wang Lifang, Zhang Jinghe. Mid—range wireless charging system for electric vehicle [J]. Transactionbs of China Electrotechnical Society,2013, 28(2):81-85.

[16] 詹姆斯·尼尔森. 电路分析基础(第二版,英文版)[M]. 王洪祥,张民,译. 北京:电子工业出版社,2018.

[17] 阿伦R.汉布利. 电工学原理与应用(第七版,英文版)[M]. 熊兰,译. 北京:电子工业出版社,2019.